EXPLORATIONS IN
NUMBER THEORY

CONTEMPORARY UNDERGRADUATE MATHEMATICS SERIES
Robert J. Wisner, Editor

MATHEMATICS FOR THE LIBERAL ARTS STUDENT, Fred Richman, Carol Walker, and Robert J. Wisner

INTERMEDIATE ALGEBRA, Edward D. Gaughan

ALGEBRA: A PRECALCULUS COURSE, James E. Hall

TRIGONOMETRY: CIRCULAR FUNCTIONS AND THEIR APPLICATIONS, James E. Hall

MODERN MATHEMATICS: AN ELEMENTARY APPROACH, SECOND EDITION, Ruric E. Wheeler

A PROGRAMMED STUDY OF NUMBER SYSTEMS, Ruric E. Wheeler and Ed R. Wheeler

FUNDAMENTAL COLLEGE MATHEMATICS: NUMBER SYSTEMS AND INTUITIVE GEOMETRY, Ruric E. Wheeler

MODERN MATHEMATICS FOR BUSINESS STUDENTS, Ruric E. Wheeler and W. D. Peeples

ANALYTIC GEOMETRY, James E. Hall

INTRODUCTORY GEOMETRY: AN INFORMAL APPROACH, SECOND EDITION, James R. Smart

MODERN GEOMETRIES, James R. Smart

AN INTUITIVE APPROACH TO ELEMENTARY GEOMETRY, Beauregard Stubblefield

GEOMETRY FOR TEACHERS, Paul B. Johnson and Carol H. Kipps

LINEAR ALGEBRA, James E. Scroggs

ESSENTIALS OF ABSTRACT ALGEBRA, Charles M. Bundrick and John J. Leeson

AN INTRODUCTION TO ABSTRACT ALGEBRA, A. Richard Mitchell and Roger W. Mitchell

INTRODUCTION TO ANALYSIS, Edward D. Gaughan

DIFFERENTIAL EQUATIONS AND RELATED TOPICS FOR SCIENCE AND ENGINEERING, Robert W. Hunt

A PRIMER OF COMPLEX VARIABLES WITH AN INTRODUCTION TO ADVANCED TECHNIQUES, Hugh J. Hamilton

CALCULUS OF SEVERAL VARIABLES, E. K. McLachlan

PROBABILITY, Donald R. Barr and Peter W. Zehna

THEORY AND EXAMPLES OF POINT-SET TOPOLOGY, John Greever

AN INTRODUCTION TO ALGEBRAIC TOPOLOGY, John W. Keesee

EXPLORATIONS IN NUMBER THEORY, Jeanne Agnew

NUMBER THEORY: AN INTRODUCTION TO ALGEBRA, Fred Richman

EXPLORATIONS IN NUMBER THEORY

JEANNE AGNEW

Oklahoma State University

BROOKS/COLE PUBLISHING COMPANY
Monterey, California
A Division of Wadsworth Publishing Company, Inc.
Belmont, California

I wish to express my gratitude to all the students who have studied number theory with me over the years. I am especially indebted to the following people, who earned Ed.D. degrees at Oklahoma State University and who provided the inspiration for the collection of much of the material presented in Units 4, 6, 7, and 8: Leon L. Palmer, "Some Analysis in a Non-Archimedean Field"; Ronald J. MacKinnon, "Elementary Approaches to Waring's Problem"; William W. Malone, "Multiplicative Number-Theoretic Functions and Their Generating Functions"; Dick R. Rogers, "Reimann p-Adic Integration"; Verbal M. Snook, "A Study of p-Adic Number Fields"; and Melvin R. Woodard, "Primes in Arithmetic Progressions."

Jeanne Agnew

This book was designed by Linda Marcetti and its production was supervised by Phyllis London. The technical illustrations were drawn by Eldon J. Hardy. It was printed and bound by Kingsport Press, Inc., Kingsport, Tennessee.

The front cover is of Ouimet Canyon, Thunder Bay, Ontario, and was photographed by P. V. Ibbetson. The back cover is of Point Sur, California, and is by Lehman J. Pinckney, Jr.

ISBN: 0-8185-0041-7
L.C. Catalog Card No: 72-189515
Printed in the United States of America

1 2 3 4 5 6 7 8 9 10—76 75 74 73 72

PREFACE

TO THE STUDENT

These pages are an invitation to you to share with me some of the thrills of contemplating the structure of the ring of integers. They are written as the result of more than ten years of acting as a guide for students like yourselves who start out with eagerness and some trepidation in this exciting pursuit. To all the students who have permitted me to share their enthusiasm and observe their difficulties, I offer my sincere thanks.

Many good books have been written in the area of number theory, some quite recently. If, after reading this one, you are prompted to explore several of the others for further enlightenment, I shall be delighted. Don't hesitate to plunge into the periodical literature to pursue some topic that excites you. Some references to guide you are suggested at the end of each of the later units. Most of all, read with a pencil in hand. Whatever inspiration comes to you, jot it down, try it out on an example, think it through.

This book is designed to help you understand the material. In the first place, examples precede the theorems. Work through the examples, notice how the definitions work in particular cases, try to prove the theorems for yourself. The numbers assigned to the theorems and definitions will help you locate them. For example, Theorem 1.6.3 designates the third theorem in Section 6 of Unit 1. More important than the number of the theorem is its name. The name will help you keep in mind what the theorem is saying. Mathematical statements in general, and theorems in particular, have important messages that can and should be put into words.

For a great many years, amateurs as well as professionals have enjoyed "pursuing the elusive integer with gun and camera." We read that the eighteenth-century mathematicians imparted their discoveries and conjectures to each other by letters. Present-day mathematicians do so also—through letters, discussions at professional meetings, summer institutes, and symposia. In addition, they invite everyone from the high school student to the retired professor to join them in the game by publishing challenges in the form of problems in many of the mathematics-oriented periodicals. Some of the problems included in this book have, as the problems indicate, been taken from the Elementary Problems section of the *American Mathematical Monthly*. The word "elementary" in this context does not necessarily mean easy—it means only that no advanced analysis is required. Many similar problems can be found in the *Monthly* or in other periodicals, such as *Mathematics Magazine*, *Mathematics Teacher*, and *Scientific American*. Several of the other problems included here have been suggested to me by students who were studying the prepublication copy of this book. The names of these students are included with the problems.

Are there any errors? Inevitably. I hope you will find them and, when you do, feel pleased with yourself more than you feel cross with me.

Above all, have fun!

TO THE INSTRUCTOR

This book is written primarily to the student, in the hope that he can and will understand it by himself. This accounts for the leisurely pace of the book, especially in the first three units. These units include the basic ideas that are needed for a study of a variety of topics in elementary number theory. Since it is obviously impossible to include every interesting avenue of pursuit, I have chosen for the last five units those areas that my students and I have enjoyed studying.

The exercises include a great many problems—too many for a student to do all of them. Therefore, each exercise set is divided into two sections. The first, called "Checks," provides enough material to allow each student to test his understanding of the section. The second section, the "Challenges," includes problems that may interest the student who wants to explore further.

My suggestion for using this book is to cover the first three units almost in their entirety. If short of time, Section 1.12 can be omitted and the last two sections of Unit 3 can be done sketchily. I then include parts of the later units or all of some of them, depending on the nature of the class. Classes with many prospective teachers seem to enjoy Unit 5 and parts of Units 4 and 6. Classes in which there are mathematics majors seem to enjoy Units 7 and 8. I have also used some of these later units quite successfully as a starting place for individual study for the undergraduate honors seminar. The last five units are independent of each other and can be studied in any order.

How you can best use the book is something you are much better able to decide than I. It is my hope that it is suitable for any student-oriented teaching technique and that it has enough flexibility to be of use in a variety of programs.

For their helpful reviews of the manuscript for this book I would like to thank William Briggs, University of Colorado, Richard Byrne, Portland State College, James H. Jordan, Washington State University, James Nyman, University of Texas at El Paso, and Robert J. Wisner, New Mexico State University.

Jeanne Agnew

CONTENTS

8   THE p-ADIC INTEGERS—JOURNEY IN SPACE

BASIC CONCEPTS AND ORIENTATION—OUR POINTS OF DEPARTURE

1.1 THE RELATION "DIVIDES"

This book is a guidebook for a tour through the set of numbers called *integers*. Intuitively, this set is very familiar to us. But, with eyes open to the guideposts, we will see properties that we have not noticed before. With persistence and some basic equipment we can uncover intriguing relationships. And with fresh and inquiring minds we may discover unsuspected properties overlooked or neglected by the simple and the great who have preceded us.

> The Road goes ever on and on
>   Down from the door where it began.
> Now far ahead the Road has gone,
>   And I must follow, if I can,
> Pursuing it with eager feet,
>   Until it joins some larger way
> Where many paths and errands meet.
>   And whither then? I cannot say.[1]

What are the integers? The set of integers is a mathematical structure called a *ring*, although such a statement by no means characterizes the integers. (See Appendix A for a formal definition of a ring.) We represent integers by lowercase letters a, b, c, ... We assume that they can be combined by two operations called addition and multiplication. The set of integers is closed under these operations so that if a and b are integers, then $a + b$ is some integer c and $a \cdot b$ is some integer d. Multiplication and addition are both commutative and associative, and multiplication is distributive over addition. The integer 0 is the additive identity, since $a + 0 = a$ for all integers a. The integer 1 is the multiplicative identity, since $a \cdot 1 = a$ for all a.

With each integer a is associated an integer, usually designated $-a$, which is the additive inverse of a; that is, $a + (-a) = 0$. Consequently, the equation $a + x = b$ has a solution in integers for every pair of integers a and b. In fact, $x = b + (-a)$ is a solution, since $a + b + (-a) = b$.

The situation in regard to multiplication is quite different. Not every integer has a multiplicative inverse. In fact, the only integers a for which the equation $a \cdot x = 1$ has a solution in integers are $a = 1$ and $a = -1$. Thus it is not surprising that equations of the form $a \cdot x = b$ do not always have a solution in integers. It is perhaps more surprising that such equations sometimes *do* have a solution.

EXAMPLE 1.1.1

$3x = 15$ has a solution $x = 5$.

$12x = -48$ has a solution $x = -4$.

$-5x = -30$ has a solution $x = 6$.

$5x = 12$ has no solution in integers.

---

[1] J. R. R. Tolkien, *The Fellowship of the Ring* (Boston: Houghton Mifflin, 1965), p. 44. Reprinted by permission.

The following definition gives a name to the property that relates a pair of integers like 3 and 15, 12 and -48, -5 and -30, but not a pair like 5 and 12.

$$\boxed{a \mid b}$$

DEFINITION 1.1.1. *Divides.* Let a and b be integers, $a \neq 0$. The statement "a divides b" means that there exists an integer x such that $ax = b$. The statement "b is a multiple of a" is equivalent to "a divides b." Those integers a such that a divides b are called the *divisors* of b. Those integers b such that b is a multiple of a are called the *multiples* of a. The notation $a \mid b$ is used to mean "a divides b."

EXAMPLE 1.1.2

Because $2 \cdot 6 = 12$, $2 \mid 12$.
Because $3 \cdot 7 = 21$, $3 \mid 21$.
The divisors of 24 are ± 1, ± 2, ± 3, ± 4, ± 6, ± 8, ± 12, ± 24.
95 and -260 are both multiples of 5.
$x \mid 0$, for every integer $x \neq 0$.

Definition 1.1.1 gives a method for determining whether or not $a \mid b$. It can be used to derive many properties involving divisors.

EXAMPLE 1.1.3. If $a \mid b$, then a divides any multiple of b.

To prove this statement on the basis of Definition 1.1.1, we can argue as follows: Because $a \mid b$, there is an integer c such that $b = ac$. A multiple of b is an integer of the form kb, where k is an integer. Since $b = ac$, $kb = kac = (kc)a = ya$, where y is the integer kc. Since we have shown that kb is of the form ya, the condition set up in Definition 1.1.1 is satisfied and $a \mid kb$.

The argument given in Example 1.1.3 contains more discussion than would usually be included in a proof. The proof in the following example is shorter but includes the essential elements.

EXAMPLE 1.1.4. If $a|b$, then $a|b + ak$ for any integer k.

Proof: $a|b$ implies $b = ac$ for some integer c.

$$b + ak = ac + ak = a(c + k).$$

Since $c + k$ is an integer y, $b + ak = ay$, which implies $a|b + ak$. ●

Definitions are fundamental equipment for exploration. Many simple properties are direct consequences of the definitions. It is well worth becoming thoroughly familiar with definitions by using them to prove as many properties as you can. Some of these properties are useful as more refined tools for deriving further information. These basic properties are singled out as theorems. A few of the more important properties of divisibility are grouped in the following theorem.

THEOREM 1.1.1. *Properties of "Divides."* Let a, b, c be integers.

1. If $a|b$ and $b|c$, then $a|c$.   (transitive property)
2. If $a|b$ and $a|c$, then $a|bx + cy$, for every choice of integers x and y.
3. If $a|b$ and $b \neq 0$, then $|a| \leq |b|$.
4. If $a|b$ and $b|a$, then $a = \pm b$.

Proof:
Property (1). $a|b$ implies $b = am$ for some integer m.
   $b|c$ implies $c = bn$ for some integer n.
   $c = bn = amn$ implies $a|c$.
Property (2). $a|b$ implies $b = am$ for some integer m.
   $a|c$ implies $c = an$ for some integer n.
   $bx + cy = amx + any = a(mx + ny)$, which implies $a|bx + cy$.
Property (3). $a|b$ implies $b = am$ for some integer m.
   $|b| = |am| = |a||m|$.
   $b \neq 0$ implies $m \neq 0$ and hence $|m| \geq 1$.
   $|m| \geq 1$ and $|a||m| = |b|$ implies $|a| \leq |b|$.
Property (4). Neither a nor b can be zero since $a|b$ and $b|a$.
   $a|b$, $b \neq 0$, implies $|a| \leq |b|$, and $b|a$, $a \neq 0$, implies $|b| \leq |a|$.
   $|a| \leq |b|$ and $|b| \leq |a|$ implies $|a| = |b|$, which implies $a = \pm b$. ●

The following examples illustrate some techniques for using the collection of properties in this theorem.

EXAMPLE 1.1.5. The only divisors of 1 are ± 1.

Proof: Since $x = 1 \cdot x$ for every integer x, $1 \mid x$ for every integer x. Let x be any divisor of 1. Then $x \mid 1$ and $1 \mid x$; therefore $x = \pm 1$. ●

EXAMPLE 1.1.6. If $a \mid 4n + 3$ and $a \mid 2n + 1$, for some n, then $a = \pm 1$.

Proof: Property (2) of the properties of "divides" implies that

$$a \mid (4n + 3) + (-2)(2n + 1).$$

But

$$(4n + 3) + (-2)(2n + 1) = 1.$$

Hence $a \mid 1$ and so $a = \pm 1$. Note that the success of this proof depends on an appropriate choice of the integers x and y in property (2). ●

EXAMPLE 1.1.7. If $c = ab$ and $a \neq \pm 1$, then $|b| < |c|$.

Proof: Property (3) of the properties of "divides" implies that $|b| \leq |c|$. But $|c| = |a| |b|$, so that if $|c| = |b|$, then $|a| = 1$ and $a = \pm 1$. Since $a \neq \pm 1$, we conclude that $|b| < |c|$. ●

Care should be taken in the use of any tool. In applying a theorem, be sure that you understand exactly what the theorem says and that you use only statements which the theorem justifies. A statement may be true but its converse false.

EXAMPLE 1.1.8. Property (3) says that if $a \mid b$, then $|a| \leq |b|$. The converse is false. It is not necessarily true that if $|a| \leq |b|$, then $a \mid b$. To prove that a statement is not necessarily true, all that is needed is an example in which the statement is false. Thus $|3| \leq |5|$, but 3 does not divide 5.

EXERCISES 1.1

Checks

1. Write the set of divisors of 39; the set of divisors of -125.

2. Prove that the set of divisors of 36 is finite. Is there an integer for which the set of divisors is an infinite set?

3. Write seven multiples of 5. Prove that the set of multiples of 5 is an infinite set. Is there an integer for which the set of multiples is a finite set?

4. Prove that the set of multiples of 12 is a subset of the set of multiples of 6. (See Example 1.1.3.)

5. Prove that the set of divisors of 6 is a subset of the set of divisors of 12.

6. Prove that for every integer a such that $a \neq 0$, $a \mid \pm a$.

7. Prove that if $a \mid b$, then $a \mid -b$.

8. If $a \mid m$ and $b \mid n$, prove that $ab \mid mn$.

9. Prove that if there exist integers x and y such that $9x + 12y = n$, then $3 \mid n$.

Challenges

10. Let $a + b = c$. Prove that if d divides any two of the integers a, b, c, then d divides all three of them. Deduce that if $d \mid c$, then d divides both a and b, or d divides neither a nor b. Give an example to show that either situation may occur.

11. If a divides $12n + 5$ and $4n + 2$ for some n, prove that $a = \pm 1$.

12. Prove that the divisors of an integer occur in pairs—that is, if $d \mid a$, there exists an integer d' such that d' divides a and $dd' = a$.

13. If $n = ab$, prove that either $|a|$ or $|b|$ is less than or equal to $\sqrt{|n|}$.

14. Let a_0, a_1, a_2 be integers. If x is an integer and $a \mid x$, prove that $a \mid a_1 x + a_2 x^2$. Show that if x is a nonzero integer such that $a_2 x^2 + a_1 x + a_0 = 0$, then $x \mid a_0$.
    What integers would you try in looking for a solution in integers of the equation $x^2 + 4x + 3 = 0$? Generalize to the case of a polynomial equation of the nth degree with integral coefficients

$$a_n x^n + a_{n-1}x^{n-1} + \cdots + a_1 x + a_0 = 0.$$

15. Because ± 1 are the divisors of the multiplicative identity 1, they are called *units* in the ring of integers. Show that the set of divisors of an integer a

is the union of two sets D_a^+ and D_a^-, where D_a^+ consists of the positive divisors of a and each element of D_a^- is obtained by multiplying some element of D_a^+ by the unit -1.

1.2   A SIEVE FOR DIVISORS

Ever since counting and arithmetic began, people have been predicting properties of the positive integers by examining tables. It is still a good idea. The following simple construction gives a basis for some information about divisors. It is a modification of an ancient technique called the Sieve of Eratosthenes (276–184 B.C.).

Write the positive integers from 1 as far as you wish (the example here stops at 60). Beside each integer place a 1. The reason for doing so is that 1 divides every integer. Beginning with 2 and by every second integer thereafter, place a 2. These integers are the multiples of 2. Identify the multiples of 3 by placing a 3 beside every third integer. Continue in this way until the end of your table. Beside each integer appears the set of positive divisors of that integer. Now count the positive divisors and obtain the third column, which is the number of positive divisors. Add the positive divisors and obtain their sum, which is recorded in the fourth column.

When we associate with a positive integer n the number of positive divisors of n, we are defining a function whose domain is the set of positive integers. Such functions are called " number-theoretic " or " arithmetic " functions in this context, but they are really just " sequences " as studied in calculus. We are here concerned with their total behavior and their relationship to each other, as well as with their limiting behavior.

$$\tau, \quad \sigma$$

DEFINITION 1.2.1. *The Functions τ and σ.* A function whose domain is the positive integers is called a number-theoretic function. The function τ is defined by $\tau(n) =$ the number of positive divisors of n. The function σ is defined by $\sigma(n) =$ the sum of the positive divisors of n.

A TABLE OF DIVISORS

n	Positive divisors of n	$\tau(n)$	$\sigma(n)$	n	Positive divisors of n	$\tau(n)$	$\sigma(n)$
1	1	1	1	31	1, 31	2	32
2	1, 2	2	3	32	1, 2, 4, 8, 16, 32	6	63
3	1, 3	2	4	33	1, 3, 11, 33	4	48
4	1, 2, 4	3	7	34	1, 2, 17, 34	4	54
5	1, 5	2	6	35	1, 5, 7, 35	4	48
6	1, 2, 3, 6	4	12	36	1, 2, 3, 4, 6, 9, 12, 18, 36	9	91
7	1, 7	2	8	37	1, 37	2	38
8	1, 2, 4, 8	4	15	38	1, 2, 19, 38	4	60
9	1, 3, 9	3	13	39	1, 3, 13, 39	4	56
10	1, 2, 5, 10	4	18	40	1, 2, 4, 5, 8, 10, 20, 40	8	90
11	1, 11	2	12	41	1, 41	2	42
12	1, 2, 3, 4, 6, 12	6	28	42	1, 2, 3, 6, 7, 14, 21, 42	8	96
13	1, 13	2	14	43	1, 43	2	44
14	1, 2, 7, 14	4	24	44	1, 2, 4, 11, 22, 44	6	84
15	1, 3, 5, 15	4	24	45	1, 3, 5, 9, 15, 45	6	78
16	1, 2, 4, 8, 16	5	31	46	1, 2, 23, 46	4	72
17	1, 17	2	18	47	1, 47	2	48
18	1, 2, 3, 6, 9, 18	6	39	48	1, 2, 3, 4, 6, 8, 12, 16, 24, 48	10	124
19	1, 19	2	20	49	1, 7, 49	3	57
20	1, 2, 4, 5, 10, 20	6	42	50	1, 2, 5, 10, 25, 50	6	93
21	1, 3, 7, 21	4	32	51	1, 3, 17, 51	4	72
22	1, 2, 11, 22	4	36	52	1, 2, 4, 13, 26, 52	6	98
23	1, 23	2	24	53	1, 53	2	54
24	1, 2, 3, 4, 6, 8, 12, 24	8	60	54	1, 2, 3, 6, 9, 18, 27, 54	8	120
25	1, 5, 25	3	31	55	1, 5, 11, 55	4	72
26	1, 2, 13, 26	4	42	56	1, 2, 4, 7, 8, 14, 28, 56	8	120
27	1, 3, 9, 27	4	40	57	1, 3, 19, 57	4	80
28	1, 2, 4, 7, 14, 28	6	56	58	1, 2, 29, 58	4	90
29	1, 29	2	30	59	1, 59	2	60
30	1, 2, 3, 5, 6, 10, 15, 30	8	72	60	1, 2, 3, 4, 5, 6, 10, 12, 15, 20, 30, 60	12	168

EXAMPLE 1.2.1

$\tau(11) = 2, \quad \sigma(11) = 12.$
$\tau(24) = 8, \quad \sigma(24) = 60.$
$\tau(1) = 1, \quad \sigma(1) = 1.$

The first impression we get from the table is that the function τ displays a quite erratic behavior. For instance, $\tau(42) = 8$, $\tau(41) = 2$, $\tau(43) = 2$. If the table were extended, would there be a point beyond which $\tau(n)$ never returns to the value 2? This is the kind of question that cannot be answered decisively by extending the table. However far we extend it, we are no nearer to the end of the integers, for there is no " end of the integers." We can, however, be led from a question to a strong feeling for what the answer must be. This makes our question become a conjecture. Then, with ingenuity and a little luck, we may be able to prove or disprove the conjecture and a theorem is born. This particular question, and the way it was answered by Euclid, will be discussed in the next section. The question is relatively easy to answer, but other questions just as simple to state have proved extremely difficult. A conjecture as yet unproved is that there is an infinite set of pairs of consecutive odd numbers for which $\tau(n) = 2$. For example, 41 and 43, or 59 and 61, are pairs with this property. This question and similar ones will be discussed in Unit 4.

The positive integers for which $\tau(n) = 2$ are of particular importance in the multiplicative structure of the integers and are given the name primes. Since an integer n is always divisible by 1 and n, the fact that $\tau(n) = 2$ implies that $n > 1$, and its only positive divisors are 1 and n.

DEFINITION 1.2.2. *Prime and Composite*. A positive integer n is called *prime* if $n > 1$ and its only positive divisors are 1 and n. In this case, $\tau(n) = 2$. A positive integer n is called *composite* if $\tau(n) > 2$.

The table on page 8 is a relatively small one. Some investigations based on it are suggested in the exercises. A table showing the distribution of the odd primes in the integers less than 2000 is found in Appendix B. Many monumental tables of primes and other related information have been prepared. The earlier ones were laboriously constructed by hand.

The use of the electronic computer makes the tables more easily prepared, hence more extensive and also more accurate. After an individual has examined enough cases to form a conjecture, the computer can be enlisted to carry on the search and either verify the conjecture in more and more cases or find a counterexample. As we proceed, we will note situations in which the computer might help us in looking for information. But before you decide to undertake the whole exploration by computer, it is wise to remember that computer time is in demand for research in many areas, that it is extremely expensive, and that, like the jet plane, the computer may get you there in a hurry but you may miss a lot on the way.

EXERCISES 1.2

1. List the primes less than 60.

2. A *prime pair* is a pair of numbers n and $n + 2$, both of which are prime. List the prime pairs less than 60.

3. Let p be a prime. Show that p, $p + 2$, $p + 4$ cannot all be prime unless $p = 3$. If p, $p + 2$, $p + 6$ are all primes, they are called a prime triple. List the prime triples less than 60.

4. If possible, find a prime quadruplet—that is, a prime p such that p, $p + 2$, $p + 6$, $p + 8$ are all primes.

5. What is the longest sequence of consecutive composite numbers in the table?

6. Find examples of primes p such that p and $p + 30$ are both prime.

7. For what integers in the table is $\tau(n)$ odd? Identify a property common to these integers.

8. It has been conjectured that there are infinitely many n such that $\tau(n) = \tau(n + 1)$. Find examples of this in the table.

9. Find instances where $\tau(n) = \tau(n + 1) = \tau(n + 2)$.

10. List the values of the following: $\tau(3)$, $\tau(16)$, $\tau(48)$; $\tau(12)$, $\tau(4)$, $\tau(48)$; $\tau(6)$, $\tau(8)$, $\tau(48)$; $\tau(3)$, $\tau(11)$, $\tau(33)$; $\tau(7)$, $\tau(5)$, $\tau(35)$; $\tau(6)$, $\tau(5)$, $\tau(30)$; $\tau(3)$, $\tau(10)$, $\tau(30)$.

11. Find the largest positive integer that divides both of the integers in each of the following pairs: 3, 16; 4, 12; 6, 8; 3, 11; 7, 5; 6, 5; 3, 10.

12. What conjecture can you make based on the information collected in problems 10 and 11?

13. Write the definition of "prime" and "composite," using the function σ in place of the function τ.

14. A positive integer is called *perfect* if $\sigma(n) = 2n$. Verify that 6 and 28 are both perfect.

15. Calculate $\sigma(n)$ for each of the integers in problem 10. Make a conjecture for σ similar to your conjecture for τ.

16. The number–theoretic function π is defined by: $\pi(n) = $ the number of primes less than or equal to n. Calculate $\pi(n)$ for $n = 60$.

17. Make calculations corresponding to those in problem 10 for $\pi(n)$. Is the behavior of π different from that of σ and τ?

18. Sketch graphs of τ, σ, π for $1 \leq n \leq 60$. What property does π possess that is different from either τ or σ?

19. Compare the behavior of π and σ as n becomes infinite with the apparent behavior of τ.

20. Check the set of divisors of the composite numbers. Are these divisors prime or composite? Make a conjecture about the divisors of composite numbers.

21. Prove that if n is composite there exist integers b and c such that $1 < b < n$, $1 < c < n$, and $bc = n$. *Hint*: Use Example 1.1.7.

1.3   WELL-ORDERING

We have looked at a table and made some conjectures about the positive integers. In trying to prove some of these conjectures, we see that we need more tools to work with. Some tools will be found in the axioms defining the integers. There are properties of the integers, such as the commutative property of addition, that are common to many mathematical systems. Whatever is proved using only the commutative property would apply to all commutative systems. Similar statements can be made concerning other ring properties of the integers. But there is a more distinctive property of the positive integers that is not possessed by either the positive rational numbers or the positive real numbers. It is called *the well-ordering principle*, but it is, as we shall see, another form of a property that you have already met in high school algebra under the name "the principle of mathematical induction."

> *The well-ordering principle.* Every nonempty set of positive elements contains a least element.

This property tells us that any collection of positive integers that we care to describe, as long as it has elements in it, must have a smallest element. It is not a property of the positive rational numbers, since we can find a set of positive rational numbers that has no smallest element. Consider the rational numbers bigger than 0 and smaller than 1. If we claim that some fraction, say a/b, is actually the smallest fraction in this set, we can always be proved wrong, since $a/2b$ is also a positive fraction and is smaller than a/b.

A little reflection about the positive integers makes the well-ordering principle seem reasonable. But what use is it? Let us see how it can be used in a proof.

THEOREM 1.3.1. *Prime Factorization.* Every integer greater than 1 is either a prime or a product of primes (some of which may be repeated).

Proof: Let S be the set of all integers greater than 1 that are not prime and that cannot be expressed as a product of primes. Assume that S is not empty. The elements of S are integers greater than 1 and hence positive. By the well-ordering principle, S must have a least element. Call this element c. Then $c > 1$, c is not prime, and c cannot be written as a product of primes.

By definition, an integer greater than 1 that is not prime must be composite. That is, $c = bd$, where $b \neq 1$ and $d \neq 1$. But $b \neq 1$ implies $d < c$ and $d \neq 1$ implies $b < c$. (See Example 1.1.7; also Exercise 1.2, problem 21.) Neither b nor d can be an element of S because c is the least element of S. So both b and d can be written as products of primes, which means that their product c can also be written as a product of primes. In this way, we have arrived at a contradiction. Our assumption that S is nonempty must be false. That is, every integer greater than 1 is either a prime or a product of primes. ●

COROLLARY 1.3.1. Every integer greater than 1 has a prime divisor.

EXAMPLE 1.3.1
$$24 = 2 \cdot 2 \cdot 2 \cdot 3 = 2^3 \cdot 3$$
$$35 = 5 \cdot 7$$
$$19 = 1 \cdot 19$$

Note that nothing is said in the theorem about whether or not the factorization into primes can be done in more than one way. Certainly the order of the factors might be different, but if some order is designated for the primes—for example, order of magnitude—is the factorization unique? The answer is yes in the integers, although there are systems that have many of the properties of the integers in which the factorization into primes is not unique. A discussion of this issue must be postponed until we have investigated the relation between the multiplicative and the additive structure of the integers. The theorem leading to the proof of unique factorization is given in Section 1.8.

The preceding theorem illustrates the use of the well-ordering principle and also the technique of proof by contradiction. Since a true hypothesis with correct logic must lead to a true conclusion, it is possible to test the truth of a hypothesis by seeing whether a false conclusion can be derived from it. If so, the hypothesis must have been false. Note that if the conclusion derived is true, nothing is proved about the truth of the hypothesis. It is quite possible, with correct logic, to derive a true conclusion from a false hypothesis. For example, assume that $1 = 2$. Equality is symmetric, so $2 = 1$. But adding the equalities $1 = 2$ and $2 = 1$ gives $3 = 3$, which is true. Therefore $1 = 2$?

The method of contradiction can be used to answer the question raised in Section 1.2 about the set of primes. Is the set of primes finite or infinite? Certainly it is large enough to generate all the positive integers greater than 1 according to the prime factorization theorem, but that still does not prove that the set is infinite. In order to get the contradiction he wanted, Euclid devised a clever construction by which, from a finite set of primes, he could create an integer that was not divisible by any of them. Let us look first at an example.

EXAMPLE 1.3.2. From the table in Section 1.2 we see that 2, 3, 5, 7, 11, 13, 17, and 19 are all prime. Consider the following:

$2 + 1 = 3$, a prime.

$2 \cdot 3 + 1 = 7$, a prime larger than 2 or 3.

$2 \cdot 3 \cdot 5 + 1 = 31$, a prime.

$2 \cdot 3 \cdot 5 \cdot 7 \cdot 11 \cdot 13 + 1 = 30{,}031 = 59 \cdot 509$, a composite number, but its smallest prime divisor is larger than 13. Thus integers of the form $p_1 \cdot p_2 \cdot p_3 \ldots p_k + 1$ are found to be either prime and greater than p_k or divisible by a prime greater than p_k. This result suggests a method of proving that the set of primes is infinite.

THEOREM 1.3.2. *The Set of Primes.* The set of primes is infinite.

Proof: Suppose that the set of primes is finite. Then we can write them all down: $p_1, p_2, p_3, \ldots, p_k$. Now consider the integer

$$b = p_1 \cdot p_2 \cdot p_3 \ldots p_k + 1.$$

The integer b cannot be a prime, since it is certainly not one of the p_i. It is a positive integer greater than 1 and thus must have a prime divisor. But $1 = b - p_1 \cdot p_2 \cdot p_3 \ldots p_k$, so that if p_i divides b, then since p_i divides $p_1 \cdot p_2 \cdot p_3 \ldots p_k$ it must divide 1, by property (2) of "divides." This is a contradiction; thus the assumption is false and the set of primes is infinite. ⬣

There are many other methods of proving this theorem. Some of them are suggested in the problems and in Unit 4.

In the next example we see how well-ordering and the method of contradiction can be used to establish a familiar formula.

EXAMPLE 1.3.3. For every positive integer n, prove that

$$1 + 2 + \cdots + n = \frac{n(n + 1)}{2}.$$

Proof: Let S be the set of positive integers for which the formula is false. Assume that S is not empty. Then there is a least integer, call it c, such that

$$1 + 2 + \cdots + c \neq \frac{c(c + 1)}{2}.$$

The integer c cannot be 1, since if $n = 1$ the formula becomes $1 = 1 \cdot 2/2$, which is true. Since $c > 1$, $c - 1$ is a positive integer. Because $c - 1 < c$, the formula must be true for $n = c - 1$. That is,

$$1 + 2 + \cdots + (c - 1) = \frac{(c - 1)c}{2}.$$

Add c to each side of this equality. We get

$$1 + 2 + \cdots + (c - 1) + c = \frac{(c - 1)c}{2} + c = \frac{c(c + 1)}{2}.$$

This result contradicts the assumption that the formula is false for $n = c$.

Thus the set of integers for which the formula is false is empty and the formula is established for all n. ●

The formula of Example 1.3.3 is a useful one. You have probably proved it already using the principle of mathematical induction. This form of the well-ordering principle is sometimes easier to use when a formula is involved, as in Example 1.3.3. The fact that well-ordering implies the principle of mathematical induction is proved in the following theorem.

THEOREM 1.3.3. *The Principle of Mathematical Induction.* Let $P(n)$ be a property defined on the positive integers. Let T be the set of integers for which P is true. If 1 belongs to T, and if $k + 1$ belongs to T whenever k belongs to T, then the set T includes the positive integers.

Proof: Let S be the set of positive integers that do not belong to T. We need to show that the hypothesis of the theorem leads to the conclusion that there are no positive integers in S. To do this we assume that S is not empty and use the well-ordering principle to arrive at a contradiction. If S is not empty, it has a least element, c. Since 1 belongs to T, $c > 1$, so that $c - 1$ is a positive integer. But c is the least element of S so that $c - 1$ must belong to T. According to the hypothesis of the theorem, if $c - 1$ belongs to T then $c - 1 + 1$ must also belong to T. But $c - 1 + 1 = c$ and does not belong to T. Thus we have arrived at a contradiction: the assumption that S is not empty must be false, and T includes all the positive integers. ●

COROLLARY 1.3.3. Let $P(n)$ be a property defined on the positive integers. Let T be the set of integers for which P is true. If 1 belongs to T, and if k belongs to T whenever all the positive integers less than k belong to T, then the set T includes all the positive integers.

Proof: Exercises 1.3, problem 3. ●

EXERCISES 1.3

Checks

1. Find a set of negative integers that has no smallest element.
2. Prove, directly from the well-ordering principle, that every integer greater than 1 has a prime divisor.

3. Prove Corollary 1.3.3.

4. Prove that 3 divides $2^{2n} - 1$, for every positive integer n.

5. A set of integers is *bounded above* if every element of the set is less than or equal to some integer K. Prove that every nonempty set of integers bounded above has a greatest element.

6. Prove that, if a and b are nonzero integers, the set of integers that are divisors of a and also divisors of b has a greatest element.

7. If $x \neq 1$, prove that for every positive integer n,

$$\frac{1 - x^n}{1 - x} = 1 + x + \cdots + x^{n-1}.$$

8. If $x \neq -1$, prove that for every positive integer n,

$$\frac{1 + x^{2n+1}}{1 + x} = 1 - x + x^2 - \cdots + x^{2n}.$$

9. Prove that the set of primes is infinite, using the integer $b = n! + 1$ to obtain a contradiction.

Challenges

10. Prove that for any positive integer k there exist sequences of k consecutive composite integers.

11. Prove that $1 + \frac{1}{2} + \frac{1}{4} + \cdots + 1/2^n < 2$ for every positive integer n.

12. Show that $2^{5n+1} + 5^{n+2}$ is divisible by 27 for $n = 0, 1, 2, \ldots$.

13. The Archimedean axiom says: If a and b are positive integers, there exists an integer N such that $aN \geq b$. Show that the well-ordering principle implies the Archimedean axiom. *Hint*: Consider the set of positive integers of the form $b - na$. Show that the assumption that $b - na > 0$ for every n leads to a contradiction.

1.4   THE DIVISION ALGORITHM

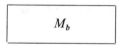

We have seen that, given an integer $b \neq 0$, there is an infinite set of integers that are multiples of b. Let us call this set M_b. Then

$$M_b = \{kb : k = 0, \pm 1, \pm 2, \ldots\}.$$

In this section we explore the relation of the integers to the set M_b.

EXAMPLE 1.4.1. Let $b = 3$. The integer 14 does not belong to M_3 but $14 = 12 + 2$—that is, 14 is 2 more than a multiple of 3. The integer $-8 = -9 + 1$—that is, -8 is 1 more than a multiple of 3. The integer -6 is a multiple of 3. What this amounts to is ordinary division. Any integer can be divided by 3 with a remainder of 0, 1, or 2. Thus $2893 \div 3$ gives a quotient of 964 and a remainder of 1, or $2893 = 964 \cdot 3 + 1$.

This arithmetic corresponds to a simple geometric picture. Think of the integers as points on the number line. Every third point is a multiple of 3. Any integer is itself a multiple of 3 or lies between two consecutive multiples of 3 and can differ from the one immediately preceding it by either 1 or 2.

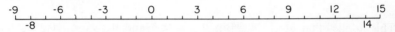

This relationship between M_b and the set of integers is one we "know" from our experience and geometric intuition. It plays a most important role in the study of the integers. We state it formally and prove it using the well-ordering principle.

THEOREM 1.4.1. *The Division Algorithm.* Let a and b be integers, $b > 0$. There exist unique integers q and r such that $a = bq + r$ and $0 \le r < b$.

Proof: We are looking for an integer r of the form $a - bq$, which has the property that $0 \le r < b$. Let S be the set of integers of this form:

$$S = \{a - bk : k = 0, \pm 1, \pm 2, \ldots\}.$$

Let S^+ be the set of positive integers in S.

First we must see whether S^+ is empty. If $a > 0$, then a belongs to S^+. If $a \le 0$, then $a - (a - 1)b$ belongs to S^+. Thus there is always at least one integer in S^+, and S^+ is nonempty. By the well-ordering principle, S^+ has a least element c. For some integer d, $c = a - bd$.

If $c > b$, then $c = b + c'$, where $c' > 0$. Since $c' = a - (d + 1)b$, c' belongs to S^+. But this result contradicts the minimum property of c, since $c' < c$. Therefore c cannot be greater than b.

If $c = b$, $a = db + b = (d + 1)b$. In this case, a is a multiple of b. We can say that $a = qb + r$, where $q = d + 1$ and $r = 0$.

If $c < b$, let $q = d$ and $c = r$. We have $a = bq + r$, where $0 < r < b$.

The required form for a is established. It remains to show that it is unique. Assume that

$$a = bq_1 + r_1, \qquad 0 \le r_1 < b;$$

also,

$$a = bq_2 + r_2, \qquad 0 \le r_2 < b.$$

Subtract the two expressions for a:

$$0 = (q_1 - q_2)b + (r_1 - r_2).$$

This implies that $b \mid r_1 - r_2$. But $0 \le r_1 < b$ and $0 \le r_2 < b$ tell us that $-b < r_1 - r_2 < b$, so that b cannot divide $r_1 - r_2$ unless $r_1 - r_2 = 0$. Thus $r_1 = r_2$. From this result it follows that $q_1 = q_2$. ⬢

It is important to note that the uniqueness of q and r is a consequence of the restriction imposed on the acceptable values of r. Some alternate forms of the division algorithm with slightly different restrictions are suggested in the problems.

The importance of the division algorithm will be realized as we proceed, for we use it again and again. The example below indicates the way in which it separates the infinite set of integers into a finite number of categories, so that statements about all the integers can be proved by considering only a finite number of cases.

EXAMPLE 1.4.2. Prove that the square of an integer is either a multiple of 4 or one more than a multiple of 4.

Proof: Let a be any integer. Use the division algorithm with $b = 4$. There are four possible cases: $a = 4q$, $a = 4q + 1$, $a = 4q + 2$, and $a = 4q + 3$. Corresponding to these cases are four possible values of a^2:

$$(4q)^2 = 16q^2 = (4q^2)4,$$
$$(4q + 1)^2 = 16q^2 + 8q + 1 = (4q^2 + 2q)4 + 1,$$
$$(4q + 2)^2 = 16q^2 + 16q + 4 = (4q^2 + 4q + 1)4,$$
$$(4q + 3)^2 = 16q^2 + 24q + 9 = (4q^2 + 6q + 2)4 + 1.$$

In the first and third cases, a^2 is a multiple of 4. In the second and fourth cases, a^2 is one more than a multiple of 4. The statement is proved. ⬢

So far we have been concerned with r. What can we say about q? It is useful to have a means of representing the integer q.

EXAMPLE 1.4.3

$$189 = 7 \cdot 25 + 14.$$
$$-245 = (-7)40 + 35.$$
$$84 = 4 \cdot 21.$$

In each case, q is the largest integer that is less than or equal to the rational number a/b. It is convenient to have a notation for this integer.

$$[x]$$

DEFINITION 1.4.1. *The Greatest Integer Function.* Let x be any real number. Then $[x]$ denotes the greatest integer less than or equal to x.

EXAMPLE 1.4.4

$[2.15] = 2, \quad [-1.876] = -2, \quad [\sqrt{15}] = 3, \quad [7] = 7,$

$\left[\dfrac{189}{25}\right] = 7, \quad \left[\dfrac{-245}{40}\right] = -7, \quad \left[\dfrac{84}{21}\right] = 4, \quad [0] = 0.$

COROLLARY 1.4.1. Let a and b be integers, $b > 0$. Then $a = bq + r$, where $0 \le r < b$, and $q = [a/b]$.

Proof: The division algorithm guarantees the existence of q and r. Since $a = bq + r$,

$$\frac{a}{b} = q + \frac{r}{b}, \qquad 0 \le \frac{r}{b} < 1.$$

If $r = 0$, a/b is an integer and $a/b = [a/b]$. If $r > 0$,

$$q < \frac{a}{b} < q + 1$$

and $q = [a/b]$.   ●

If $a > 0$, the integer $[a/b]$ can be thought of as counting the number of positive multiples of b that are less than or equal to a. For example, since $7 \cdot 25 < 189 < 8 \cdot 25$, there are seven multiples of 25 which are less than 189.

EXAMPLE 1.4.5. The number of multiples of 3 that are less than or equal to 50 is $[50/3]$. The number of multiples of 2 that are less than or equal to 50 is $[50/2]$. The number of multiples of 6 that are less than or equal to 50 is $[50/6]$.

The greatest integer function is useful in many counting situations. Some of the properties of this function are suggested in Exercises 1.4, problem 11. In proving statements regarding $[x]$, it is frequently useful to write the real number x in the form $[x] + \theta$, where $0 \le \theta < 1$.

EXAMPLE 1.4.6. Prove that $0 \le [x] - 2[x/2] \le 1$, for every real number x. If x is an integer, this is simply a restatement of Corollary 1.4.1, with $x = a$ and $b = 2$. For any real number x, we can write

$$x = [x] + \theta, \qquad 0 \le \theta < 1.$$

Since $[x]$ is an integer, it must be either of the form $2k$ or of the form $2k + 1$, according to the division algorithm. If $[x] = 2k$,

$$x = 2k + \theta \quad \text{and} \quad \frac{x}{2} = k + \frac{\theta}{2},$$

so that $[x/2] = k$. In this case,

$$[x] - 2\left[\frac{x}{2}\right] = 2k - 2k = 0.$$

If $[x] = 2k + 1$,

$$x = 2k + 1 + \theta \quad \text{and} \quad \frac{x}{2} = k + \frac{1 + \theta}{2}.$$

Since $0 \leq \theta < 1$, we have $0 < (1 + \theta)/2 < 1$, so that $[x/2]$ is again equal to k. Now, however,

$$[x] - 2\left[\frac{x}{2}\right] = 2k + 1 - 2k = 1.$$

Can you write a similar inequality for $[x] - 3[x/3]$ or, in general, for $[x] - k[x/k]$, where k is any positive integer?

EXERCISES 1.4

Checks

1. Find q and r if $a = -35$ and $b = 3$; if $a = 178$ and $b = 15$.

2. If p is a prime greater than 4, prove that p has the form $4k + r$, where $r = 1$ or 3.

3. If $a = 4q_1 + 3$ and $b = 4q_2 + 3$, prove that $ab = 4q_3 + 1$, where q_1, q_2, and q_3 represent integers.

4. Let $a = 6q + r$, where $0 \leq r < 6$ and $q \geq 1$. For what values of r might a be a prime?

5. Show that $0 \leq r_1 < b$ and $0 \leq r_2 < b$ imply $-b < r_1 - r_2 < b$.

6. Calculate $[-3.777]$, $[3.918]$, $[4]$, $[-7/2]$.

7. Sketch the graph of the function $[x]$.

Challenges

8. (a) Prove that of any three consecutive integers, exactly one is a multiple of 3.
   (b) Prove that of any three consecutive even integers, exactly one is a multiple of 3.
   (c) Prove that of any three consecutive odd integers, exactly one is a multiple of 3.

9. Prove that 3, 7, 11 is the only set of three consecutive primes of the form c, $c + 4$, $c + 8$.

10. Extend the division algorithm to include negative b. Does the formula $q = [a/b]$ hold in this case? State a form of the division algorithm in which $-b/2 < r \leq b/2$. A form in which $1 \leq r \leq b$.

11. (a) If n is an integer and x a real number, prove that $[x+n]=[x]+n$.
     (b) For any real numbers x and y, prove that $[x+y]\geq[x]+[y]$. Under what conditions does $[2x]=2[x]$?
     (c) If a and b are positive integers, prove that

$$\left[\frac{1}{b}\left[\frac{n}{a}\right]\right]=\left[\frac{n}{ab}\right].$$

     Is this relation true if a or b is negative?

12. If a product of primes is of the form $4q+3$, prove that at least one of the primes must have this form. *Hint*: Use problem 2.

1.5   LEAST COMMON MULTIPLE AND THE SET M_a+M_b

In the preceding section we considered the set of multiples of b and its relation to the set of integers. Let us now experiment with combining sets of multiples in various ways.

What are the ways of combining sets? Let X and Y be two sets. By the *union* of the sets, we mean the set of elements belonging either to set X or to set Y or to both. This is written $X\cup Y$. By the *intersection* of the sets X and Y, we mean the set of elements belonging to both X and Y. This is written $X\cap Y$. The diagram illustrates these operations.

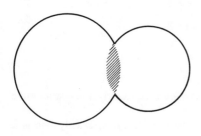

$X=$ the set of points inside the left-hand circle.

$Y=$ the set of points inside the right-hand circle.

$X\cap Y=$ the set of points in the shaded area.

$X\cup Y=$ the set of points in the area enclosed by the solid black line.

The operations of union and intersection can be applied to sets whose elements are perfectly general: points, functions, apples, people. If the elements of the sets belong to some mathematical structure of a specialized nature, such as the ring of integers, in which it is possible to add elements of the set and to multiply elements of the sets, we can define products and sums of sets.

The *set product*, $X \cdot Y$, means the set consisting of all possible products of the form $x \cdot y$, where x is an element of X and y is an element of Y. The *set sum*, $X + Y$, is the set consisting of all possible sums of the form $x + y$, where x is an element of X and y is an element of Y. Note that the sign $\cdot$ or $+$ written between two sets refers to multiplication or addition of sets, and the sign $\cdot$ or $+$ written between two elements refers to multiplication or addition of the elements. Unless the elements of the sets belong to a mathematical system where multiplication and addition are possible, the definitions do not make sense. We are interested in sets of integers, so there is no problem.

EXAMPLE 1.5.1

$$M_8 = \{0, \pm 8, \pm 16, \pm 24, \pm 32, \pm 40, \ldots\}$$
$$M_6 = \{0, \pm 6, \pm 12, \pm 18, \pm 24, \pm 30, \pm 36, \ldots\}$$
$$M_8 \cup M_6 = \{0, \pm 6, \pm 8, \pm 12, \pm 16, \pm 18, \pm 24, \ldots\}$$
$$M_8 \cap M_6 = \{0, \pm 24, \pm 48, \pm 72, \ldots\}$$
$$M_8 \cdot M_6 = \{0, \pm 48, \pm 96, \pm 144, \pm 192, \ldots\}$$
$$M_8 + M_6 = \{0, \pm 2, \pm 4, \pm 6, \pm 8, \pm 10, \ldots\}$$

Note that

$$M_8 \cap M_6 = M_{24}, \quad M_8 \cdot M_6 = M_{48}, \quad M_8 + M_6 = M_2.$$

In the preceding example we see that the operations of inter-section, set product, or set sum, when applied to two sets of multiples, give a set of multiples. Will this always happen? Why? Is it surprising? Is it important? This section begins to answer these questions.

First, look at $M_a \cdot M_b$. The set $M_a = \{ka: k \text{ is an integer}\}$ and the set $M_b = \{k'b: k' \text{ is an integer}\}$; hence $M_a \cdot M_b = \{kk'ab: k, k' \text{ are integers}\}$. Thus each element of $M_a \cdot M_b$ is an element of M_{ab}; that is, $M_a \cdot M_b \subset M_{ab}$. Conversely, $M_{ab} = \{k''ab: k'' \text{ an integer}\}$. But $(k''a)$ is an element of M_a and b is an element of M_b, so each element of M_{ab} is an element of $M_a \cdot M_b$. That is, $M_{ab} \subset M_a \cdot M_b$. Since each is included in the other, $M_{ab} = M_a \cdot M_b$. This result uses nothing more than the definition of divides and the definition of set product. It is not too surprising or especially deep.

The situation regarding $M_a \cap M_b$ and $M_a + M_b$ is more interesting. We will pay special attention to the set $M_a + M_b$. What is its structure? Since $M_a = \{ax: x \text{ is an integer}\}$ and $M_b = \{by: y \text{ is an integer}\}$, then $M_a + M_b = \{ax + by: x \text{ and } y \text{ are integers}\}$. Sums of the form $ax + by$ are called *linear combinations* of a and b. The word linear refers to the fact that a and b occur to the first degree. The set $M_a + M_b$ is the set of linear combinations of a and b with integer coefficients.

THEOREM 1.5.1. *The Set* $M_a + M_b$. Let a and b be integers, not both zero. Let d be the least positive integer in the set $M_a + M_b$. Then $M_a + M_b = M_d$.

Proof: Since $|a|$ and $|b|$ belong to $M_a + M_b$, and a and b are not both zero, the set $M_a + M_b$ has some positive elements. By the well-ordering principle, the set of positive integers in $M_a + M_b$ has a least element. Thus d is well defined. Since d is an element of $M_a + M_b$, there exist integers x_0 and y_0 such that $d = ax_0 + by_0$.

Let kd be any multiple of d. Then

$$kd = k(ax_0 + by_0) = (ak)x_0 + (bk)y_0.$$

This result shows that kd is an element of $M_a + M_b$ and $M_d \subset M_a + M_b$.

Conversely, any element of $M_a + M_b$ has the form $ax + by$ for some integers x and y. By the division algorithm,

$$ax + by = qd + r, \qquad 0 \le r < d.$$

Set $d = ax_0 + by_0$ and solve for r;

$$r = ax + by - qd = a(x - qx_0) + b(y - qy_0).$$

This shows that r is an element of $M_a + M_b$. But since $r < d$ and d is the least positive element in $M_a + M_b$, then r must be zero—that is, $d \mid ax + by$. Thus $M_a + M_b \subset M_d$. Hence $M_a + M_b = M_d$. ◆

COROLLARY 1.5.1a. The smallest positive element of $M_a + M_b$ is a divisor of a and a divisor of b.

COROLLARY 1.5.1b. If there exist integers x_0 and y_0 such that $ax_0 + by_0 = 1$, then $M_a + M_b$ is the set of all integers.

COROLLARY 1.5.1c. If there exist integers x_0 and y_0 such that $ax_0 + by_0 = 1$, then the only positive integer that divides both a and b is 1.

EXAMPLE 1.5.2

$M_{24} + M_{18} = \{24x + 18y: x \text{ and } y \text{ are integers}\} = M_6$.
$M_5 + M_2 = $ the integers, since $1 \cdot 5 + (-2)2 = 1$.

THEOREM 1.5.2. *The Set $M_a \cap M_b$*. Let a and b be nonzero integers. Let m be the smallest positive integer in the set $M_a \cap M_b$. Then $M_a \cap M_b = M_m$.

Proof: The proof of this theorem is very similar to the proof of Theorem 1.5.1 and is left as an exercise. ●

$$\boxed{[a, b]}$$

DEFINITION 1.5.1. *Least Common Multiple*. Let a and b be integers, neither of which is zero. The least positive integer that is both a multiple of a and a multiple of b is called the *least common multiple* of a and b and is designated $[a, b]$. If either a or b is zero, the least common multiple of a and b is zero.

EXAMPLE 1.5.3

$[8, 6] = 24$.   (See Example 1.5.1 for the set $M_8 \cap M_6$.)
$[24, 18] = 72, [5, 2] = 10, [6, 0] = 0, [0, 0] = 0$.

THEOREM 1.5.3. *Least Common Multiple*. Let a and b be nonzero integers. Then $[a, b] = m$ if and only if the following conditions hold: $m > 0$; $a \mid m$, $b \mid m$; and m divides every integer M that is a common multiple of a and b.

Proof: Let $[a, b] = m$. By definition, $a \mid m$ and $b \mid m$, and m is the least positive element of $M_a \cap M_b$. Since every element of the set $M_a \cap M_b$ is a multiple of m (Theorem 1.5.2), m divides every integer M that is a common multiple of a and b.

Conversely, suppose that $a \mid m$ and $b \mid m$, $m > 0$. Then m is a positive element in $M_a \cap M_b$. Since m divides all other elements in $M_a \cap M_b$, $m \leq |M|$ for every nonzero element in $M_a \cap M_b$, by property (3) of "divides." This means that m is the least positive element in $M_a \cap M_b$; that is, $m = [a, b]$. ●

EXERCISES 1.5

Checks

1. Write several elements of $M_4 \cap M_{10}$. What is [4, 10]? Is the set $M_4 \cap M_{10}$ closed under multiplication and subtraction?

2. Write several elements of $M_4 + M_{10}$. What is the least positive element of this set? Write it in the form $4x + 10y$. Is the set $M_4 + M_{10}$ closed under multiplication and subtraction?

3. Show that $M_{12} \subset M_6$ (see Exercises 1.1 problem 4). What is [6, 12]?

4. Prove that $M_4 \subset M_4 + M_{10}$ but that M_4 and $M_4 + \{1\}$ have no elements in common.

5. Let Z represent the ring of integers. Use the division algorithm to prove that

$$Z = M_4 \cup [M_4 + \{1\}] \cup [M_4 + \{2\}] \cup [M_4 + \{3\}].$$

Challenges

6. If $a|b$, prove that $M_b \subset M_a$ and that $[a, b] = |b|$.

7. If $k > 0$, prove that $[ka, kb] = k[a, b]$.

8. Carry out the proof of Theorem 1.5.2.

9. Let a, b, c be positive integers. Define d to be the least positive element in the set $M_a + M_b + M_c$. Prove that the set $M_a + M_b + M_c = M_d$.

10. Let a, b, c be nonzero integers. Define $[a, b, c]$ to be the least positive element in the set $M_a \cap M_b \cap M_c$. Prove that $[a, b, c] = [[a, b], c]$.

11. Prove that the set $M_a \cap M_b$ and the set $M_a + M_b$ are closed under multiplication and subtraction. Prove that $M_a \cup M_b$ is closed under multiplication but not necessarily closed under subtraction. Show that $M_a + M_b$ is the smallest subset of the integers that is closed under subtraction and multiplication and contains $M_a \cup M_b$.

1.6   GREATEST COMMON DIVISOR
AND THE SET $M_a + M_b$

In the study of the structure of the set $M_a + M_b$, we were led to statements regarding the integers that divide both a and b. Such integers are called common divisors of a and b. To study them, we would naturally look at the intersection of the set of divisors of a and the set of divisors of b.

$$\boxed{D_a}$$

Let D_a denote the set of divisors of a. If a is not zero, D_a is a finite set, since the integers in D_a lie between $-a$ and a, according to the third property of "divides" (Theorem 1.1.1). D_0 contains all the integers except 0.

EXAMPLE 1.6.1

$$D_{24} = \{\pm1, \pm2, \pm3, \pm4, \pm6, \pm8, \pm12, \pm24\}.$$
$$D_{18} = \{\pm1, \pm2, \pm3, \pm6, \pm9, \pm18\}.$$
$$D_{45} = \{\pm1, \pm3, \pm5, \pm9, \pm15, \pm45\}.$$
$$D_{24} \cap D_{18} = \{\pm1, \pm2, \pm3, \pm6\}.$$
$$D_{24} \cap D_{45} = \{\pm1, \pm3\}.$$
$$D_{18} \cap D_{45} = \{\pm1, \pm3, \pm9\}.$$
$$D_{18} \cap D_{24} \cap D_{45} = \{\pm1, \pm3\}.$$

Each of the sets listed in Example 1.6.1 is finite and nonempty and therefore has a greatest element. Will this always be the case? Since $\pm1 \,|\, a$ and $\pm1 \,|\, b$, $D_a \cap D_b$ will always be nonempty. Since $D_a \cap D_b \subset D_a$ and $D_a \cap D_b \subset D_b$, this intersection will be finite as long as D_a or D_b is finite; that is, as long as a and b are not both zero. A finite set has a greatest element. This means that the following definition makes sense.

$$\boxed{(a, b)}$$

DEFINITION 1.6.1. *Greatest Common Divisor*. Let a and b be integers, not both zero. The largest integer that is a common divisor of a and b is called the *greatest common divisor* of a and b and is denoted by (a, b).

EXAMPLE 1.6.2

$(24, 18) = 6$. Note that $D_{24} \cap D_{18} = D_6$.
$(24, 45) = 3$. Note that $D_{24} \cap D_{45} = D_3$.
$(18, 45) = 9$. Note that $D_{18} \cap D_{45} = D_9$.

Definition 1.6.1 could be extended to apply to a set of integers not all zero. Thus $(24, 18, 45) = 3$. Note that $D_{24} \cap D_{18} \cap D_{45} = D_3$. In Example 1.5.2 we saw that $M_{24} + M_{18} = M_6$. Is this related to the fact that $(24, 18) = 6$?

Consider $(24/d, 18/d)$, where d is a common divisor of 24 and 18. If $d = 2$, we get $(12, 9) = 3$. If $d = 3$, we get $(8, 6) = 2$. If $d = 6$, we get $(4, 3) = 1$. Consider $(45/d, 18/d)$. For what d will $(45/d, 18/d) = 1$?

This example illustrates some properties of the greatest common divisor. Could one of these properties be set up as an alternate definition of (a, b)? Definition 1.6.1 is a reasonable and descriptive one, but it is not easy to use it in manipulative arguments. It is helpful to be equipped with as many alternate forms of a definition as possible. These "if and only if" statements, which could replace the definition, are called *characterizations*.

THEOREM 1.6.1. *Characterizations of the Greatest Common Divisor.* Let a and b be integers, not both zero.

1. $d = (a, b)$ if and only if d is the least positive element in $M_a + M_b$.

2. $d = (a, b)$ if and only if d is a positive common divisor of a and b and every common divisor of a and b is a divisor of d.

3. $d = (a, b)$ if and only if d is a positive common divisor of a and b and $(a/d, b/d) = 1$.

Proof:

1. Let $ax_0 + by_0$ be the least positive element in $M_a + M_b$ and let d' be a common divisor of a and b. By property (2) of divides, d' divides $ax_0 + by_0$, so $d' \le ax_0 + by_0$. By the properties of the set $M_a + M_b$ (Theorem 1.5.1), we know that $ax_0 + by_0$ divides every element of the set. Since a and b belong to the set, $ax_0 + by_0$ is a divisor of a and b. Thus $ax_0 + by_0$ is a common divisor of a and b and is greater than or equal to every common divisor. By Definition 1.6.1, $ax_0 + by_0$ must be the greatest common divisor, (a, b).

Conversely, let $d = (a, b)$. Then $d \mid ax_0 + by_0$, so that, by property (3) of divides, $d \le ax_0 + by_0$. But $ax_0 + by_0$ is a common divisor of a and b; hence $ax_0 + by_0 \le d$. This implies $d = ax_0 + by_0$.

2. Let d be a positive common divisor of a and b. If every common divisor of a and b divides d, then certainly d is the greatest common divisor. Conversely, if $d = (a, b)$, then, from characterization (1), there exist integers x_0 and y_0 such that $d = ax_0 + by_0$. Property (2) of divides then implies that every divisor of a and b is also a divisor of d.

3. Let $d = (a, b)$. Then, for some x_0 and y_0, $d = ax_0 + by_0$. Thus $(a/d)x_0 + (b/d)y_0 = 1$. Any integer that divides a/d and b/d must divide 1. This implies that $(a/d, b/d) = 1$.

Conversely, suppose that $(a/d, b/d) = 1$ and let $(a, b) = n$, where $n > d$. But $d|n$ by characterization (2), so that $n = kd$, where $k > 1$. Since $kd|a$ and $kd|b$ imply $k|a/d$ and $k|b/d$, this contradicts the hypothesis that $(a/d, b/d) = 1$. ●

COROLLARY 1.6.1. $(a, b) = 1$ if and only if there exist integers x_0 and y_0 such that $ax_0 + by_0 = 1$.

Proof: If 1 is an element of $M_a + M_b$, it must be the least positive element. ●

EXAMPLE 1.6.3. Prove that $(a, b) = (a + bk, b)$ for every integer k.

Proof: Let $d = (a, b)$ and let $f = (a + bk, b)$ for some integer k. We wish to prove $d = f$. Since $d|a$ and $d|b$, $d|a + bk$ and $d|b$. By characterization (2), $d|f$. On the other hand, since $f|a + bk$ and $f|b$, then $f|a + bk - bk$ by property (2) of divides—that is, $f|a$. But $f|a$ and $f|b$ imply $f|d$. Now since $d|f$ and $f|d$, we conclude that $d = f$. ●

EXAMPLE 1.6.4. For any integer x, prove that

$$(2x + 9, x + 4) = 1.$$

Proof: Since

$$(2x + 9) + (-2)(x + 4) = 1,$$

then

$$(2x + 9, x + 4) = 1 \qquad \text{(Corollary 1.6.1).} ●$$

EXAMPLE 1.6.5. Prove that if $d = (a, b)$ and $d = ax_0 + by_0$, then $(x_0, y_0) = 1$.

Proof: Since $d|a$, $a = da'$, and since $d|b$, $b = db'$. This means that

$$d = da'x_0 + db'y_0.$$

That is,

$$a'x_0 + b'y_0 = 1,$$

which implies $(x_0, y_0) = 1$. ●

EXAMPLE 1.6.6. Prove that if $a = bq + r$, then $(a, b) = (b, r)$.

Proof: $(b, r)|b$ and $(b, r)|r$. Thus $(b, r)|a$, since $a = bq + r$. This means that (b, r) is a common divisor of a and b; hence $(b, r)|(a, b)$ by characterization (2). Now $r = a - bq$; therefore $(a, b)|r$. Since $(a, b)|b$ and $(a, b)|r$, we have $(a, b)|(b, r)$. Therefore $(a, b) = (b, r)$. ●

The foregoing examples illustrate how to use the information contained in Theorem 1.6.1. These results are also convenient in their own right. It is clear that the case when the greatest common divisor is 1 is an important one. It is singled out for special terminology.

$$\boxed{(a, b) = 1}$$

DEFINITION 1.6.2. *Relatively Prime.* Let a and b be integers. If the greatest common divisor of a and b is 1, the integers a and b are said to be *relatively prime*.

Corollary 1.6.1 says that a and b are relatively prime if and only if there exist integers x_0 and y_0 such that $ax_0 + by_0 = 1$—that is, if and only if the set $M_a + M_b$ is the set of all integers. We have seen also that if $(a, b =)d$, then a/d and b/d are relatively prime. We now establish an important property of relatively prime pairs.

EXAMPLE 1.6.7. $(12, 5) = 1$ and $(12, 35) = 1$. What about $(12, 175)$? We have $(12, 175) = 1 = (12, 5)(12, 35)$. Is this true for pairs that are not relatively prime? $(12, 8) = 4$ and $(12, 18) = 6$. Is $(12, 144) = 24$?

THEOREM 1.6.2. *Multiplicative Property of Relatively Prime Pairs.* If $(a, b) = 1$ and $(a, c) = 1$, then $(a, bc) = 1$, and conversely if $(a, bc) = 1$ then $(a, b) = (a, c) = 1$.

Proof: Since $(a, b) = 1$, there exist integers x_0 and y_0 such that $ax_0 + by_0 = 1$. Since $(a, c) = 1$, there exist integers x_1 and y_1 such that $ax_1 + cy_1 = 1$. Multiply these equalities:

$$(ax_0 + by_0)(ax_1 + cy_1) = 1.$$

If we rearrange the left side, we get

$$a(ax_0 x_1 + by_0 x_1 + cx_0 y_1) + bc(y_0 y_1) = 1.$$

Thus there exist integers

$$x_2 = ax_0 x_1 + by_0 x_1 + cx_0 y_1 \quad \text{and} \quad y_2 = y_0 y_1$$

such that $ax_2 + bcy_2 = 1$. This proves that $(a, bc) = 1$.

The proof of the converse is immediate, using Corollary 1.6.1. ●

COROLLARY 1.6.2. If $(a, n_i) = 1$ for $i = 1, 2, 3, \ldots, k$, then $(a, n_1 n_2 n_3 \ldots n_k) = 1$.

The preceding theorem is an important result and also illustrates the use of the criterion for relatively prime pairs. Note that in order to prove $(a, b) = 1$, the important thing is the *existence* of integers having a certain property. The actual value of these integers, and even whether or not there is more than one pair with the desired property, is not important here.

EXERCISES 1.6

Checks

1. Find $(48, 60)$, $(24, 30)$, $(12, 15)$, $(16, 20)$, $(4, 5)$, $(28, 35)$.
2. If p is a prime, prove that $(p, a) = 1$ or $(p, a) = p$.

3. For any integer x, prove that $(15x + 17, 10x + 11) = 1$.

4. If $(a, b) = 1$ and $d' | a$ and $d'' | b$, prove that $(d', d'') = 1$.

5. If $(a, b) = d$, $(a, c) = f$, and $(b, c) = 1$, prove that $(d, f) = 1$.

6. Prove that if $(a, b) = d$ and $k > 0$, then $(\pm ka, \pm kb) = kd$, and, conversely, if $(\pm ka, \pm kb) = kd$ and $k > 0$, then $(a, b) = d$.

Challenges

7. If $a | c$ and $b | c$ and $(a, b) = 1$, prove that $ab | c$. Give an example to show that the assumption $(a, b) = 1$ is needed.

8. If $(a, b) = d$ and $(a, c) = f$ and $(b, c) = 1$, prove that $(a, bc) = df$. Give an example to show that this may not be true if $(b, c) \neq 1$. *Hint*: Let $(a, bc) = k$. Use problems 5 and 7, plus characterization (2) of the greatest common divisor, to prove that $k | df$ and $df | k$.

9. If $(a, b) = d$ and $(a, c) = f$, prove that $(a, bc) = (a, df)$.

10. Define (a, b, c) to be the greatest integer that is a divisor of the three integers a, b, and c. Prove that $(a, b, c) = (d, c)$, where $d = (a, b)$.

11. Prove that $(a, b, c) = 1$ if and only if there exist integers x, y, z such that $ax + by + cz = 1$.

12. Give an example to show that it is possible to have $(a, b, c) = 1$ but no one of (a, b), (a, c), or (b, c) equal to 1. Prove that if $(a, b, c) = 1$, the integers (a, b), (a, c), (b, c) are relatively prime in pairs.

1.7   THE EUCLIDEAN ALGORITHM

What about a method of calculating the greatest common divisor? We cannot write down the sets D_a and D_b unless the integers a and b are rather small. In this section we look at an algorithm or systematic rule of procedure by which the greatest common divisor d can be found. The investigations in this section are largely numerical. The change of pace will give us a better insight into the concept of the greatest common divisor. The algorithm developed here is found in Euclid's *Elements* and therefore is attributed to him. Although he did not, as far as we know, have a computer available, the technique can easily be arranged as a computer program.

EXAMPLE 1.7.1. Let $a = 324$ and $b = 126$. In Example 1.6.6 we discovered that if $a = bq + r$, then $(a, b) = (b, r)$. This suggests that the division algorithm can be used to change the problem to a similar one involving smaller numbers which might therefore be easier. Thus $324 = 2 \cdot 126 + 72$ so that $(324, 126) = (126, 72)$. If we repeat the process, $126 = 72 + 54$ and $(126, 72) = (72, 54)$. Again, $72 = 54 + 18$ and $(72, 54) = (54, 18)$. Since $54 = 3 \cdot 18$ we see that $(54, 18) = (18, 0) = 18$. Since $(324, 126) = (126, 72) = (72, 54) = (54, 18) = 18$, we obtain $(324, 126) = 18$. Will the process always terminate? At each application of the division algorithm the new integer r is strictly less than the r of the preceding step; that is, $18 < 54 < 72$. A strictly decreasing sequence of positive integers must be finite.

The repeated applications of the division algorithm illustrated in the preceding example are set up as a formal procedure in the following theorem.

THEOREM 1.7.1. *The Euclidean Algorithm.* Let a and b be integers, $b > 0$. To find (a, b), apply the division algorithm as follows:

$$a = bq_1 + r_1; \quad b = r_1 q_2 + r_2; \quad r_1 = r_2 q_3 + r_3; \ldots;$$

$$r_{i-1} = r_i q_{i+1} + r_{i+1}; \ldots; \quad r_{k-1} = r_k q_{k+1} + 0.$$

The remainder $r_k = (a, b)$.

Proof: Since $b > 0$, the first application of the division algorithm is valid and $0 \le r_1 < b$. If $r_1 = 0$, $k = 0$, then $b | a$ and hence $(a, b) = b$. If $r_1 > 0$, the second application is justified, and $0 \le r_2 < r_1$. At each step, if $r_i > 0$, the application of the division algorithm is justified and $0 \le r_{i+1} < r_i$. For some k, the remainder r_{k+1} is zero, since

$$b > r_1 > r_2 > \cdots > r_i > r_{i+1} > \cdots \ge 0.$$

But the defining equations imply $(a, b) = (b, r_1)$, as well as

$$(b, r_1) = (r_1, r_2), \ldots, (r_{i-1}, r_i) = (r_i, r_{i+1}), \ldots, (r_{k-1}, r_k) = (r_k, 0)$$

(see Example 1.6.6). Hence $(a, b) = (r_k, 0) = r_k$.   ◆

If one wishes to obtain an expression for r_k in the form $ax + by$, this can be done by repeated substitution. The first equation gives $r_1 = a - bq_1$. Substitute for r_1 in the equation $r_2 = b - r_1 q_2$ and collect the terms involving a and b. In this way, r_2 is found as a linear combination of

a and b. The procedure can be continued. Attempting to obtain a formula in this way for r_k would be quite complicated and of little use. Algorithms are rules of procedure and rarely yield useful formulas. The following example should make the procedure clear.

EXAMPLE 1.7.2. Find the greatest common divisor of 612 and 84.

$$612 = 7 \cdot 84 + 24.$$
$$84 = 3 \cdot 24 + 12.$$
$$24 = 2 \cdot 12 + 0.$$

Hence $(612, 84) = 12$.

Find integers x_0 and y_0 such that $612x_0 + 84y_0 = 12$.

$24 = 612 - 7 \cdot 84.$

$12 = 84 - 3 \cdot 24 = 84 - 3[612 - 7 \cdot 84] = -3 \cdot 612 + 22 \cdot 84.$

Thus $x_0 = -3$ and $y_0 = 22$.

Are these the only possible such integers? Try $x_0 = -3 + 84 = 81$ and $y_0 = 22 - 612 = -590$. Since

$$81 \cdot 612 = 49572 \quad \text{and} \quad -590 \cdot 84 = -49560,$$

then

$$81 \cdot 612 + (-590) \cdot 84 = 12.$$

Simple substitution will show that

$$x = -3 + 84k \quad \text{and} \quad y = 22 - 612k$$

will be a possible choice for any integer k.

The Euclidean algorithm enables us to find solutions in integers for an equation of the form $ax + by = n$. This is an example of a problem in *Diophantine Equations* studied extensively in number theory. This study refers to the problem of finding solutions in integers for various classes of algebraic equations. When we find integers x and y such that $ax + by = d$, we are finding solutions in integers of a linear equation in two variables. Do all such equations have solutions in integers? By what method can a solution be found? The Euclidean algorithm, together with what we know about the structure of the set $M_a + M_b$, provides answers to these questions. We will postpone formal discussion of these equations until Unit 5, but if you wish, you can answer these questions for yourself. The following example may help.

EXAMPLE 1.7.3. Find integers x and y such that $612x + 84y = 18$. Since $(612, 84) = 12$, every integer of the form must be a multiple of 12. Since 18 is not a multiple of 12, there are no integers x and y such that $612x + 84y = 18$.

EXAMPLE 1.7.4. Find integers x and y such that $612x + 84y = 156$. In Example 1.7.2. we found that $(-3)612 + (22)84 = 12$. Since $156 = 13 \cdot 12$, we get

$$(-39)612 + (286)84 = 13 \cdot 12.$$

Thus

$$x = -39 \quad \text{and} \quad y = 286$$

is a solution of the equation $612x + 84y = 156$. Are there any other solutions? Suppose that $612x + 84y = 156$ and $612x_0 + 84y_0 = 156$. Subtract these equations:

$$612(x - x_0) + 84(y - y_0) = 0$$

or

$$51(x - x_0) = -7(y - y_0).$$

This relationship is satisfied if

$$x - x_0 = 7k \quad \text{and} \quad y - y_0 = -51k.$$

Substitution shows that

$$x = -39 + 7k \quad \text{and} \quad y = 286 - 51k$$

is a solution of $612x + 84y = 156$ for every integer k.

Are these the only solutions? To answer, we need an important property of divisors, one that we investigate in the next section.

EXERCISES 1.7

Checks

1. Calculate the greatest common divisor of 300 and 222.

2. Prove that $(28, 75) = 1$ and find integers x and y such that $28x + 75y = 1$.

3. Calculate $(300, 222, 416)$.

4. Find integral solutions of $9x - 6y = 57$.

5. Find integral solutions of $3x + 5y = 47$. Are there any in which both x and y are positive?

6. Show that in order to find the greatest common divisor of 144 and 89, one needs ten steps of the Euclidean algorithm.

Challenges

7. The integers 144 and 89 are the twelfth and eleventh entries in a remarkable sequence called the *Fibonacci sequence*. This is defined by

    $$u_1 = 1, \quad u_2 = 1, \quad u_n = u_{n-1} + u_{n-2}, \qquad \text{for } n = 3, 4, \ldots .$$

    Prove that in finding (u_{n+2}, u_{n+1}) by the Euclidean algorithm, n divisions are necessary.

8. Prove that the equation $ax + by = n$ has a solution in integers when $(a, b) = 1$. State a necessary condition for the existence of a solution in integers if $(a, b) = d \neq 1$.

9. A student received a prize of $200, all of which he must spend on mathematics and English books. If the average cost is $13 for mathematics books and $9 for English books, how shall he divide his purchases in order to receive the largest number of books? How shall he divide his purchases if he needs to own six mathematics books?

10. The system of inequalities obtained in the Euclidean algorithm can be written as

    $$\frac{a}{b} = q_1 + \frac{r_1}{b} = q_1 + \frac{1}{b/r_1},$$

    but

    $$\frac{b}{r_1} = q_2 + \frac{r_2}{r_1} = q_2 + \frac{1}{r_1/r_2}.$$

This procedure leads to a fraction representing a/b in which the denominator is a succession of fractions:

$$\frac{a}{b} = q_1 + \cfrac{1}{q_2 + \cfrac{1}{q_3 + \cfrac{1}{q_4 + \cdots \cfrac{1}{q_k}}}}$$

A fraction of this type is called a *simple continued fraction*. Write 612/84 as a simple continued fraction. Write 75/28 in this way.

1.8   THE FUNDAMENTAL THEOREM OF FACTORIZATION

In the preceding section we have obtained an algorithm for calculating (a, b). How can we calculate the least common multiple $[a, b]$? Certainly ab is a common multiple, but it is not always the least common multiple.

EXAMPLE 1.8.1. $M_8 \cap M_6 = M_{24}$ (see Example 1.5.1). In this case, $[a, b] = 24$ and $ab = 48$. On the other hand, $M_5 \cap M_6 = M_{30}$. In this case, $[a, b] = 30$ and $ab = 30$. Where do the situations differ? Are there multiples of 6 such that $8 \mid 6n$ but $8 \nmid n$? Yes, 24 is such a multiple. Are there multiples of 6 such that $5 \mid 6n$ but $5 \nmid n$? No. Every multiple of 6 that is also a multiple of 5 must be a multiple of 30. The difference lies in the fact that $(8, 6) = 2$ whereas $(5, 6) = 1$.

The following theorem is a basic property of the structure of the integers. It is this theorem that is the key to the fact that the factorization of an integer into products of primes is unique, apart from the order of the factors. This property is so familiar to us that, although it is involved in much of our arithmetic, we take it for granted. For this reason, Theorem 1.8.1. is referred to as the Fundamental Theorem of Factorization.

THEOREM 1.8.1. *The Fundamental Theorem of Factorization.* If c divides ab, and $(c, a) = 1$, then c divides b.

Proof: Since $(c, a) = 1$, there exist integers x_0 and y_0 such that $cx_0 + ay_0 = 1$. Multiply both sides of this equation by b:

$$bcx_0 + aby_0 = b.$$

But $c|ab$, so $c|aby_0$. Also, $c|bcx_0$. By the second property of divides,

$$c|bcx_0 + aby_0.$$

That is, $c|b$. ●

COROLLARY 1.8.1. If p is a prime and $p|ab$, then $p|a$ or $p|b$.

Proof: By definition, the divisors of p are 1 and p. Therefore $(p, a) = 1$ or p. If $(p, a) = p$, then $p|a$. If $(p, a) = 1$, then $p|b$. ●

EXAMPLE 1.8.2. $6|11n$ implies that $6|n$. $6|35n$ implies that $6|n$. $6|34n$ does not imply that $6|n$, since $(6, 34) = 2$.

$7|11n$ implies that $7|n$. $7|34n$ implies that $7|n$. $7|35n$ does not imply that $7|n$, since $(7, 35) = 7$.

EXAMPLE 1.8.3. If $(a, b) = 1$, prove that $(a + b, a - b) = 1$ or 2. Let $(a + b, a - b) = d$. By the second property of divides, $d|a + b$ and $d|a - b$ imply $d|2a$ and $d|2b$. If $(d, 2) = 1$, the fundamental theorem implies that $d|a$ and $d|b$. Therefore $d|(a, b)$. But $(a, b) = 1$, so $d = 1$. If $(d, 2) = 2$, $d = 2d'$, where $d'|a$ and $d'|b$, so $d'|(a, b)$, and $d' = 1$ so $d = 2$.

In Example 1.8.1 we saw that $[a, b] \leq ab$ and that equality appears to hold in case $(a, b) = 1$. The following theorem makes precise the relationship between (a, b) and $[a, b]$. Again the fundamental theorem of factorization is a necessary part of the argument.

THEOREM 1.8.2. *Connection between* (a, b) *and* $[a, b]$. Let a and b be positive integers. Then

$$[a, b] \cdot (a, b) = ab.$$

Proof: Let $(a, b) = d$. By the third characterization of (a, b), we can write

$$a = da', \quad b = db', \quad \text{where} \quad (a', b') = 1.$$

We wish to prove that $[a, b] = a'b'd$. To do this, we show that

(1) $a \mid a'b'd$ and $b \mid a'b'd$ and

(2) if $a \mid n$ and $b \mid n$ then $a'b'd \mid n$.

It is clear that $a \mid a'b'd$, since $a'b'd = ab'$. Also, $b \mid a'b'd$, since $a'b'd = a'b$. Let n be a common multiple of a and b. Since $a \mid n$, $n = ak$ for some integer k. By hypothesis, $b \mid ak$; that is, $b'd \mid a'dk$, which implies $b' \mid a'k$. But $(b', a') = 1$. By the fundamental theorem, $b' \mid k$; that is, $k = b'k'$ for some integer k'. This implies that $n = ab'k' = a'b'dk'$, so that n is a multiple of $a'b'd$. This result, together with the fact that $a'b'd$ is a common multiple of a and b, proves that $[a, b] = a'b'd$. ●

COROLLARY 1.8.2. If $(a, b) = 1$, then $[a, b] = ab$.

EXAMPLE 1.8.4. If a, b, c are relatively prime in pairs, prove that $[a, b, c] = abc$.

$$[a, b, c] = [[a, b], c]$$

(see Exercises 1.5, problem 10). Since $(a, b) = 1$, $[a, b] = ab$. The multiplicative property of relatively prime pairs tells us that $(a, c) = 1$ and $(b, c) = 1$ imply $(ab, c) = 1$. Therefore $[ab, c] = abc$.

With the help of Corollary 1.8.2, we can extend the counting process begun in Example 1.4.6. There the greatest integer function was used to count the number of positive multiples of the integer a that are less than or equal to n. There are $[n/a]$ integers in this set. Eventually we would like to determine how many of the first n positive integers are prime by counting those that are composite. As a first step in this direction, consider the set of positive integers less than or equal to n and multiples of either a or b. To count these it is necessary to identify those integers that are common to M_a and M_b. If $(a, b) = 1$, these integers are in M_{ab}.

EXAMPLE 1.8.5. How many positive integers less than or equal to 50 are multiples of 2 or 3? There are $[50/2]$ multiples of 2 and $[50/3]$ multiples of 3. The integers that are multiples of both 2 and 3 are included in each of these sets. These integers are all multiples of 6. Thus the count with duplications eliminated is $[50/2] + [50/3] - [50/6]$. How many positive integers less than or equal to 50 are multiples of 2, 3, or 5? In this case, the possibilities of duplication are increased. The multiples of 6, the multiples of 10, and the multiples of 15 each occur in two of the sets of multiples of 2, 3, or 5, whereas the multiples of all three—that is, the multiples of 30—occur in M_2, M_3, M_5, M_6, M_{10}, and M_{15}. The count, with duplications eliminated, is

$$\left[\frac{50}{2}\right] + \left[\frac{50}{3}\right] + \left[\frac{50}{5}\right] - \left[\frac{50}{6}\right] - \left[\frac{50}{10}\right] - \left[\frac{50}{15}\right] + \left[\frac{50}{30}\right].$$

EXERCISES 1.8

Checks

1. Calculate $[300, 222]$. (See Exercises 1.7, problem 1.)
2. Calculate $[28, 320, 55]$. (See Exercises 1.5, problem 10.)
3. If $(a, b) = 1$ prove that $(4a + b, a - b) = 1$ or 5.
4. If $(m, n) = 1$, prove that $(m + n, mn) = 1$.
5. How many positive integers less than or equal to 50 are multiples of 3, 5, or 7? How many are multiples of 2, 3, 5, or 7?
6. How many positive integers less than or equal to 50 are multiples of 6 or 14?

Challenges

7. If $d_1 e_1 = d_2 e_2$ and $(e_1, e_2) = 1$, prove that $[d_1, d_2] = d_1 e_1 = d_2 e_2$.
8. If $(a, b, c) = 1$, prove that $[ab, bc, ca] = abc$, where a, b, c are positive integers. *Hint*: Use Exercises 1.5, problem 7, and the fact that $(b, (a, c)) = 1$ to calculate $[[ab, bc], ca]$.
9. If a, b, c are positive integers, prove that $abc = (a, b, c)[ab, bc, ca]$.
10. If a, b, c are positive integers, prove that

$$abc = (a, b, c)[(a, b), (b, c), (c, a)][a, b, c].$$

Hint: Reduce to the case $(a, b, c) = 1$. Use Exercises 1.6, problem 12.

11. If $(a, n) = d$ and $(r, n) = 1$, prove that $(r - a, d) = 1$.
12. Find two positive integers such that their sum will be a factor of their product. Problem E1452, *American Mathematical Monthly*, Vol. 68 (1961), p. 177.

1.9 THE DIVISORS OF A PRODUCT

What information about divisors can be obtained by an application of the fundamental theorem of factorization proved in the preceding section? We look first at the divisors of p^n, where p is a prime, and assure ourselves that our intuition is correct.

THEOREM 1.9.1. *The Divisors of p^n*. Let p be a prime. Then p^n has $n + 1$ divisors: $\{1, p, p^2, \ldots, p^n\}$.

Proof: We use mathematical induction. By definition, a prime has two positive divisors: $\{1, p\}$. The statement is thus true for $n = 1$.

Suppose that the set of divisors of p^k is $\{1, p, p^2, \ldots, p^k\}$. We wish to find the divisors of p^{k+1}. Let $d \,|\, p^{k+1}$, then $p^{k+1} = dm$, for some integer m. If $p \,|\, d$, then

$$d = pd' \quad \text{and} \quad p^{k+1} = pd'm.$$

In this case, $p^k = d'm$, so that $d' \,|\, p^k$. This means that d' is one of the elements of the set $\{1, p, p^2, \ldots, p^k\}$, which implies that d belongs to the set $\{p, p^2, p^3, \ldots, p^{k+1}\}$.

Now suppose that p does not divide d. Since p is prime, this means that $(p, d) = 1$. By the multiplicative property of relatively prime pairs, this implies that $(p^{k+1}, d) = 1$. But since $d \,|\, p^{k+1}$, we must have $d = 1$. Thus the set of divisors of p^{k+1} is $\{1, p, p^2, \ldots, p^{k+1}\}$. By the principle of mathematical induction, the statement is true for every n. ◆

COROLLARY 1.9.1. If p is a prime, then $p^a \,|\, p^b$ if and only if $0 \le a \le b$.

What about the divisors of a product? What information do we have about the divisors of ab if the divisors of a and the divisors of b are known?

EXAMPLE 1.9.1

$$D_4 = \{\pm 1, \pm 2, \pm 4\}.$$
$$D_6 = \{\pm 1, \pm 2, \pm 3, \pm 6\}.$$
$$D_4 \cdot D_6 = \{\pm 1, \pm 2, \pm 3, \pm 4, \pm 6, \pm 8, \pm 12, \pm 24\} = D_{24}.$$
$$D_3 = \{\pm 1, \pm 3\}.$$
$$D_8 = \{\pm 1, \pm 2, \pm 4, \pm 8\}.$$
$$D_3 \cdot D_8 = \{\pm 1, \pm 2, \pm 3, \pm 4, \pm 6, \pm 8, \pm 12, \pm 24\} = D_{24}.$$

It appears that $D_a \cdot D_b = D_{ab}$. Because of the symmetry of the sets of divisors, we need to look only at the positive divisors. The three positive divisors of 4 and the six positive divisors of 6 multiplied pairwise yield the eight positive divisors of 24. In this case, there must be some duplicates among the 12 pairwise products. If we look closer, we see that $2 \cdot 6 = 4 \cdot 3$, $2 \cdot 1 = 1 \cdot 2$, $4 \cdot 1 = 2 \cdot 2$, and $1 \cdot 6 = 2 \cdot 3$. Now consider the pairwise products obtained from the two positive divisors of 3 and the four positive divisors of 8. There are exactly eight, and they correspond to the positive divisors of 24. It is not surprising that the difference lies in the fact that $(4, 6) = 2$ while $(3, 8) = 1$.

THEOREM 1.9.2. *The Divisors of ab.* The integer d divides ab if and only if d can be written in the form $d = d'd''$, where $d' | a$ and $d'' | b$. If $(a, b) = 1$, a positive divisor d can be written as $d'd''$ for exactly one pair of positive divisors d' and d''.

Proof: Let $d' | a$ and $d'' | b$. Then $d'd'' | ab$. Hence if $d = d'd''$, then $d | ab$.

Now suppose that $d | ab$. Let $(d, a) = d'$. Then

$$d = d'k \quad \text{and} \quad a = d'a'$$

where $(k, a') = 1$. Since $d | ab$, then $d'k | d'a'b$; that is, $k | a'b$. But $(k, a') = 1$, and, by the fundamental theorem of factorization, $k | b$. Thus $d = d'd''$, where $d'' = k$ and is a divisor of b. This proves the first part of the theorem.

Is there a unique pair of positive divisors such that $d = d'd''$? Yes, provided that $(a, b) = 1$. Suppose that $d = d'd''$ and, also, $d = f'f''$, where d' and f' are positive divisors of a and d'' and f'' are positive divisors of b. Since $(a, b) = 1$, we have

$$(d', f'') = 1 \quad \text{and} \quad (f', d'') = 1.$$

(This result is an immediate consequence of Corollary 1.6.1. See Exercises 1.6, problem 4.) But $d'd'' = f'f''$ implies that $d' | f'f''$. Therefore since $(d', f'') = 1$, then $d' | f'$. Also, $f' | d'd''$; and since $(f', d'') = 1$, $f' | d'$. This result implies $f' = d'$. Similarly, $f'' = d''$, and the divisors of ab can be written in the form $d'd''$ in exactly one way. ◆

EXAMPLE 1.9.2. In Example 1.9.1 it was seen that if $a = 4$ and $b = 6$, duplicate representations occur for the divisors 2, 4, 6, and 12.

Can you relate this duplication to the fact that $(4, 6) = 2$? (See Exercises 1.9, problems 10 and 11.)

The preceding theorem gives some interesting information about the number-theoretic functions τ and σ defined in Section 1.1. In Example 1.9.1 we saw that $\tau(24) = \tau(3)\tau(8)$. The theorem tells us that the positive divisors of ab are in one-to-one correspondence with the pairwise products of the positive divisors of a and the positive divisors of b, provided that $(a, b) = 1$. That is, $\tau(ab) = \tau(a)\tau(b)$, provided that $(a, b) = 1$. Number-theoretic functions possessing this property occur very frequently in number theory. They are called *multiplicative functions*. A function whose definition is associated with divisibility is likely to be multiplicative.

DEFINITION 1.9.1. *Multiplicative Function.* A number-theoretic function is *multiplicative* if for every pair of integers m, n such that $(m, n) = 1$, $f(mn) = f(m)f(n)$. The function f is *completely multiplicative* if $f(mn) = f(m)f(n)$ for every pair m and n.

EXAMPLE 1.9.3. Let $f(n) = n^2$. Since $f(mn) = (mn)^2 = m^2 n^2 = f(m)f(n)$, f is completely multiplicative. Since $\tau(4)\tau(6) = 12 \neq \tau(24)$, τ is not completely multiplicative.

THEOREM 1.9.3. *Multiplicative Property of τ and σ.* The number-theoretic functions τ and σ are multiplicative.

Proof: Let $(m, n) = 1$. Let d_i', $i = 1, 2, \ldots, \tau(m)$ be the positive divisors of m and d'', $j = 1, 2, \ldots, \tau(n)$ be the positive divisors of n. The set of positive divisors of mn is the set of products $d_i' d_j''$ where $i = 1, 2, \ldots, \tau(m)$ and $j = 1, 2, \ldots, \tau(n)$. Since there are exactly $\tau(m)\tau(n)$ distinct integers in this set (by the preceding theorem), $\tau(mn) = \tau(m)\tau(n)$ when $(m, n) = 1$; that is, τ is multiplicative.

Also

$$\sigma(m) = (d_1' + d_2' + \cdots + d_{\tau(m)}')$$

and

$$\sigma(n) = (d_1'' + d_2'' + \cdots + d_{\tau(n)}'').$$

The product of these two expressions is the sum of the products obtained by choosing d_i' from the first and d_j'' from the second—that is, the sum of the $\tau(m)\tau(n)$ expressions $d_i'd_j''$. This is exactly $\sigma(mn)$. Thus when $(m, n) = 1$, $\sigma(m)\sigma(n) = \sigma(mn)$ and σ is multiplicative. ⬢

EXAMPLE 1.9.4

Let $m = 6$ and $n = 5$.
$\sigma(6) = 1 + 2 + 3 + 6 = 12$.
$\sigma(5) = 1 + 5 = 6$.
$$(1 + 2 + 3 + 6)(1 + 5) = 1 + 2 + 3 + 6 + 5 + 10 + 15 + 30$$
$$= \sigma(30) = 72.$$

EXERCISES 1.9

Checks

1. Construct the set of divisors of 56 from the divisors of 7 and the divisors of 8. Construct these also from the divisors of 4 and the divisors of 14. For which divisors do duplicate representations occur?

2. Using the table in Section 1.2, calculate $\tau(110)$, $\tau(114)$, $\tau(390)$, $\sigma(75)$, and $\sigma(96)$.

3. Prove Corollary 1.9.1.

4. If p and q are prime and $p \mid q^k$, prove that $p = q$.

5. Let $f(n) = n^k$, for some integer k. Prove that f is completely multiplicative.

6. The function π is defined by $\pi(n) = $ the number of primes less than or equal to n. Is π multiplicative?

Challenges

7. If f is multiplicative and $f(a) \neq 0$ for some $a \neq 0$, prove that $f(1) = 1$.

8. If a, b, c are integers that are relatively prime in pairs, prove that
$$\tau(abc) = \tau(a)\tau(b)\tau(c).$$

Show that $(a, b, c) = 1$ is not sufficient to imply that $\tau(abc) = \tau(a)\tau(b)\tau(c)$.

9. If $(m, n) = 1$, show that
$$\sum_{d \mid mn} g(d) = \sum_{d' \mid m} \left\{ \sum_{d'' \mid n} g(d'd'') \right\}$$

for any function g.

10. Let $d = d'd'' = f'f''$, where $d' \neq f'$. Prove that at least one of (d', f'') and (d'', f') is not equal to 1.

11. In the notation of Theorem 1.9.2, prove that if at least one of $(d', b/d'')$ and $(a/d', d'')$ is not equal to 1, then d is a divisor of ab, for which duplicate representations occur. *Hint*: If $(d', b/d'') = k > 1$, let $f' = d'/k$.

12. Let a be a fixed integer. Define $f(n) = (a, n)$. Prove that f is multiplicative.

1.10 THE FUNDAMENTAL THEOREM OF ARITHMETIC AND THE STANDARD FORM OF n

We have seen some of the advantages of writing an integer as a product of factors that are relatively prime. If p and q are distinct primes, then p^a and q^b are relatively prime integers, for any positive integral exponent a and b. This suggests that factoring an integer into powers of distinct primes would be a good idea. In considering factorization, we can restrict ourselves to integers greater than 1. The integers ± 1 are units and have no divisors other than ± 1. The negative integers are the products of the unit -1 and the positive integers. The integer 0 is divisible by every nonzero integer and thus has too many factorizations to be interesting.

In the factorization theorem (Theorem 1.3.1), we proved that every integer greater than 1 is either a prime or a product of primes. Let us set up a standard way of representing this product of primes.

DEFINITION 1.10.1. *Standard Form of n.* The integer $n > 1$ is said to be written in *standard form* when it is written as a product of powers of primes, where the primes are arranged in ascending order of magnitude and the exponents are all greater than or equal to 1:

$$n = p_1^{a_1} p_2^{a_2} \cdots p_k^{a_k},$$

where

$$p_1 < p_2 < \cdots < p_k, \quad a_i \geq 1, \quad i = 1, 2, \ldots, k.$$

$$\boxed{\prod_{i=1}^{k} p_i^{a_i}.}$$

Since the products are apt to become cumbersome, we write them in a shorter form, making use of the greek letter $\prod$ to indicate that a product is to be taken. Thus

$$p_1^{a_1} p_2^{a_2} \cdots p_k^{a_k} = \prod_{i=1}^{k} p_i^{a_i}.$$

Here $p_i^{a_i}$ tells us that the terms to be multiplied are all primes to some power, and the primes to be used are identified by stating that i takes in turn the values 1, 2, 3, ..., k. In some cases, it may be desirable to indicate the primes to be included by writing

$$\prod_{p|n} p^{a_p}.$$

The notation $p|n$ means that exactly one term is to be included for each prime p that divides n, and the exponent a_p is the highest power of the prime p which divides n.

EXAMPLE 1.10.1

$60 = 2 \cdot 2 \cdot 3 \cdot 5 = 2^2 \cdot 3 \cdot 5.$
$2520 = 2^3 \cdot 3^2 \cdot 5 \cdot 7.$
$17 = 17.$

Exactly three distinct primes divide 360, namely 2, 3, 5. The highest power of 2 which divides 360 is 3; thus $a_2 = 3$. The highest power of 3 which divides 360 is 2 so that $a_3 = 2$. The highest power of 5 which divides 360 is 1 and $a_5 = 1$. Therefore

$$\prod_{p|360} p^{a_p} = 2^3 \cdot 3^2 \cdot 5.$$

Our numerical experience leads us to feel confident that the expression of an integer in standard form can be done in only one way. Such is indeed true. It should not be assumed, however, that the unique expression of an element as a product of primes is a property of rings in general. Rather simple examples of rings can be given in which the representation of an element as a product of primes is not unique. (See Exercises 1.10, problem 12.) Theorem 1.8.1, the fundamental theorem of factorization, is the property that enables us to derive this result. The theorem which follows is sometimes called the unique factorization theorem or, more frequently, the fundamental theorem of arithmetic.

THEOREM 1.10.1. *The Fundamental Theorem of Arithmetic.* An integer can be written in standard form in exactly one way.

Proof: As we have seen earlier, we need consider only $n > 1$. Suppose that there is some $n > 1$ which can be written in standard form in more than one way. Let S be the set of such integers. By the well-ordering principle, S has a smallest element c. That is,

$$c = p_1^{a_1} p_2^{a_2} \cdots p_k^{a_k} = q_1^{b_1} q_2^{b_2} \cdots q_s^{b_s}$$

where both expressions are in standard form. There is no loss of generality in assuming that $p_1 \le q_1$. Since $p_1 | c$,

$$p_1 | q_1^{b_1} q_2^{b_2} \cdots q_s^{b_s}.$$

By the fundamental theorem of factorization, this implies that $p_1 | q_i^{b_i}$ for some i—that is, $p_1 = q_i$. Because of the ordering in increasing magnitude, $p_1 = q_1$.

Now consider the integer c/p_1. Since this integer is smaller than c, it is either 1 or has a unique representation in standard form. If $c/p_1 = 1$, $c = p_1$ and can have no other prime factors, so it has a unique representation in standard form. If $c/p_1 > 1$, then

$$p_1^{a_1-1} p_2^{a_2} \cdots p_k^{a_k} \quad \text{and} \quad q_1^{b_1-1} q_2^{b_2} \cdots q_s^{b_s}$$

are identical. But this is impossible unless the original representations for c were identical. Thus the assumption that n can be written in standard form in more than one way leads to a contradiction. ●

EXAMPLE 1.10.2. Written in standard form, $200772 = 2^2 \cdot 3^3 \cdot 11 \cdot 13^2$. What are the divisors of 200772? Consider any prime p. If $p \ne 2$, 3, 11, or 13, $(p, 200772) = 1$ and p^k does not divide 200772 for any $k \ge 1$. Since $(2^k, 3^3 \cdot 11 \cdot 13^2) = 1$, the fundamental theorem implies that $2^k | 200772$ if and only if $2^k | 2^2$; that is, $k = 0, 1, 2$. Similarly, $3^k | 200772$ if and only if $k = 0, 1, 2, 3$; $11^k | 200772$ if and only if $k = 0, 1$; and $13^k | 200772$ if and only if $k = 0, 1, 2$. The divisors of 200772 must be integers of the form $2^a \cdot 3^b \cdot 11^c \cdot 13^d$, where $a = 0, 1, 2$; $b = 0, 1, 2, 3$; $c = 0, 1$; $d = 0, 1, 2$. Since there are 3 choices for a, 4 choices for b, 2 choices for c, and 3 choices for d, there are $3 \cdot 4 \cdot 2 \cdot 3 = 72$ integers that can be created in this way. These integers are exactly the divisors of 200772. In this way we see that $\tau(200772) = 72$.

THEOREM 1.10.2. *The Divisors of n.* If $n = p_1^{a_1} p_2^{a_2} \ldots p_k^{a_k}$, then the positive divisors of n are the integers $\{p_1^{b_1} p_2^{b_2} \ldots p_k^{b_k}$, such that $0 \le b_i \le a$, $i = 1, 2, \ldots, k\}$. The number of positive divisors of n is

$$\tau(n) = (a_1 + 1)(a_2 + 1) \ldots (a_k + 1).$$

Proof: Let $d | n$. The only prime divisors of d are $p_1, p_2, \ldots, p_k$, since $p | d$ and $d | n$ imply that $p | n$. Therefore

$$d = p_1^{b_1} p_2^{b_2} \cdots p_k^{b_k},$$

where we know that $b_i \ge 0$, $i = 1, 2, \ldots, k$.

Since $d | n$, $p_i^{b_i} | n$. This implies that $p_i^{b_i} | p_i^{a_i}$, by the fundamental theorem of factorization—that is, $b_i \le a_i$, $i = 1, 2, \ldots, k$. Therefore the divisors of n have the form stated.

There are $a_i + 1$ choices for the exponent of p_i. The total number of positive divisors is therefore $(a_1 + 1)(a_2 + 1) \cdots (a_k + 1)$. ◆

Factorization of integers into powers of distinct primes leads to a representation for the greatest common divisor. In this case, it is helpful to choose a form in which the primes shown are chosen so that the integers will be comparable. Any prime that divides either integer is included in the expression of the integers as products of prime powers, and the exponent zero is used where necessary.

EXAMPLE 1.10.3. $24 = 2^3 \cdot 3$ and $45 = 3^2 \cdot 5$. We would write 24 and 45 in comparable form as follows:

$$24 = 2^3 \cdot 3 \cdot 5^0 \quad \text{and} \quad 45 = 2^0 \cdot 3^2 \cdot 5.$$

In this way it is easy to pick the greatest common divisor:

$$(24, 45) = 2^0 \cdot 3 \cdot 5^0 = 3.$$

We can also identify the least common multiple:

$$[24, 45] = 2^3 \cdot 3^2 \cdot 5 = 450.$$

COROLLARY 1.10.2. If $n = p_1^{a_1} p_2^{a_2} \ldots p_k^{a_k}$ and $m = p_1^{b_1} p_2^{b_2} \ldots p_k^{b_k}$, $a_i \ge 0$, $b_i \ge 0$, are comparable representations for m and n, then

$$(m, n) = p_1^{c_1} p_2^{c_2} \cdots p_k^{c_k}$$

where c_i = the smaller of a_i and b_i, and

$$[m, n] = p_1^{d_1} p_2^{d_2} \cdots p_k^{d_k}$$

where d_i = the larger of a_i and b_i.

Proof: The proof is left as an exercise (Exercises 1.10, problem 4). ●

The following examples indicate some of the questions that can be answered using the idea of factorization into prime powers. The first concerns identifying integers that have a certain number of positive divisors.

EXAMPLE 1.10.4. Since $45 = 3^2 \cdot 5$,

$$\tau(45) = (2 + 1)(1 + 1) = 6.$$

Are there any other n for which $\tau(n) = 6$? Since

$$\tau(n) = \prod_{i=1}^{k}(a_i + 1) \qquad \text{when} \qquad n = \prod_{i=1}^{k} p_i^{a_i},$$

we must choose a_i so that $\prod_{i=1}^{k}(a_i + 1) = 6$. The only factorizations of 6 are $1 \cdot 6$ and $2 \cdot 3$. This means that $n = p^5$, for any prime p, and $n = p^1 q^2$, for any distinct primes p and q, are the only possibilities. Check this result against the table in Section 1.2.

EXAMPLE 1.10.5. If $uv = n^2$ and $(u, v) = 1$, prove that there exist integers r and s such that $u = r^2$ and $v = s^2$. Let $n = \prod p_i^{a_i}$. Then $n^2 = \prod p_i^{2a_i}$. Since $(u, v) = 1$, no prime can divide both u and v. For each i, either $p_i | u$ or $(p_i, u) = 1$. If $(p_i, u) = 1$, by the multiplicative property of relatively prime pairs, $(p_i^{2a_i}, u) = 1$. Therefore, since $p_i^{2a_i} | n^2$, it follows from the fundamental theorem of factorization that $p_i^{2a_i} | v$. Similarly, if $p_i | u$, $(p_i^{2a_i}, v) = 1$ and $p_i^{2a_i} | u$. Thus

$$u = \prod_{p_i | u} p_i^{2a_i} \quad \text{and} \quad v = \prod_{p_i | v} p_i^{2a_i}.$$

The desired result follows if we choose

$$r = \prod_{p_i | u} p_i^{a_i} \quad \text{and} \quad s = \prod_{p_i | v} p_i^{a_i}.$$

The fact that n has a unique representation as a product of primes enables us to obtain an expression for the number of primes less than or equal to n. Each composite number less than or equal to n is divisible by a prime less than $\sqrt{n}$. In Example 1.8.5 we saw how to use the greatest integer function to count positive composite numbers.

EXAMPLE 1.10.6. How many primes are less than 50? There are four primes less than $\sqrt{50}$; namely, 2, 3, 5, 7. The number of positive multiples of 2, 3, 5, 7 that are less than or equal to 50 is $[50/2] + [50/3] + [50/5] + [50/7] - [50/6] - [50/10] - [50/14] - [50/15] - [50/21] - [50/35] + [50/30] + [50/42] + [50/70] + [50/105] = 38$. The primes 2, 3, 5, 7 are themselves included in this set of multiples, so that the number of composite positive integers less than or equal to 50 is $38 - 4 = 34$, and the number of primes less than or equal to 50 is $49 - 34 = 15$.

EXERCISES 1.10

Checks

1. Write in standard form: 286, 390, 1278, 842.

2. Write the product represented by $\prod_{p|1260} p^{a_p}$.

3. Write in comparable form and obtain the following:

    $(286, 390), (286, 1278), (286, 842).$

4. Prove Corollary 1.10.2. Extend it to obtain (a, b, c). Illustrate this technique by calculating $(286, 390, 1278)$.

5. Use the formula for $\tau(n)$ obtained in Theorem 1.10.2 to prove that τ is multiplicative.

6. Write the set of divisors of $p^2 q^5$ where p and q are distinct primes.

7. A unitary divisor is a divisor d having the property that $(d, n/d) = 1$. Write the unitary divisors of $p^2 q^5$.

Challenges

8. Characterize the values of n for which $\tau(n) = 14$, $\tau(n) = 15$, $\tau(n) = 18$.

9. Prove that n is a square if and only if $\tau(n)$ is odd.

10. Prove that the product of the divisors of n is $n^{\tau(n)/2}$. *Hint*: Remember that the divisors of n occur in pairs whose product is n.

11. Prove that n is a kth power if and only if k divides each of the exponents in the standard representation of n.

12. If $uv = n^k$ and $(u, v) = 1$, prove that there exist integers r and s such that $u = r^k$ and $v = s^k$.

13. Prove that every positive integer has a unique representation in the form m^2n, where n is square-free—that is, not divisible by the square of a prime.

14. Let R consist of the set of elements of the form $\alpha = a + b\sqrt{-5}$, where a and b are integers, and addition and multiplication are defined by

$$(a + b\sqrt{-5}) + (c + d\sqrt{-5}) = (a + c) + (b + d)\sqrt{-5}$$

and

$$(a + b\sqrt{-5})(c + d\sqrt{-5}) = (ac - 5bd) + (bc + ad)\sqrt{-5}.$$

An element is a *unit* if it divides 1. An element is a *prime* if it is not a unit and if every factorization $\pi = \alpha\beta$ has either α or β a unit. Define

$$N(a + b\sqrt{-5}) = a^2 + 5b^2.$$

(a) Prove that $N(\alpha)N(\beta) = N(\alpha\beta)$.

(b) If α is a unit—that is, $\alpha\beta = 1$ for some β—prove that $N(\alpha) = 1$.

(c) From part (b), prove that ± 1 are the only units in the system.

(d) Prove that 3 is a prime by showing that $3 = \alpha\beta$, where neither α nor β is a unit, implies $a^2 + 5b^2 = 3$, which is impossible for integers a and b.

(e) Prove that 7 is a prime.

(f) Prove that $1 + 2\sqrt{-5}$ and $1 - 2\sqrt{-5}$ are primes, by showing that $1 \pm 2\sqrt{-5} = \alpha\beta$, where neither α nor β is a unit, implies that either $N(\alpha)$ or $N(\beta) = 3$.

(g) Prove that in this system factorization into primes is not unique by considering $3 \cdot 7$ and $(1 + 2\sqrt{-5})(1 - 2\sqrt{-5})$. [See Hardy and Wright, *An Introduction to the Theory of Numbers* (London: Oxford University Press, 1938), Chapter 12, for more detail.]

1.11   FORMULAS FOR MULTIPLICATIVE FUNCTIONS

In the preceding section a formula was developed for $\tau(n)$ that was based on the standard form of n. Although the multiplicative property of τ was not used explicitly in deriving this formula, it was implied in the counting process used. What about formulas for other multiplicative

functions? The standard form of n plays an important role in this connection. For a list of some important number-theoretic functions and
their definitions, see Appendix C.

Recall that the function f is multiplicative if whenever $(m, n) = 1$,
$f(m)f(n) = f(mn)$. We first use mathematical induction to extend this
property to a property of f over products of k integers relatively prime in
pairs.

THEOREM 1.11.1. *Formula for Multiplicative Functions.* If f is
multiplicative and n_1, n_2, $\ldots$, n_k are relatively prime in pairs, then
$f(n_1 n_2 n_3 \ldots n_k) = f(n_1)f(n_2)f(n_3) \ldots f(n_k)$. In particular, if $n = p_1^{a_1} p_2^{a_2} p_3^{a_3} \ldots p_k^{a_k}$
then

$$f(n) = f(p_1^{a_1})f(p_2^{a_2})f(p_3^{a_3}) \ldots f(p_k^{a_k}).$$

Proof: The statement $f(n_1 n_2 \ldots n_k) = f(n_1)f(n_2) \ldots . f(n_k)$ is an
identity if $k = 1$, and it is the defining property of multiplicative functions
when $k = 2$. Suppose that this property is true for fixed k. Consider
$n = n_1 n_2 \ldots n_k n_{k+1}$. Since $(n_{k+1}, n_i) = 1$ for $i = 1, 2, \ldots, k$, the multiplicative property of prime pairs (Corollary 1.5.4) implies that

$$(n_{k+1}, n_1 n_2 \ldots n_k) = 1.$$

Since f is multiplicative,

$$f(n_{k+1} n_1 n_2 \ldots n_k) = f(n_{k+1})f(n_1 n_2 \ldots n_k).$$

But by the induction hypothesis,

$$f(n_1 n_2 \ldots n_k) = f(n_1)f(n_2) \ldots f(n_k)$$

and hence

$$f(n_1 n_2 \ldots n_k n_{k+1}) = f(n_1)f(n_2) \ldots f(n_k)f(n_{k+1}).$$

The truth of the first statement of the theorem follows by induction.

Since the primes p_1, p_2, $\ldots, p_k$ are all distinct, $(p_i^{a_i}, p_j^{a_j}) = 1$
if $i \neq j$. Thus when n is written in standard form it can be thought of as a
product of k factors, $p_i^{a_i}$, which are relatively prime in pairs. Let $n_i = p_i^{a_i}$.
Then

$$f(n) = f(p_1^{a_1})f(p_2^{a_2}) \ldots f(p_k^{a_k}). \quad \blacklozenge$$

If we use the shorter notation suggested for the standard form of n, the formula becomes:

$$f(n) = \prod_{i=1}^{k} f(p_i^{a_i}), \qquad \text{when } n = \prod_{i=1}^{k} p_i^{a_i}.$$

The importance of this theorem is that it reduces the problem of deriving a formula for $f(n)$ to that of deriving a formula for $f(p^{a_i})$, a much easier job. Of course, it is necessary to know that f is multiplicative before this theorem can be used. Since we have already shown that the function σ is multiplicative, it is now a simple problem to derive a formula for $\sigma(n)$.

THEOREM 1.11.2.   *A Formula for $\sigma(n)$*

$$\sigma(n) = \prod_{i=1}^{k}(1 + p_i + \cdots + p_i^{a_i}) = \prod_{i=1}^{k} \frac{p_i^{a_i+1} - 1}{p_i - 1}.$$

Proof: Let $n = \prod_{i=1}^{k} p_i^{a_i}$ be the standard form of n. Then $\sigma(n) = \prod_{i=1}^{k} \sigma(p_i^{a_i})$. The divisors of p^a are $1, p, p^2, \ldots, p^a$. Therefore, by definition, $\sigma(p^a) = 1 + p + \cdots + p^a$. This establishes the first form of $\sigma(n)$.

Since the sum $1 + p + \cdots + p^a = (p^{a+1} - 1)/(p - 1)$ (see Exercises 1.3, problem 7), the second expression is immediate. ●

EXAMPLE 1.11.1. $1008 = 2^4 \cdot 3^2 \cdot 7$. Hence

$$\sigma(1008) = \sigma(2^4)\sigma(3^2)\sigma(7)$$

$$= \frac{2^{4+1} - 1}{2 - 1} \cdot \frac{3^{2+1} - 1}{3 - 1} \cdot \frac{7^{1+1} - 1}{7 - 1}$$

$$= 31 \cdot 13 \cdot 8 = 3224.$$

The other form for σ gives

$$\sigma(1008) = (1 + 2 + 4 + 8 + 16)(1 + 3 + 9)(1 + 7) = 31 \cdot 13 \cdot 8 = 3224.$$

The formulas in Theorem 1.11.2 can be used for deriving properties of the function σ. Which form of the expression is more suitable depends on the situation and on your own preference.

EXAMPLE 1.11.2. For what n is $\sigma(n) = 12$? This requires

$$\prod_{p|n} (1 + p + \cdots + p^{a_p}) = 12.$$

If only one prime divides n, $p = 11$ and $a_p = 1$ satisfy the equation. If there are two prime divisors, we need $p = 3$, $a_p = 1$, along with $p = 2$, $a_p = 1$, so that $n = 3 \cdot 2$. These are the only possibilities. Note that we need to check only primes less than 12.

EXAMPLE 1.11.3. Prove that

$$\sigma(n) < n \prod_{p|n} \frac{p}{p-1}.$$

Proof:

$$\sigma(n) = \prod_{p|n} \frac{p^{a_p+1} - 1}{p - 1} < \prod_{p|n} \frac{p^{a_p+1}}{p - 1} = \prod_{p|n} p^{a_p} \prod_{p|n} \frac{p}{p - 1} = n \prod_{p|n} \frac{p}{p - 1}. \quad \bullet$$

Since multiplicative functions are so naturally related to the multiplicative structure of the integers, it is not surprising that a great many multiplicative functions exist and that they are encountered frequently when exploring the integers. An important multiplicative function will be introduced in Unit 2. In Unit 7 the multiplicative functions are discussed in more detail, along with the structure of the set of number-theoretic functions. At present we will look briefly at a way in which sums of multiplicative functions yield new multiplicative functions.

EXAMPLE 1.11.4. The constant function $u(n) = 1$ is certainly multiplicative; in fact, it is completely multiplicative. Consider the function f such that $f(n)$ is the sum formed by adding a term $u(d)$ for each divisor d of n. Thus

$$f(6) = u(1) + u(2) + u(3) + u(6) + 1 + 1 + 1 + 1 = 4.$$

Since $u(d) = 1$ for every d, the sum has a term 1 for each divisor of n and hence is equal to the number of divisors of n. Thus the function f defined by $f(n) = \sum_{d|n} u(d)$ is actually the function τ.

THEOREM 1.11.3. *The Function* $F(n) = \sum_{d|n} f(d)$. The function F, such that $F(n) = \sum_{d|n} f(d)$, is multiplicative if f is multiplicative.

Proof: Let $(a, b) = 1$.

$$F(ab) = \sum_{d|ab} f(d).$$

But the divisors of ab can be expressed as products of the form $d'd''$, where $d'|a$ and $d''|b$. Therefore

$$\sum_{d|ab} f(d) = \sum_{d'|a} \sum_{d''|b} f(d'd'') = \sum_{d'|a} \sum_{d''|b} f(d')f(d''),$$

since $(d', d'') = 1$ and f is multiplicative. But $\sum_{d'|a} f(d') = F(a)$ and $\sum_{d''|b} f(d'') = F(b)$. Thus $F(ab) = F(a)F(b)$ and F is multiplicative. ●

EXAMPLE 1.11.5. Find an expression for $\sum_{d|n} d^2$. Let $F(n) = \sum_{d|n} f(d)$, where f is the function defined by $f(n) = n^2$. Since f is multiplicative, F is multiplicative, and we can find its value for any n if we calculate $F(p^a)$ for p a prime and use Theorem 1.11.1. By definition,

$$F(p^a) = \sum_{d|p^a} d^2 = 1 + p^2 + p^4 + \cdots + p^{2a}.$$

This is a geometric series whose sum is $(p^{2a+2} - 1)/(p^2 - 1)$. If

$$n = \prod_{i=1}^{k} p_i^{a_i},$$

we obtain in this way

$$F(n) = \prod_{i=1}^{k} \frac{p_i^{2a_i+2} - 1}{p_i^2 - 1}.$$

In Example 1.11.4 we saw that $\sum_{d|n} u(d) = \tau(n)$, where u is the constant function $u(n) = 1$. Does there exist any other function g such that $\sum_{d|n} g(d) = \tau(n)$? If the function F has the property that $F(n) = \sum_{d|n} f(d)$, is this representation unique?

EXAMPLE 1.11.6. Let $\sum_{d|n} g(d) = \tau(n)$. Prove that $g(n) = 1$ for every n; that is, $g = u$. Let $n = 1$. The equation becomes $g(1) = \tau(1)$, so that $g(1) = 1$. Since $\tau(2) = g(1) + g(2)$, $g(2) = \tau(2) - g(1) = 2 - 1 = 1$. Suppose that $g(k) = 1$ for every $k < n$. Since $\sum_{d|n} g(d) = \tau(n)$, we can write

$$g(n) = \tau(n) - \sum_{\substack{d|n \\ d<n}} g(d).$$

Now, for every d in the summation, $g(d) = 1$ by hypothesis. Since all the divisors of n are included except n itself, there are $\tau(n) - 1$ terms in the sum. Therefore $g(n) = \tau(n) - \tau(n) + 1 = 1$. By induction, we conclude that $g(n) = 1$ for every n. See Exercises 1.11, problem 13.

EXERCISES 1.11

Checks

1. Calculate $\sigma(72)$, $\sigma(250)$, $\sigma(8000)$.

2. Prove that there is no positive integer n such that $\sigma(n) = 5$. *Hint*: Remember that $\sigma(n) > n$ if $n > 1$.

3. If f and g are multiplicative, prove that the pointwise product, fg, is also multiplicative. $[fg(n) = f(n)g(n).]$

4. Prove that $\sigma(m)/m$ is the sum of the reciprocals of the positive divisors of m.

5. Let u_1 be the identity function, $u_1(n) = n$. Prove that

$$\sum_{d|n} u_1(d) = \sigma(n).$$

6. A positive integer is called *perfect* if $\sigma(n) = 2n$. Prove that if $2^k - 1$ is a prime, then $2^{k-1}(2^k - 1)$ is a perfect number.

Challenges

7. For what values of n does $\sigma(n) = b$ if $b = 14$; $b = 15$; $b = 16$; $b = 18$?

8. Prove that if n is an even number, there exists a prime p such that $p = 2^k - 1$ and $n = 2^{k-1}p$. *Hint*: Set $n = 2^{k-1}n_0$. Show that $\sigma(n) = 2n$ implies that $n_0 = (2^k - 1)u$, where $u = 1$ and $2^k - 1$ is prime.

9. If the number of distinct prime factors of n is less than or equal to 5, prove that $\sigma(n) < 5n$.

10. Find an expression for

$$\sum_{d|n} \tau(d).$$

11. The function $\sigma_k(n)$ is defined to be the sum of the kth powers of the positive divisors of n. Find an expression for $\sigma_k(n)$. Note that $\tau = \sigma_0$.

12.   Define f as follows: $f(1) = 1$, $f(n) = (-1)^k$, where k is the total number of prime factors of n counting repetitions. For example, if $n = 12$, $k = 3$. Let $F(n) = \sum_{d|n} f(d)$. Prove that $F(n)$ is either 0 or 1. For what integers is $F(n) = 0$?

13.   Let $F(n) = \sum_{d|n} f(d)$ and $F(n) = \sum_{d|n} g(d)$. Prove that $f(n) = g(n)$ for every positive integer n.

1.12   THE p-VALUE OF AN INTEGER

Let us look again at the fundamental theorem of arithmetic and this time focus attention on a particular prime. This theorem implies that the factorization of a nonzero integer n determines for a fixed prime p a unique integer k such that p^k divides n and p^{k+1} does not divide n.

EXAMPLE 1.12.1

Let $p = 5$, $n = 35$. Since $35 = 5 \cdot 7$, $k = 1$.
Let $p = 5$, $n = 1875$. Since $1875 = 5^4 \cdot 3$, $k = 4$.
Let $p = 5$, $n = -105$. Since $-105 = (-1)3 \cdot 5 \cdot 7$, $k = 1$.
Let $p = 5$, $n = 48$. Since $48 = 2^4 \cdot 3$, $k = 0$.

THEOREM 1.12.1. *p-Factorization.* Every nonzero integer n can be written uniquely in the form $p^k n_0$, where p is a fixed prime, k is a nonnegative integer, and n_0 is an integer such that $(n_0, p) = 1$.

Proof: Let the standard form of $|n|$ be $p_1^{a_1} p_2^{a_2} \ldots p_r^{a_r}$. Then $n = \pm p_1^{a_1} p_2^{a_2} \ldots p_r^{a_r}$. If p does not divide n, then $k = 0$ and $n_0 = n$. If $p|n$, then $p = p_i$ for some i, $i = 1, 2, \ldots, r$. In this case, $n = p_i^{a_i} n_0$, where

$$n_0 = \pm \prod_{\substack{j=1 \\ j \neq i}}^{r} p_j^{a_j} \quad \text{and} \quad k = a_i. \quad \bullet$$

EXAMPLE 1.12.2.   Let $p = 5$. We write $35 = 5 \cdot 7$, $1875 = 5^4 \cdot 3$, $-105 = 5(-21)$, $48 = 5^0 \cdot 48$, $-25 = 5^2(-1)$.

In general, we can separate integers into classes according to the highest power of p which divides that integer. An expression for the highest power of p that divides $n!$ can be written using the greatest integer function.

EXAMPLE 1.12.3. Find the highest power of 3 that divides 50! How many integers in the set 1, 2, 3, ..., 50 are multiples of 3? There are [50/3]. Some of these are also multiples of 9 and contribute a factor of 3^2 to the product 50!. There are [50/9] multiples of 9. Of these [50/27] are multiples of 27 and contribute a factor of 3^3 to the product. Since $3^4 = 81 > 50$, there are no multiples of 81 less than 50. The sum

$$\left[\frac{50}{3}\right] + \left[\frac{50}{9}\right] + \left[\frac{50}{27}\right] = 16 + 5 + 1$$

gives the exact number of factors of 3 that occur in the product 50!. Note that $[50/3^n] = 0$ for all $n \geq 4$. These terms could have been included without altering the value of the sum. The result could then be written as

$$\left[\frac{50}{3}\right] + \left[\frac{50}{3^2}\right] + \left[\frac{50}{3^3}\right] + \cdots + \left[\frac{50}{3^n}\right] + \cdots ;$$

that is, $\sum_{n=1}^{\infty}[50/3^n]$, where the sum has only a finite number of nonzero terms.

The following theorem gives the result in the general case.

THEOREM 1.12.2. *The p-Factorization of n!.*

$$n! = p^e n_0, \quad \text{where} \quad (n_0, p) = 1 \quad \text{and} \quad e = \sum_{k=1}^{\infty} \left[\frac{n}{p^k}\right].$$

Proof: The proof can be carried out on the lines of the preceding example, or by induction on n, and is left as an exercise. ●

EXAMPLE 1.12.4. Prove that the binomial coefficient $\binom{27}{13}$ is divisible by 5.

$$\binom{27}{13} = \frac{27!}{13!\,14!} = \frac{5^6 a}{5^2 b 5^2 c} = 5^2 \frac{a}{bc}.$$

If we assume that we already know that the binomial coefficient is an integer, this proves that $5 | \binom{27}{13}$. How could you use Theorem 1.12.1 to prove the binomial coefficient is an integer?

The p-factorization of an integer allows us to define an interesting number-theoretic function called the *p-value of n*. If $n \neq 0$, n has a unique representation in the form $p^k n_0$, where $(n_0, p) = 1$. In this case, the p-value of n is defined to be $1/p^k$. If $n = 0$, this definition is not useful, since k is not uniquely determined. In fact, $p^k | 0$ for every k. Since $1/p^k$ approaches zero as k becomes large, it is reasonable to define the p-value of 0 to be 0. The remainder of this section is concerned with a discussion of the properties of the p-value function. This function is of major importance in Unit 8.

DEFINITION 1.12.1. *p-Value, p a Fixed Prime.* The p-value of an integer is a function from the integers onto the set of numbers $\{0, 1, 1/p, 1/p^2, \ldots, 1/p^k, \ldots\}$ which is defined as follows: The p-value of n is $1/p^k$ if $n = p^k n_0$, where $(n_0, p) = 1$. The p-value of 0 is 0.

$$\boxed{|n|_p}$$

The notation $|n|_p$ is used to represent the p-value of n.

EXAMPLE 1.12.5

$$|35|_5 = \frac{1}{5}, \quad |100|_5 = \frac{1}{25}, \quad |48|_5 = 1, \quad |-60|_5 = \frac{1}{5}.$$

$$|35|_3 = 1, \quad |105|_3 = \frac{1}{3}, \quad |3500|_7 = \frac{1}{7}, \quad |3500|_5 = \frac{1}{5^3}, \quad |3500|_2 = \frac{1}{2^2}.$$

Is there an integer whose p-value is the same for all p? Consider $|a \cdot b|_p$ and $|a|_p \cdot |b|_p$. Is $|n|_p$ a multiplicative function? Is it completely multiplicative? Try some examples of $|a|_p$, $|b|_p$ and $|a + b|_p$. Can you make a guess about the relationship between $|a + b|_p$ and the pair of numbers $|a|_p$ and $|b|_p$?

The notation used for p-value is very similar to the notation used for absolute value. The reason is that the p-value function has many properties similar to the properties of absolute value. Remember from elementary algebra the following: $|a| \geq 0$ and $|a| = 0$ implies $a = 0$; $|ab| = |a| |b|$; $|a + b| \leq |a| + |b|$. The following theorem shows that the p-value function has these properties and, in addition, satisfies an inequality that is "stronger" than the triangle inequality.

THEOREM 1.12.3. *Basic Properties of p-Value.*

1. $|a|_p \geq 0$ and equals zero only if $a = 0$.
2. $|ab|_p = |a|_p |b|_p$.
3. $|a + b|_p \leq \max(|a|_p, |b|_p)$, non-Archimedean property.
4. $|a + b|_p \leq |a|_p + |b|_p$, triangle inequality.

Proof: 1. This statement follows directly from the definition of p-value.

2. If either a or b is zero, property (2) is trivial. Let $a = p^\alpha a_0$ and $b = p^\beta b_0$, where $(a_0, p) = 1$ and $(b_0, p) = 1$. The multiplicative property of relatively prime pairs implies $(a_0 b_0, p) = 1$. Therefore $ab = p^{\alpha+\beta} a_0 b_0 = p^{\alpha+\beta} c_0$, where $(c_0, p) = 1$, and

$$|ab|_p = \frac{1}{p^{\alpha+\beta}} = \left(\frac{1}{p^\alpha}\right)\left(\frac{1}{p^\beta}\right) = |a|_p |b|_p.$$

3. Again the statement is trivial if either a or b is 0. Let $a = p^\alpha a_0$ and $b = p^\beta b_0$. There is no loss of generality in assuming $\alpha \geq \beta$, so that $1/p^\alpha \leq 1/p^\beta$. Then

$$a + b = p^\beta(a_0 p^{\alpha-\beta} + b_0).$$

If $(a_0 p^{\alpha-\beta} + b_0, p) = 1$, let $c_0 = a_0 p^{\alpha-\beta} + b_0$. Then $a + b = p^\beta c_0$, where $(c_0, p) = 1$, so that

$$|a + b|_p = \frac{1}{p^\beta} = \max(|a|_p, |b|_p).$$

If $p \mid a_0 p^{\alpha-\beta} + b_0$, there exists a $\gamma > \beta$ such that

$$p^\beta(a_0 p^{\alpha-\beta} + b_0) = p^\gamma c_0,$$

where $(c_0, p) = 1$. In this case,

$$|a + b|_p = \frac{1}{p^\gamma} < \frac{1}{p^\beta}$$

so that

$$|a + b|_p < \max(|a|_p, |b|_p).$$

4. Since

$$\max \left(|a|_p, |b|_p \right) \le |a|_p + |b|_p,$$

property (4) is an immediate consequence of property (3). ●

COROLLARY 1.12.2. If $|a|_p \ne |b|_p$, then

$$|a + b|_p = \max \left(|a|_p, |b|_p \right).$$

A value function that satisfies the inequality in property (3) of Theorem 1.12.3—that is, $|a + b|_p \le \max\{|a|_p, |b|_p\}$—is classified as *non-Archimedean*. The ring of integers, with the p-value function defined above, is a non-Archimedean valued ring.

EXERCISES 1.12

Checks

1. Find the 5-value of each of the integers: 45, 70, −250, −55, 135, 99.

2. Find the 3-value of the integers in problem 1. Find the 7-value of these integers.

3. Write an integer whose 5-value is $\frac{8}{9}$ times its 3-value. What property characterizes such integers?

4. Write an integer that has p-value 1 for each of the primes 3, 5, 11. Can you give the general description of such an integer?

5. Find the exponent of 7 in the factorization of 82!. Prove that 7 does not divide the binomial coefficient $\binom{82}{12}$ but does divide the binomial coefficient $\binom{82}{36}$. Remember that

$$\binom{n}{k} = \frac{n!}{k!\,(n-k)!}.$$

6. Use Theorem 1.12.2 to write 20! in standard form. (Note that if $p > 20$, p does not divide 20!.)

7. Give examples to show that equality and inequality are each possible in property (3), Theorem 1.12.3.

8. Prove that $|1|_p = 1$ for every p; $|-a|_p = |a|_p$ for every p and for every a.

Challenges

9. Prove that $|1 + x|_p = 1$ if $|x|_p < 1$.

10. Prove that if $a | b$, then $|a|_p \geq |b|_p$ for every p.

11. Prove that if $|a|_p \geq |b|_p$ for every p, then $a | b$. Contrast the relation between p-value and divisibility (as established in problems 10 and 11) with the relation between absolute value and divisibility.

12. Prove that $|(a, b)|_p = \max(|a|_p, |b|_p)$. What can you say about $|[a, b]|_p$?

CONGRUENCE— A WELL-MARKED PATH

2.1 RESIDUE CLASSES

The division algorithm, introduced in Section 1.4, was put to use on several occasions in Unit 1. For our present purposes, this theorem is a basic and most essential tool. So much is this the case that before starting, we will find it useful to take a new look at the division algorithm and set it up in a notation that makes it easier to use.

EXAMPLE 2.1.1. For any integer a, there exist unique integers q and r such that $a = 6q + r$, where $0 \leq r < 6$. This statement reworded says: a belongs to exactly one of the sets M_6, $M_6 + \{1\}$, $M_6 + \{2\}$,

$M_6 + \{3\}$, $M_6 + \{4\}$, $M_6 + \{5\}$. (Recall that $M_6 + \{r\}$ is the set of integers of the form $6q + r$, where q is an integer.) If a_1 and a_2 are both in $M_6 + \{r\}$, then $a_1 = 6q_1 + r$ and $a_2 = 6q_2 + r$, so that $a_1 - a_2 = 6(q_1 - q_2)$ and $6\,|\,a_1 - a_2$. The condition $6\,|\,a_1 - a_2$ identifies the fact that a_1 and a_2 are in the same set $M_6 + \{r\}$, although it does not tell us the value of r. For example, $6\,|\,17 - 5$, also, $6\,|\,250 - (-2)$. Here 17 and 5 belong to $M_6 + \{5\}$, whereas 250 and -2 belong to $M_6 + \{4\}$. Two integers that fall in the same set $M_6 + \{r\}$—that is, whose difference is divisible by 6—are said to be *congruent modulo* 6.

$$\boxed{a \equiv b \bmod m}$$

DEFINITION 2.1.1. $a \equiv b \bmod m$. Let a and b be integers, and m an integer greater than zero. If $m\,|\,a - b$, then a and b are said to be *congruent modulo* m. This is written $a \equiv b \bmod m$. The integer m is called the *modulus*.

THEOREM 2.1.1. *Characterizations of* $a \equiv b \bmod m$. Let a and b be integers and let m be a fixed positive integer. The following statements are equivalent:

  1. $a \equiv b \bmod m$.
  2. a and b belong to the same set $M_m + \{r\}$ for some r, $0 \le r < m$.
  3. There exist integers q_1 and q_2 such that $a = q_1 m + r$ and $b = q_2 m + r$, for some r.
  4. There exists an integer q such that $a = qm + b$.

Proof: Statement (1) implies statement (2). Since $a \equiv b \bmod m$, Definition 2.1.1 implies that $m\,|\,a - b$; that is, $a - b = km$ for some k. By the division algorithm,

$$b = qm + r, \quad 0 \le r < m.$$

Hence

$$a = (q + k)m + r.$$

Thus a and b both belong to $M_m + \{r\}$.

   Statement (2) implies statement (3), from the definition of $M_m + \{r\}$.

Statement (3) implies statement (4). If

$$a = q_1 m + r \quad \text{and} \quad b = q_2 m + r,$$

then

$$a = b = (q_1 - q_2)m = qm \quad \text{and} \quad a = qm + b.$$

Statement (4) implies statement (1), since if $a - b = qm$, $m \mid a - b$.

Since (1) implies (2), (2) implies (3), (3) implies (4), and (4) implies (1), the four statements are equivalent. ⬢

Congruence modulo m is a relation defined on the set of integers. It is a special type of relation known as an *equivalence relation*, which means that it has three properties—reflexive, symmetric, and transitive. A relation is *reflexive* if each element is related to itself. In our case, since $m \mid a - a$, then $a \equiv a \bmod m$. A relation is *symmetric* if whenever a is related to b, b is related to a. In our case, if $a \equiv b \bmod m$, then $m \mid a - b$, or $m \mid b - a$, which implies that $b \equiv a \bmod m$. The *transitive* property means that if a is related to b and b is related to c, then a is related to c. In our case, if $a \equiv b \bmod m$ and $b \equiv c \bmod m$, then $m \mid a - b$ and $m \mid b - c$. This implies $m \mid a - b + b - c$, or $a \equiv c \bmod m$. An equivalence relation separates the set on which it is defined into disjoint subsets called *equivalence classes*. In our case, the set in question is the set of integers. The equivalence classes are the sets $M_m + \{r\}$, $0 \le r < m$.

DEFINITION 2.1.2. *Residues and Residue Classes.* The sets

$$M_m + \{r\} = \{qm + r : m \text{ and } r \text{ fixed integers}, q = 0, \pm 1, \pm 2, \ldots\}$$

are called *residue classes* modulo m. The elements of these sets are called *residues* mod m. If an integer a belongs to the set $M_m + \{r\}$, where $0 \le r < m$, then r is the *least non-negative residue congruent to* a mod m.

EXAMPLE 2.1.2. Let $m = 6$. There are six residue classes mod 6: M_6, $M_6 + \{1\}$, $M_6 + \{2\}$, $M_6 + \{3\}$, $M_6 + \{4\}$, $M_6 + \{5\}$. The least non-negative residue congruent to 250 mod 6 is 4. The least non-negative residue congruent to -31 mod 6 is 5. Any integer congruent to 5 mod 6 belongs to the set $M_6 + \{5\}$. The set $M_6 + \{17\} =$ the set $M_6 + \{5\}$.

THEOREM 2.1.2. *The Division Algorithm Restated.* Every integer belongs to exactly one of the m residue classes $M_m + \{r\}$, $0 \le r < m$.

Proof: Restatement of Theorem 1.4.1. ●

Given a fixed positive integer m, the set of integers is the union of m distinct residue classes, the sets $M_m + \{r\}$, $0 \le r < m$. An interesting question to consider is the relationship between residue classes for different values of m.

EXAMPLE 2.1.3. $M_6 + \{5\} \subset M_3 + \{5\}$, since every multiple of 6 is also a multiple of 3. If a is in $M_2 + \{5\}$, then $2 | a - 5$. If a is in $M_3 + \{5\}$, then $3 | a - 5$. Since $(2, 3) = 1$, this implies $6 | a - 5$. These statements imply

$$(M_2 + \{5\}) \cap (M_3 + \{5\}) \subset (M_6 + \{5\}).$$

Is this a proper inclusion? Since $6 | a - 5$ implies that both 2 and 3 divide $a - 5$, we have

$$(M_6 + \{5\}) \subset (M_2 + \{5\}) \cap (M_3 + \{5\}).$$

These two results imply that

$$(M_6 + \{5\}) = (M_2 + \{5\}) \cap (M_3 + \{5\}).$$

Because the ideas of congruence and divisibility are so closely related, the theorems we have proved about divisors can be used to obtain many results similar to those in Example 2.1.3. Some of these results are suggested in the exercises. You can undoubtedly create more for yourself. The following theorem gives a more general form of the relation in Example 2.1.3.

THEOREM 2.1.3. *Multiplicative Property of the Modulus.* If $a \equiv b \bmod m$, and $a \equiv b \bmod n$, then $a \equiv b \bmod [m, n]$.

Proof: Since $a \equiv b \bmod m$ implies that $m | a - b$, and $a \equiv b \bmod n$ implies that $n | a - b$, we see that $a - b$ is a common multiple of m and n. By the definition of least common multiple, it follows that $[m, n] | a - b$; that is, $a \equiv b \bmod [m, n]$. ●

COROLLARY 2.1.3a. If $a \equiv b \bmod m$, and $a \equiv b \bmod n$, and $(m, n) = 1$, then $a \equiv b \bmod mn$.

COROLLARY 2.1.3b. Let $m_1, m_2, \ldots, m_k$ be a set of integers relatively prime in pairs. If $a \equiv b \bmod m_i$, for $i = 1, 2, \ldots, k$, then $a \equiv b \bmod m_1 m_2 m_3 \cdots m_k$.

Proof: Use induction on k. ●

The converse of Theorem 2.1.3 is an immediate consequence of the properties of "divides." Since $m \mid a - b$ and $m' \mid m$ implies $m' \mid a - b$, there is a relationship between residue classes mod m and residue classes mod m' where $m = m'd$. This situation is discussed in the following theorem.

THEOREM 2.1.4. *Nested Residue Classes.* Let $m = m'd$, where m and m' are integers greater than zero. Each residue class mod m' is the union of d distinct residue classes mod m; that is, $a \equiv b \bmod m'$ if and only if, for some i, $0 \le i < d$, $a \equiv b + im' \bmod m$. The value of i is unique.

Proof: If $a \equiv b + im' \bmod m$, then $a = b + im' + km$ for some k—that is, $a = b + (i + kd)m'$, which implies that $a \equiv b \bmod m'$.

Suppose that $a \equiv b \bmod m'$. If $a = b$, then $a \equiv b + 0 \cdot m' \bmod m$. If $a \ne b$, then $a = b + km'$, where $k \ne 0$. By the division algorithm there is a unique i such that $k = dq + i$, $0 \le i < d$. Thus

$$a = b + im' + qdm' = b + im' + qm.$$

This result implies that $a \equiv b + im' \bmod m$.

The theorem follows by the definition of union of sets. ●

COROLLARY 2.1.4. $a \equiv b \bmod p^k$ if and only if, for some i, $0 \le i < p$, $a \equiv b + ip^k \bmod p^{k+1}$. That is, each residue class mod p^k is the union of p residue classes mod p^{k+1}.

EXAMPLE 2.1.4. Let $p = 3$. The class $M_3 + \{1\} = \{\cdots -11, -8, -5, -2, 1, 4, \cdots\}$. In this class are integers like $-8, 1, 10$ which are congruent to 1 mod 9. Also, there are integers such as $-5, 4, 13$ which are congruent to 4 mod 9, and integers such as $-11, -2, 7$ which are congruent to 7 mod 9. There are no integers congruent to 2, 3, 5, 6, 8, 9 mod 9, since these numbers are not congruent to 1 mod 3. Thus

$$M_3 + \{1\} = [M_9 + \{1\}] \cup [M_9 + \{4\}] \cup [M_9 + \{7\}].$$

Each of the residue classes mod 9 is, in turn, the union of three residue classes mod 27. For example,

$$M_9 + \{4\} = [M_{27} + \{4\}] \cup [M_{27} + \{13\}] \cup [M_{27} + \{22\}].$$

A geometric picture of the situation is shown in the figure. Each residue class is represented by a finite interval.

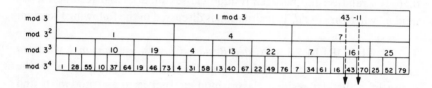

The integers 43 and -11 are both congruent to 1 mod 3, to 7 mod 3^2, and to 16 mod 3^3. But $43 \equiv 43$ mod 3^4 and $-11 \equiv 70$ mod 3^4.

A sequence of sets A_n with the property that $A_{n+1} \subset A_n$ is called a nested sequence. Thus the sequence of residue classes $M_{p^n} + \{r\}$ is a nested sequence of residue classes.

In Section 2.2 we shall see how the nature of the residue classes makes it possible for us to carry out our exploration by using only representatives of these classes and the notation $\equiv$. Nevertheless, it is important for us to have a basic understanding of the relationships expressed in Theorems 2.1.3 and 2.1.4 lest we jump at conclusions too quickly or overlook important information not set out explicitly in the statement $a \equiv b$ mod m.

EXERCISES 2.1

Checks

1. Rephrase the statement $5 \equiv 17$ mod 3 in the three alternate ways suggested in Theorem 2.1.1.

2. List five integers in $M_{10} + \{2\}$; in $M_{13} + \{5\}$; in $M_{11} + \{1\}$. For what r, $0 \leq r < 11$, do the following integers belong to $M_{11} + \{r\}$: 81, -3, 142, -75, 0.

3. Prove that

$$M_{10} + \{2\} \subset M_5 + \{2\}.$$

Prove that

$$[M_{10} + \{2\}] \cdot [M_{10} + \{3\}] \subset [M_{10} + \{6\}].$$

Can $\subset$ be replaced by $=$?

4. Prove that $[M_{10} + \{2\}] + [M_{10} + \{5\}] = M_{10} + \{7\}$.

5. Show that the geometric concept of similar triangles in the plane is an equivalence relation but that the relation $a < b$ defined on the integers is not an equivalence relation.

6. Express the residue class $M_4 + \{2\}$ as the union of residue classes mod 12.

Challenges

7. Prove that if $a \mid b$, $M_a + \{r\} \supset M_b + \{r\}$.

8. Prove that $[M_a + \{r\}] \cap [M_b + \{r\}] = M_{[a,\, b]} + \{r\}$. How is $[M_a + \{r\}] \cup [M_b + \{r\}]$ related to $M_{[a,\, b]} + \{r\}$?

9. Prove that $M_m + \{r_1\}$ and $M_m + \{r_2\}$ are either disjoint or identical.

10. The concept of nested residue classes is closely related to the expansion of an integer in base p. Prove that every positive integer can be written uniquely in the form

$$n = a_0 + a_1 p + a_2 p^2 + \cdots + a_k p^k, \qquad 0 \leq a_i < p,$$

where k is the unique integer such that $p^k \leq n < p^{k+1}$. This statement is equivalent to the statement: If $n \equiv a_0 + a_1 p + \cdots + a_r p^r \bmod p^{r+1}$, then for some $a_{r+1}, 0 \leq a_{r+1} < p, n \equiv a_0 + a_1 p + \cdots + a_r p^r + a_{r+1} p^{r+1} \bmod p^{r+2}$.

11. Prove that every negative integer can be written in the form

$$n = a_0 + a_1 p + a_2 p^2 + \cdots + a_k p^k - p^{k+1}, \quad 0 \leq a_i < p$$

where k is the unique integer such that $-p^{k+1} \leq n < -p^k$.

2.2 ARITHMETIC OF CONGRUENCE AND THE RING OF RESIDUES

The many relations among the residue classes suggest that it must be possible to combine statements about congruence in a variety of ways. Since congruence is much like equality, we are tempted to hope that we could manipulate with congruence in much the same way as we do with equality. We have nothing to lose by trying.

EXAMPLE 2.2.1. $22 \equiv 12 \bmod 5$ and $-1 \equiv 14 \bmod 5$. Can we add? $21 \equiv 26 \bmod 5$? Yes. Can we multiply? $-22 \equiv 168 \bmod 5$? Yes. Division is a bit trickier. We must stay within the integers, but $22 \equiv 12 \bmod 5$ ought to allow us to say $11 \equiv 6 \bmod 5$. True. Try another example. $8 \equiv 20 \bmod 12$, and $3 \equiv -9 \bmod 12$. Can we add? $11 \equiv 11 \bmod 12$ is true. Can we multiply? $24 \equiv -180 \bmod 12$ is true. Can we divide the common factor 4 from $8 \equiv 20 \bmod 12$? $2 \equiv 5 \bmod 12$ is false! Try again. $80 \equiv 20 \bmod 12$. Can we divide the common factor 5? $16 \equiv 4 \bmod 12$ is true. Can we divide the common factor 10? $8 \equiv 2 \bmod 12$ is false! If we write the congruence in terms of divisibility, the problem is soon identified. $12 | 8 - 20$ means $12 | 4(2 - 5)$. This does not imply that $12 | (2 - 5)$, since $(12, 4) \neq 1$. However, $12 | 5(16 - 4)$ does imply $12 | (16 - 4)$, since $(12, 5) = 1$. [Fundamental Theorem of Factorization, 1.8.1.] Why did no problem arise when the modulus was 5?

THEOREM 2.2.1. *The Arithmetic of Congruence.*

1. If $a \equiv b \bmod m$ and $c \equiv d \bmod m$, then $a + c \equiv b + d \bmod m$.
2. If $a \equiv b \bmod m$ and k is an integer, then $ka \equiv kb \bmod m$.
3. If $a \equiv b \bmod m$ and $c \equiv d \bmod m$, then $ac \equiv bd \bmod m$.
4. If $a \equiv b \bmod m$, then $a^k \equiv b^k \bmod m$, for every integer $k > 0$.
5. If $ka \equiv kb \bmod m$, and $k \neq 0$, then $a \equiv b \bmod m/(m, k)$.
6. If $ka \equiv kb \bmod m$, and $(k, m) = 1$, then $a \equiv b \bmod m$.

Proof:

1. Since $m | a - b$ and $m | c - d$, then $m | a - b + c - d$; that is, $m | (a + c) - (b + d)$, so that

$$a + c \equiv b + d \bmod m.$$

2. If $a \equiv b \bmod m$, then $m | a - b$ and therefore $m | ka - kb$, so that $ka \equiv kb \bmod m$.

3. Since $a \equiv b \bmod m$, property (2) implies that $ac \equiv bc \bmod m$. But $c \equiv d \bmod m$ implies $bc \equiv bd \bmod m$. From the transitive property of congruence, $ac \equiv bd \bmod m$.

4. This follows from (3) for $k = 2$ by multiplying the congruence $a \equiv b \bmod m$ with itself. The statement $a^k \equiv b^k \bmod m$ can be established by induction.

5. Let $(k, m) = d$. Then $k = k'd$ and $m = m'd$, where $(k', m') = 1$ [characterization of the greatest common divisor, Theorem 1.6.1]. Now

$m|ka - kb$ implies that $m'd|k'd(a - b)$—that is, $m'|k'(a - b)$. Since $(m', k') = 1$, this means that $m'|a - b$, or $a \equiv b \bmod m'$. This is statement (5), since $m' = m/(m, k)$.

6. This is an immediate consequence of statement (5) and is stated separately because it is often used in this form. ●

COROLLARY 2.2.1. Let $f(x)$ be a polynomial of degree n with integral coefficients:

$$f(x) = a_n x^n + a_{n-1} x^{n-1} + \cdots + a_2 x^2 + a_1 x + a_0.$$

If $a \equiv b \bmod m$, then $f(a) \equiv f(b) \bmod m$.

Proof: The proof of the corollary involves combining the statements of the theorem and is left as an exercise. ●

The arithmetic of congruence suggests that if a and b belong to the same residue class mod m, they can be treated in some ways as if they were equal. If the integers in a residue class are in some sense "practically equal," we ought to choose one of them to represent the whole class. This choice could be made in many ways.

EXAMPLE 2.2.2. Let $m = 6$. The set $\{0, 1, 2, 3, 4, 5\}$ consists of exactly one integer from each residue class mod 6. So does the set $\{0, \pm 1, \pm 2, 3\}$. So does the set $\{0, 5, 10, 15, 20, 25\}$. Any set of six integers, exactly one from each residue class mod 6, is called a complete residue system mod 6. In the first set, the smallest possible non-negative integers are chosen as representatives. The second chooses the representatives so that they will be as small as possible in absolute value. The third chooses multiples of 5. Can you find a complete residue system mod 6 in which each residue is a multiple of 4? As we proceed, we will have occasion to use residue systems chosen in each of these ways. The best choice depends on the situation.

DEFINITION 2.2.1. *Complete Residue System.* A set of m integers that consists of exactly one integer from each of the residue classes mod m is called a complete residue system mod m. The set $\{0, 1, 2, \ldots, m - 1\}$ is the least non-negative complete residue system mod m.

EXAMPLE 2.2.3. The least non-negative complete residue system mod 6 is $\{0, 1, 2, 3, 4, 5\}$. Here 2 represents the class $M_6 + \{2\}$ and 5 represents the class $M_6 + \{5\}$. What about the integer $2 + 5 = 7$? Certainly 7 belongs to some residue class mod 6, but it is not the representative chosen for its class. Since $7 \equiv 1 \bmod 6$, 7 belongs to $M_6 + \{1\}$. When we are considering residue classes mod 6, we can write $2 + 5 = 1$.

Similarly, the product of 2 and 5 is 10, which belongs to $M_6 + \{4\}$, so that when we are considering residue classes mod 6, $2 \cdot 5 = 4$. Let us put a circle around the $+$ and $\cdot$ to indicate that we are adding and multiplying in terms of residue classes mod 6. Define $a \oplus b$ to be the integer c in the least non-negative complete residue system which represents the residue class to which $a + b$ belongs. Multiplication is defined by: $a \odot b = c$ if and only if c is a least non-negative residue such that $a \cdot b \equiv c \bmod 6$. From these definitions, we can construct the addition and multiplication tables shown:

$\oplus$ mod 6	0	1	2	3	4	5
0	0	1	2	3	4	5
1	1	2	3	4	5	0
2	2	3	4	5	0	1
3	3	4	5	0	1	2
4	4	5	0	1	2	3
5	5	0	1	2	3	4

$\odot$ mod 6	0	1	2	3	4	5
0	0	0	0	0	0	0
1	0	1	2	3	4	5
2	0	2	4	0	2	4
3	0	3	0	3	0	3
4	0	4	2	0	4	2
5	0	5	4	3	2	1

The structure of the tables indicates that the complete residue system mod 6, with the operations of addition and multiplication mod 6, constitutes a commutative ring and thus has many properties similar to those of the integers.

THEOREM 2.2.2. *The Ring of Residues.* Let $\{a_1, a_2, \ldots, a_m\}$ be the least non-negative complete residue system mod m. Define addition by $a_i \oplus a_j = a_k$ if and only if $a_i + a_j \equiv a_k \bmod m$. Define multiplication by $a_i \odot a_j = a_k$ if and only if $a_i \cdot a_j \equiv a_k \bmod m$. The set $\{a_1, a_2, \ldots, a_m\}$ with these operations is a commutative ring, denoted by Z_m.

Proof: The commutative and associative properties of addition and multiplication, as well as the distributive property of multiplication over addition, are immediate consequences of the corresponding properties of the integers. For example, to prove

$$a_i \odot (a_j \oplus a_k) = (a_i \odot a_j) \oplus (a_i \odot a_k),$$

we argue as follows: $a_j \oplus a_k = a_r$ implies $a_j + a_k \equiv a_r \bmod m$. Also, $a_i \odot a_r = a_n$ implies

$$a_n \equiv a_i \cdot a_r \equiv a_i \cdot a_j + a_i \cdot a_k \bmod m.$$

But

$$a_i \cdot a_j \equiv a_i \odot a_j \bmod m \quad \text{and} \quad a_i \cdot a_k \equiv a_i \odot a_k \bmod m,$$

so that

$$a_n \equiv (a_i \odot a_j) + (a_i \odot a_k) \bmod m;$$

that is,

$$a_n = (a_i \odot a_j) \oplus (a_i \odot a_k).$$

Exactly one of the integers $a_i = 0$. Let a_m designate this integer. Then a_m is the additive identity, since $a_m + a_i \equiv 0 + a_i \bmod m$; that is, $a_m \oplus a_i = a_i$ for every a_i. Every element has an additive inverse: if $a_m - a_i \equiv a_s \bmod m$, then $a_i \oplus a_s = a_m$. This inverse is unique, since $a_i \oplus a_s = a_i \oplus a_{s'} = a_m$ implies that $a_i + a_s \equiv a_i + a_{s'}$ which implies that $a_s = a_{s'}$. ◆

COROLLARY 2.2.2a. The residue classes mod m constitute a ring with the sum of two classes defined to be the residue class containing the sum of the elements of the original classes, and the product of two classes defined to be the class containing the products of elements in the original classes.

COROLLARY 2.2.2b. Any complete residue system mod m, with operations defined as in the theorem, is a ring.

For purposes of investigation, it is wise to construct addition and multiplication tables in the ring of residues for several different values of m. We include for future reference the tables mod 7 and mod 8 in the following example.

EXAMPLE 2.2.4. The rings Z_7 and Z_8. Use the complete residue system 0, 1, 2, 3, 4, 5, 6, mod 7.

$\oplus$	0	1	2	3	4	5	6
0	0	1	2	3	4	5	6
1	1	2	3	4	5	6	0
2	2	3	4	5	6	0	1
3	3	4	5	6	0	1	2
4	4	5	6	0	1	2	3
5	5	6	0	1	2	3	4
6	6	0	1	2	3	4	5

$\odot$	0	1	2	3	4	5	6
0	0	0	0	0	0	0	0
1	0	1	2	3	4	5	6
2	0	2	4	6	1	3	5
3	0	3	6	2	5	1	4
4	0	4	1	5	2	6	3
5	0	5	3	1	6	4	2
6	0	6	5	4	3	2	1

Use the complete residue system 0, 1, 2, 3, 4, 5, 6, 7, mod 8.

$\oplus$	0	1	2	3	4	5	6	7
0	0	1	2	3	4	5	6	7
1	1	2	3	4	5	6	7	0
2	2	3	4	5	6	7	0	1
3	3	4	5	6	7	0	1	2
4	4	5	6	7	0	1	2	3
5	5	6	7	0	1	2	3	4
6	6	7	0	1	2	3	4	5
7	7	0	1	2	3	4	5	6

$\odot$	0	1	2	3	4	5	6	7
0	0	0	0	0	0	0	0	0
1	0	1	2	3	4	5	6	7
2	0	2	4	6	0	2	4	6
3	0	3	6	1	4	7	2	5
4	0	4	0	4	0	4	0	4
5	0	5	2	7	4	1	6	3
6	0	6	4	2	0	6	4	2
7	0	7	6	5	4	3	2	1

EXERCISES 2.2

Checks

1. If $a \equiv b$ mod m, and $c \equiv d$ mod m, prove that $ax + cy \equiv bx + dy$ mod m for any integers x and y.

2. Prove that $10^k \equiv 1$ mod 9, for every integer $k > 0$.

3. Prove the remaining properties necessary to show Z_m is a ring.

4. Construct addition and multiplication tables mod 2, 3, 4, 5, 9, 10, 11, 12.

5. Compare the addition tables mod m. Does the additive identity occur in each row? What does this imply about additive inverses? Does each element occur in each row? What does this imply about the equation $x + a = b$? Is the table symmetric? What property does this imply about addition mod m? Find the additive inverse of 3 in Z_7. Find the additive inverse of 4 in Z_6, of 5 in Z_{11}, of 3 in Z_{12}. State and prove a theorem about the additive inverse of a in Z_m.

6. Compare the multiplication tables. Does the multiplicative identity occur in each row? What does this imply about multiplicative inverses? Is the answer different for different values of m? Compare the second and last row in each of the multiplication tables. Is the situation the same regardless of the modulus? State and prove the property exhibited in the last row of the tables.

7. Solve the following equations in Z_8 :

$$3x = 7, \quad 2x = 6, \quad 5x = 2, \quad 6x = 4.$$

In each case, how many solutions exist? For which a does $ax = 2$ have a unique solution? For which a does $ax = 2$ have two solutions? Is there an a for which $ax = 2$ has more than two solutions? Is there an a such that $ax = 2$ has no solutions? Answer the same questions for $ax = 7$. For what b does $2x = b$ have one solution? Two solutions? No solutions?

8. Compare the number of solutions of the equation $ax = b$ for different a and b in Z_7 with the solutions in Z_8. Consider the same equation in Z_{11}.

Challenges

9. In ordinary arithmetic, if $a^2 = b^2$, then $a = \pm b$. Is an analogous statement true in the ring of residues mod m?

10. Try other statements of ordinary arithmetic and check their analogs in various rings of residues.

11. If p is a prime, prove that the binomial coefficient $\binom{p}{r} \equiv 0 \mod p$ for $r = 1, 2, 3, \ldots, p - 1$. Derive from this the conclusion

$$(a + b)^p \equiv a^p + b^p \mod p.$$

12. Prove that an integer is congruent, mod 10, to its units digit. Use this result to prove that a fourth power must have 0, 1, 5, or 6 for its units digit.

13. In decimal notation (that is, base 10), the integer

$$a_n a_{n-1} a_{n-2} \cdots a_0 = a_n 10^n + a_{n-1} 10^{n-1} + \cdots + a_1 10 + a_0.$$

Prove that an integer in decimal notation is congruent mod 9 to the sum of its digits. This implies a well-known test for divisibility by 9—namely, an integer is divisible by 9 if and only if the sum of its digits is divisible by 9.

14. In elementary arithmetic, a process known as "casting out nines" is frequently taught as a means of checking multiplication. Suppose that $ab = c$. Add the digits of a; if the sum has more than one digit, add the digits of the sum and continue until an integer a' that is less than 10 is reached. Proceed in the same way to obtain b' from b and c' from c. The

product $a'b'$ reduced in the same way ought to be c'. Example: $a = 358$, $b = 219$, $c = 78402$. $3 + 5 + 8 = 16$, $1 + 6 = 7$, so $a' = 7$; $2 + 1 + 9 = 12$, $1 + 2 = 3$, so $b' = 3$; $7 + 8 + 4 + 0 + 2 = 21$, so $c' = 3$. Now $a'b' = 7 \cdot 3 = 21$, and $2 + 1 = 3$. Prove that $a \equiv a'$ mod 9, $b \equiv b'$ mod 9, and $c \equiv c'$ mod 9, so that if the multiplication is correct, $a'b' \equiv c'$ mod 9. What type of error would this method fail to detect?

15. Construct a test for divisibility by 11.

2.3   UNITS IN THE RING OF RESIDUES

If you have done the exercises in the preceding section faithfully, you will have already guessed much of what is said in this section. The structure of the addition table is much the same for any modulus. The structure of the multiplication table, however, for a prime modulus is quite different from the structure of the multiplication table for a composite modulus.

DEFINITION 2.3.1. *Units.* An element of a ring is called a unit if it has a multiplicative inverse.

EXAMPLE 2.3.1. In the ring of integers, Z, the only units are 1 and -1. How can we identify the units in Z_m? The table shown collects the information in the multiplication tables already constructed.

Ring	Units	Ring	Units
Z_2	1	Z_8	1, 3, 5, 7
Z_3	1, 2	Z_9	1, 2, 4, 5, 7, 8
Z_4	1, 3	Z_{10}	1, 3, 7, 9
Z_5	1, 2, 3, 4	Z_{11}	1, 2, 3, 4, 5, 6, 7, 8, 9, 10
Z_6	1, 5	Z_{12}	1, 5, 7, 11
Z_7	1, 2, 3, 4, 5, 6		

It seems a reasonable guess that the units in Z_m are the residues that are relatively prime to the modulus.

THEOREM 2.3.1. *Identification of Units.* The element a in Z_m is a unit if and only if $(a, m) = 1$. The multiplicative inverse of a unit is unique.

Proof: Suppose that $(a, m) = 1$. By the characterization of relatively prime pairs, there exist integers x_0, y_0 such that $ax_0 + my_0 = 1$. But $ax_0 = 1 - my_0$ implies $ax_0 \equiv 1 \bmod m$. The residue that has been chosen to represent the class containing x_0 is the inverse of a in Z_m.

Conversely, if a is a unit, there is an integer b such that $ab = 1$ in Z_m. This means that $ab - 1 = my$ for some integer y; that is, there exist integers b and $-y$ such that $ab + (-y)m = 1$. Therefore $(a, m) = 1$.

To see that the inverse is unique, we suppose that $ax_1 = 1$ and $ax_2 = 1$ in Z_m. Then $ax_1 \equiv ax_2 \bmod m$, and the arithmetic of congruences implies that $x_1 \equiv x_2 \bmod m$, since $(a, m) = 1$. Because x_1 and x_2 belong to the same complete residue system, the fact that they are congruent implies that they are equal. ●

Can we predict the value of the inverse if it exists? Certainly if $1 \cdot x = 1$, then $x = 1$, so the inverse of 1 is 1. Also, $(m - 1)^2 \equiv 1 \bmod m$. Therefore $(m - 1)x = 1$ implies that $x = m - 1$, and $m - 1$ is its own inverse. Are there any other residues in Z_m such that $a^2 = 1$?

EXAMPLE 2.3.2.

In Z_7, $a^2 = 1$ if and only if $a = 1$ or $a = 6$.
In Z_8, $a^2 = 1$ if and only if $a = 1, 3, 5, 7$.
In Z_9, $a^2 = 1$ if and only if $a = 1, 8$.
In Z_{10}, $a^2 = 1$ if and only if $a = 1, 9$.
In Z_{11}, $a^2 = 1$ if and only if $a = 1, 10$.
In Z_{12}, $a^2 = 1$ if and only if $a = 1, 5, 7, 11$.

THEOREM 2.3.2. *The Square Roots of* 1, *mod p*. Let p be a prime. Then $a^2 = 1$ in Z_p if and only if $a = 1$ or $a = p - 1$.

Proof: Let $a^2 = 1$ in Z_p. Then $p \mid a^2 - 1$—that is, $p \mid (a - 1) \times (a + 1)$. Since p is prime, this implies $p \mid a + 1$, or $p \mid a - 1$. But a is a unit in Z_p so that $1 \le a \le p - 1$. Therefore $p \mid a + 1$ implies $a = p - 1$ and $p \mid a - 1$ implies $a = 1$. Conversely, direct substitution shows that $a^2 = 1$ if $a = 1$ and if $a = p - 1$. ●

The fact that there are only two square roots of 1 in Z_p, where p is prime, allows us to prove a famous theorem, one leading to a characterization of a prime. It is based on evaluating $(p - 1)! \bmod p$.

EXAMPLE 2.3.3. $6! = 1 \cdot 2 \cdot 3 \cdot 4 \cdot 5 \cdot 6$. In Z_7, each element except 1 and 6 has a unique inverse different from itself. Thus the integers are divided into pairs whose product is congruent to 1 mod 7: $2 \cdot 4 = 1$, $3 \cdot 5 = 1$ in Z_7 and hence $6! = 6$ in Z_7; that is, $6! \equiv 6$ mod 7. Similarly,

$$10! = 1 \cdot 2 \cdot 3 \cdot 4 \cdot 5 \cdot 6 \cdot 7 \cdot 8 \cdot 9 \cdot 10$$
$$= 1 \cdot (2 \cdot 6) \cdot (3 \cdot 4) \cdot (5 \cdot 9) \cdot (7 \cdot 8) \cdot 10$$
$$\equiv 10 \text{ mod } 11.$$

Calculate 12! mod 13, and 16! mod 17.

THEOREM 2.3.3.   *Wilson's Theorem.* If p is a prime, $(p - 1)! \equiv p - 1$ mod p.

Proof: For $p = 2$ and 3, the congruence can be established by simple calculation. Assume that $p \geq 5$ and consider $(p - 1)!$. The set of $p - 3$ integers $\{2, 3, \ldots, p - 2\}$ can be grouped into $(p - 3)/2$ pairs i, j such that $i \cdot j \equiv 1$ mod p. This result is guaranteed by the two preceding theorems, since each i has an inverse that is unique and different from itself. Thus $(p - 1)! \equiv p - 1$ mod p.   ◆

The converse of Wilson's theorem is also true. That is, if m is an integer greater than 1, and $(m - 1)! \equiv m - 1$ mod m, then m is a prime. The proof is outlined in the exercises.

EXERCISES 2.3

Checks

1. Find the inverse of each of the units listed in Example 2.3.1.

2. Prove that the inverse of a unit is a unit and the product of two units is a unit. Thus the units form a group with the operation multiplication.

3. Use the appropriate multiplication tables to solve the following equations:

$$5x = 3 \text{ in } Z_6; \quad 4x = 2 \text{ in } Z_7; \quad 7x = 5 \text{ in } Z_{12}; \quad 7x = 2 \text{ in } Z_9.$$

4. Calculate $(m - 1)!$ mod m, for $m = 4, 5, 8, 9, 10, 12, 14, 15$. Remember to use the arithmetic of congruences to shorten the work.

5. Find the product of the units mod m, for each m in the table in Example 2.3.1.

Challenges

6. If a is a unit in Z_m, prove that $m - a$ is also a unit in Z_m.

7. Prove that $ab \equiv 1$ mod m implies $(a, m) = 1$ and $(b, m) = 1$.

8. Prove that if m is composite and greater than 4, then $(m - 1)! \equiv 0$ mod m. Use this and problem 4 to prove the converse of Wilson's theorem.

9. Prove that if $a^2 = 1$ in Z_m, then $(m - a)^2 = 1$ in Z_m. Hence show that if $m > 2$, then the units mod m can be grouped in pairs such that $r_i r_j \equiv \pm 1$ mod m.

10. Prove that the product of the units mod m is congruent to ± 1 mod m.

11. If p is a prime, prove that $(p - 2)! \equiv 1$ mod p.

12. Prove that $13!/5^2 \equiv 2! \, 3!$ mod 5. In general, prove that

$$\frac{n!}{p^k} \equiv (-1)^k k! \, (n - pk)! \text{ mod } p, \qquad k = \left[\frac{n}{p} \right].$$

13. Let p be an odd prime. Then $a^2 = 1$ in the ring of residues mod p^k if and only if $a = 1$ or $a = p^k - 1$.

14. Let p and q be odd primes. What can you say about the values of a such that $a^2 = 1$ in the ring of residues mod pq?

15. If p is a prime not less than n, where n is a given positive integer, show that $(n - 1)! \, (p - n)! \equiv (-1)^n$ mod p. This is E1603, *American Mathematical Monthly*, Vol. 70 (1963), p. 668.

2.4 THE EULER ϕ FUNCTION

The table in Example 2.3.1 lists the units in Z_m for small values of m. How many units are there for a particular modulus? There are two in Z_3, two in Z_4, and four in Z_{12}. Is the number of units a multiplicative function of the modulus? Before exploring these questions further, we introduce a name for the set of units and for the function that counts them. The units are identified (Theorem 2.3.1) as the residues that are relatively prime to m. It is these residues that we wish to single out.

DEFINITION 2.4.1. *Reduced Residue System.* A reduced residue system mod m is a set of integers consisting of exactly one integer from each residue class mod m whose elements are relatively prime to m.

EXAMPLE 2.4.1. Let $m = 12$. Four of the residue classes mod 12 consist of integers relatively prime to 12. A reduced residue system mod 12 consists of four integers, one from each of these classes. For example, $\{1, 5, 7, 11\}$ or $\{-11, 17, 31, -1\}$ or $\{5, 25, 35, 55\}$.

$$\boxed{\phi(n)}$$

DEFINITION 2.4.2 *Euler ϕ Function*. The function ϕ is a function from the positive integers to the positive integers, defined as follows: $\phi(1) = 1$, $\phi(n) = $ the number of elements in a reduced residue system mod n.

EXAMPLE 2.4.2.

$$\phi(2) = 1, \quad \phi(3) = 2, \quad \phi(4) = 2, \quad \phi(5) = 4, \quad \phi(6) = 2.$$

THEOREM 2.4.1 *Characterizations of ϕ*.

1. $\phi(n)$ is the number of units in Z_n, if $n > 1$.

2. $\phi(n)$ is the number of integers in the set $\{1, 2, 3, \ldots, n - 1, n\}$ that are relatively prime to n.

Proof. Definition 2.4.2 is equivalent to statement (1) since the units are the residues that are relatively prime to n. To prove that statement (2) is also equivalent, consider the complete residue system $\{1, 2, 3, \ldots, n\}$, and apply Definition 2.4.1. ●

The second statement of the theorem gives the best characterization to use in guessing the value of $\phi(n)$ for a particular n. If n is prime, then $(a, n) = 1$ for every a such that $1 \le a \le n - 1$. In this case, $\phi(n) = n - 1$. It is almost as easy to calculate $\phi(n)$ if n is a power of a prime.

EXAMPLE 2.4.3. $n = 3^2$. Consider $\{1, 2, 3, 4, 5, 6, 7, 8, 9\}$. Since $(3^2, n) = 1$ if and only if $(3, n) = 1$, we need to eliminate from this set the multiples of 3. There are $\frac{9}{3} = 3$ multiples—namely, 3, 6, 9. Hence $\phi(9) = 9 - 3 = 6$.

THEOREM 2.4.2 $\phi(p^k)$. If p is prime, for any integer $k \geq 1$,

$$\phi(p^k) = p^k - p^{k-1} = p^{k-1}(p-1) = p^k\left(1 - \frac{1}{p}\right).$$

Proof: Consider the set $\{1, 2, 3, \ldots, p^k\}$. For a in this set, $(a, p^k) = 1$ if and only if $(a, p) = 1$. There are $p^k/p = p^{k-1}$ multiples of p in this set. The number of a such that $(a, p^k) = 1$ is therefore $p^k - p^{k-1}$. ●

The determination of $\phi(m)$ in case m is a composite number that is not a prime power is somewhat more complicated. Consider, first, a direct approach.

EXAMPLE 2.4.4. $n = 24$. Consider the set $\{1, 2, 3, \ldots, 23, 24\}$. The primes that divide 24 are 2 and 3; thus $(a, 24) \neq 1$ if and only if 2 or 3 divides a. There are twelve multiples of 2 in the set and eight multiples of 3. But there are duplications involved, since both sets contain multiples of 6. There are four multiples of 6. Thus

$$\phi(24) = 24 - 8 - 12 + 4 = 8.$$

Note that $\phi(3) = 2$ and $\phi(8) = 4$, so that $\phi(24) = \phi(3)\phi(8)$. This result leads us to hope that ϕ is a multiplicative function. Let us investigate $\phi(24)$ by another method that might help us prove that ϕ is multiplicative.

EXAMPLE 2.4.5. Write the integers $1, 2, 3, \ldots, 23, 24$ in three rows as follows:

①	2	3	4	⑤	6	⑦	8
9	10	⑪	12	⑬	14	15	16
⑰	18	⑲	20	21	22	㉓	24

Each column in this array belongs to the same residue class mod 8. The elements in each column form a complete residue system mod 3. Since $(3, 8) = 1$, $(a, 24) = 1$ if and only if $(a, 3) = 1$ and $(a, 8) = 1$. We wish to consider the columns that are relatively prime to 8 and in these columns pick those elements that are relatively prime to 3. There are $\phi(8) = 4$ such columns, and in each there are $\phi(3) = 2$ suitable elements. The eight values of a are circled as shown.

The technique outlined in this example is one way of proving that the function ϕ is in fact a multiplicative function. Another method of proof is outlined in the exercises.

THEOREM 2.4.3. $\phi(n)$. The function ϕ is multiplicative.

Proof: Let m and n be positive integers such that $(m, n) = 1$. Write the integers $1, 2, 3, \ldots, mn$ in n rows and m columns.

$$
\begin{array}{ccccc}
1 & 2 & 3 & \cdots & m \\
m + 1 & m + 2 & m + 3 & \cdots & 2m \\
2m + 1 & 2m + 2 & 2m + 3 & \cdots & 3m \\
\vdots & \vdots & \vdots & & \vdots \\
(n-1)m + 1 & (n-1)m + 2 & (n-1)m + 3 & \cdots & nm
\end{array}
$$

Each column belongs to the same residue class mod m. Each column contains n elements, no two of which are congruent mod n; hence each column is a complete residue system mod n. Thus there are $\phi(m)$ columns in which each element is relatively prime to m and in each of these columns there are $\phi(n)$ elements relatively prime to n. But $(a, mn) = 1$ if and only if both $(a, m) = 1$ and $(a, n) = 1$. Hence $\phi(mn) = \phi(m)\phi(n)$. ●

COROLLARY 2.4.3. Let $n = \prod_{i=1}^{r} p_i^{a_i}$. Then $\phi(n)$ is given by each of the following expressions:

$$\phi(n) = \prod_{i=1}^{r} (p_i^{a_i} - p_i^{a_i - 1}),$$

$$\phi(n) = n \prod_{p|n} \left(1 - \frac{1}{p}\right),$$

$$\phi(n) = n \prod_{p|n} (p - 1) / \prod_{p|n} p.$$

Proof: The formula for multiplicative functions (Theorem 1.11.1) gives $\phi(n) = \prod_{i=1}^{k} \phi(p_i^{a_i})$. Each of the preceding forms arises when $\phi(p_i^{a_i})$ is replaced by its value in one of the forms listed in Theorem 2.4.2. ●

EXAMPLE 2.4.6.

$$\phi(210) = \phi(3)\phi(5)\phi(7) = 2 \cdot 4 \cdot 6 = 48.$$
$$\phi(2^3 \cdot 5^2 \cdot 11^4) = 2^3 \cdot 5^2 \cdot 11^4 \cdot \tfrac{1}{2} \cdot \tfrac{4}{5} \cdot \tfrac{10}{11} = 2^5 \cdot 5^2 \cdot 11^3.$$

EXAMPLE 2.4.7. For what x is $\phi(x) = 12$? Since $\prod p^{k-1}$ $\times (p-1) = 12$, the largest prime that divides x must be less than or equal to 13. If only one prime divides x, we get $\phi(13) = 12$. Since $\phi(2) = 1$, $\phi(26) = 12$ also. Since $10 \nmid 12$, the prime $11 \nmid x$. With $p = 7$, a number of combinations are possible: $x = 7 \cdot 4$, $x = 7 \cdot 3$, $x = 7 \cdot 6$. The prime 5 cannot be used. Although $\phi(5) = 4$ and $4 \mid 12$, there is no integer with $\phi(n) = 3$. The prime 3 must be considered since $\phi(3) = 2$, which divides 12. It may also occur with $k = 2$, since $3^{2-1} = 3$, which divides 12. In this way we find $x = 3^2 \cdot 4$, and $\phi(x) = 12$.

There are still many unsolved problems related to the finding of x such that $\phi(x) = n$. Some computational results are announced by K. W. Wegner and S. R. Savitzky, Carleton College, Minnesota, in the article "Solutions of $\phi(x) = n$, where ϕ is Euler's ϕ Function," in *American Mathematical Monthly*, Vol. 77 (1970), p. 287.

EXERCISES 2.4

Checks

1. Calculate $\phi(n)$ for $n = 48, 96, 121, 3025$.

2. Prove that $\phi(n)$ is even if $n > 2$.

3. Prove that a belongs to a reduced residue system mod m if and only if $m - a$ belongs to a reduced residue system mod m.

4. Prove that the sum of the elements of a reduced residue system mod m is congruent to 0 mod m if $m > 2$.

5. Use the method of Example 2.4.4 to show that if p and q are distinct primes, then

$$\phi(p^k q^m) = p^k q^m - p^{k-1} q^m - p^k q^{m-1} + p^{k-1} q^{m-1}.$$

Challenges

6. Prove that if $a \mid b$, then $\phi(a) \mid \phi(b)$.

7. Prove that if n is odd, then $\phi(2n) = \phi(n)$, and if n is even, $\phi(2n) = 2\phi(n)$.

8. If p is prime, prove that if $p^2 \mid n$, then $\phi(n) = p\phi(n/p)$, and if $p \mid n$ but $p^2 \nmid n$, then $\phi(n) = (p-1)\phi(n/p)$.

9. If n has k distinct odd prime divisors, prove that $2^k \mid \phi(n)$.

10. For what n is $\phi(3n) = 3\phi(n)$? For what n is $\phi(3n) = 2\phi(n)$?

11. Let m and n be integers, $(m, n) = 1$. Let $r_1, r_2, \ldots, r_{\phi(m)}$ be a reduced residue system mod m and $s_1, s_2, \ldots, s_{\phi(n)}$ be a reduced residue system mod n. Show that the $\phi(m)\phi(n)$ integers $nr_i + ms_j$, $i = 1, 2, \ldots, \phi(m)$ and $j = 1, 2, \ldots, \phi(n)$, constitute a reduced residue system mod mn. *Hint*: Show that $nr_i + ms_j$ is relatively prime to both m and n and that no two of these integers are congruent mod mn.

12. For what integers x is $\phi(x) = 4$?

13. For what integers x is $\phi(x) = 18$?

14. For what integers x is $\phi(x) = 70$?

15. For any pair of positive integers a and b, show that there are infinitely many pairs of positive integers A and B such that: $\phi(A) \equiv 0$ mod a, $\phi(B) \equiv 0$ mod b and $\phi(A + B + AB) \equiv 0$ mod $(a + b)$. This is E1483, *American Mathematical Monthly*, Vol. 68 (1961), p. 803.

2.5   SUMS INVOLVING ϕ

In Section 1.11 we found that, given a multiplicative function f, the function F defined by

$$F(n) = \sum_{d \mid n} f(d)$$

is also multiplicative. Suppose that the ϕ function is the given multiplicative function f. What does F look like?

EXAMPLE 2.5.1.

$n = 12$.

$$F(12) = \sum_{d \mid 12} \phi(d) = \phi(1) + \phi(2) + \phi(3) + \phi(4) + \phi(6) + \phi(12)$$

$$= 1 + 1 + 2 + 2 + 2 + 4 = 12.$$

$n = 15$.

$$F(15) = \sum_{d \mid 15} \phi(d) = \phi(1) + \phi(3) + \phi(5) + \phi(15)$$

$$= 1 + 2 + 4 + 8 = 15.$$

We can guess that $\sum_{d \mid n} \phi(d) = n$. This result can be proved by evaluating $F(p^k)$ directly and using the formula for a multiplicative function. This method is suggested as an exercise. We prove it here directly without using any theorems about multiplicative functions and without using the multiplicative property of ϕ. The proof illustrates a "counting" method that is useful in explorations in number theory. First a numerical case.

EXAMPLE 2.5.2. Let $n = 12$. Consider the set $\{1, 2, 3, 4, 5, 6, 7, 8, 9, 10, 11, 12\}$.

Divide this set into classes C_d, where C_d is the integers a in the set such that $(a, 12) = d$. There will be a class for each divisor of 12.

$C_1 = \{1, 5, 7, 11\}$. $(a, 12) = 1$. There are $\phi(12)$ integers in C_1.

$C_2 = \{2, 10\}$. $(a, 12) = 2$ if and only if $a = 2k$, where $(k, 6) = 1$. There are $\phi(6)$ possible k; hence $\phi(6)$ integers in C_2.

$C_3 = \{3, 9\}$. $(a, 12) = 3$ if and only if $a = 3k$, where $(k, 4) = 1$. There are $\phi(4)$ possible k; hence $\phi(4)$ integers in C_3.

$C_4 = \{4, 8\}$. $a = 4k$, where $(k, 3) = 1$; hence $\phi(3)$ integers in C_4.

$C_6 = \{6\}$. $a = 6k$, where $(k, 2) = 1$; hence $\phi(2)$ integers in C_6.

$C_{12} = \{12\}$. $a = 12k$, one possible k; hence $\phi(1)$ integers in C_{12}.

Thus

$$12 = \phi(12) + \phi(6) + \phi(4) + \phi(3) + \phi(2) + \phi(1) = \sum_{d|12} \phi(d).$$

Note that we could just as well have written $\sum_{d|n} \phi(n/d)$, since if $d|n$ so does n/d and the set $\{d: d|n\} = \{n/d: d|n\}$.

THEOREM 2.5.1. *A Sum of ϕ Values.* Let n be a positive integer. Then

$$n = \sum_{d|n} \phi(d).$$

Proof: Separate the set $\{1, 2, 3, \ldots, n\}$ into classes

$$C_d = \{a: 1 \leq a \leq n, (a, n) = d\}.$$

Each integer $1 \leq a \leq n$ falls into exactly one of the classes C_d.

Let $(a, n) = d$. Then $a = dk$, $n = dn'$ and $(k, n') = 1$ (characterization of the greatest common divisor, Theorem 1.6.1). There are $\phi(n')$ possible values of k and hence $\phi(n')$ possible values of a. Each class C_d has $\phi(n') = \phi(n/d)$ integers. The total number of integers is

$$n = \sum_{d|n} \phi\left(\frac{n}{d}\right) = \sum_{d|n} \phi(d). \quad \blacklozenge$$

A simple formula like this one suggests some questions. Is there any other function f such that $n = \sum_{d|n} f(d)$, or is ϕ the only one? Could this formula be used as a definition of ϕ? $1 = \phi(1)$; hence $\phi(1) = 1$. $2 = \phi(1) + \phi(2)$; hence $\phi(2) = 2 - \phi(1) = 1$. $3 = \phi(1) + \phi(3)$; hence $\phi(3) = 3 - 1$. $4 = \phi(1) + \phi(2) + \phi(4)$; hence

$$\phi(4) = 4 - \phi(2) - \phi(1) = 4 - 1 - 1 = 2.$$

In general, $n = \sum_{d|n} \phi(n)$; hence

$$\phi(n) = n - \sum_{\substack{d|n \\ d<n}} \phi(d).$$

Thus if the value of ϕ has been determined for $k < n$, the value of $\phi(n)$ can be uniquely determined. The formulas $\phi(n) = 1$ and $n = \sum_{d|n} \phi(d)$ define ϕ uniquely by induction.

This situation is a particular case of a more general question. Suppose that $F(n) = \sum_{d|n} f(d)$. Is the function f uniquely determined? Can we solve for f in terms of F? To do this nicely, we need another number-theoretic function. An example may suggest how to define it.

EXAMPLE 2.5.3. Let $F(n) = \sum_{d|n} f(d)$ and consider $n = p$, $p^2, p^3, \ldots$.

$F(p) = f(1) + f(p)$; hence $f(p) = F(p) - f(1) = F(p) - F(1)$.

Then $F(p^2) = f(1) + f(p) + f(p^2)$; hence $f(p^2) = F(p^2) - f(1) - f(p) = F(p^2) - F(p)$. The divisors of p^k are the divisors of p^{k-1} with one added—namely, p^k—so that

$$F(p^k) = \sum_{d|p^k} f(d) = \sum_{d|p^{k-1}} f(d) + f(p^k) = F(p^{k-1}) + f(p^k).$$

Thus

$$f(p^k) = F(p^k) - F(p^{k-1}).$$

Now some composite n. Let p and q be distinct primes. We have $F(pq) = f(1) + f(p) + f(q) + f(pq)$ which implies

$$f(pq) = F(pq) - F(q) - F(p) - F(1).$$

Then $F(p^2q) = f(1) + f(p) + f(q) + f(pq) + f(p^2) + f(p^2q)$.

Hence

$$f(p^2q) = F(p^2q) - F(pq) - F(p^2) + F(p).$$

If we could find the right coefficients $g(d)$, this sum would have the form

$$f(n) = \sum_{d|n} g(d)F\left(\frac{n}{d}\right).$$

This would mean that

$$f(p^2q) = g(1)F(p^2q) + g(p)F(pq) + g(q)F(p^2) + g(pq)F(p)$$
$$+ g(p^2)F(q) + g(p^2q)F(1).$$

Compare with $f(p^2q)$ above. We want $g(1) = 1$, a pleasant situation. Also, $g(p) = -1$ and $g(q) = -1$; surely, then, $g(p) = -1$ for every prime. Now $g(pq) = 1$. This will be consistent if g is multiplicative. But $g(p^2) = 0$. A multiplicative g with $g(p^2) = 0$ would make $g(p^2q) = 0$. This is exactly what we want! Check out $f(p^k)$ in the light of the suggested g. How should we define $g(p^3)$?

The number-theoretic function g that we have searched out is called the Möbius function and is denoted by the greek letter μ. It is clear from the definition that μ is multiplicative.

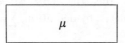

DEFINITION 2.5.1. *The Möbius Function.* The function μ is defined as follows:

$$\mu(1) = 1, \quad \mu(p) = -1, \quad \mu(p^k) = 0 \quad \text{if } k > 1,$$

$$\mu(n) = \prod_{i=1}^{r} \mu(p_i^{a_i}) \quad \text{if } n = \prod_{i=1}^{r} p_i^{a_i},$$

where p is a prime.

EXAMPLE 2.5.4

$\mu(2) = \mu(3) = \mu(5) = -1.$
$\mu(6) = 1.$
$\mu(210) = (-1)^4 = 1.$
$\mu(1) + \mu(2) + \mu(3) + \mu(6) = 1 - 1 - 1 + 1 = 0.$
$\mu(1) + \mu(2) + \mu(4) + \mu(8) + \mu(16) = 1 - 1 + 0 + 0 + 0 = 0.$

THEOREM 2.5.2. *A Property of μ.* Let n be an integer greater than 1. Then

$$\sum_{d\mid n} \mu(d) = 0.$$

Proof: Since μ is multiplicative, the function F, defined by $F(n) = \sum_{d\mid n} \mu(d)$, is also multiplicative. Hence we need only evaluate $F(p^k)$. But

$$F(p^k) = \mu(1) + \mu(p) + \mu(p^2) + \cdots + \mu(p^k) = 1 - 1 + 0 = 0.$$

Thus

$$F(n) = \prod_{p\mid n} F(p^k) = 0 \qquad \text{if } n > 1. \quad \blacksquare$$

THEOREM 2.5.3. *The Möbius Inversion Formula.* If f is any number-theoretic function and F is defined by $F(n) = \sum_{d\mid n} f(d)$, then

$$f(n) = \sum_{d\mid n} \mu(d) F\left(\frac{n}{d}\right).$$

Proof: To prove this theorem, we evaluate the summation $\sum_{d\mid n} \mu(d) F(n/d)$. Since $F(n/d)$ is, by definition, $\sum_{d'\mid n/d} f(d')$, we can write the summation in the form

$$\sum_{d\mid n} \mu(d) \sum_{d'\mid n/d} f(d') = \sum_{d\mid n} \sum_{d'\mid n/d} \mu(d) f(d').$$

This means that the summation is to be extended over all pairs d and d' such that $dd'\mid n$. We can perform this summation by considering d' fixed and, for each d', taking all the d such that $dd'\mid n$—that is, such that $d\mid n/d'$. Thus the sum can be written

$$\sum_{d'\mid n} f(d') \sum_{d\mid n/d'} \mu(d).$$

But the special property of μ proved in the previous theorem tells us that the sum $\sum_{d\mid n/d'} \mu(d)$ will be zero except in the special case that $n/d' = 1$. Thus the only nonzero term in the sum is the term obtained when $d' = n$, which tells us that

$$\sum_{d\mid n} \mu(d)F\left(\frac{n}{d}\right) = f(n). \quad \bullet$$

COROLLARY 2.5.3

$$\phi(n) = n \sum_{d\mid n} \frac{\mu(d)}{d}$$

Proof: Apply the theorem to the equation $n = \sum_{d\mid n} \phi(d)$. $\quad \bullet$

EXERCISES 2.5

Checks

1. Evaluate $\mu(n)$ for each of the first 30 integers.

2. Prove that $\mu(n) = (-1)^k$ if n is the product of k distinct primes, no one of which occurs to a power greater than 1.

3. Use the formula in Corollary 2.5.3 to evaluate $\phi(p^k)$.

4. Calculate $\sum_{d\mid p^k} \phi(d)$. Let $F(n) = \sum_{d\mid n} \phi(d)$. Assume that ϕ is known to be multiplicative and use the calculated value of $F(p^k)$ to determine $F(n)$.

5. If p is prime, show that $\sigma(p) + \phi(p) = p\tau(p)$.

Challenges

6. Apply the Möbius inversion theorem to the expressions $\sigma(n) = \sum_{d\mid n} d$ and $\tau(n) = \sum_{d\mid n} 1$.

7. Use the formulas $\mu(1) = 1$, $\sum_{d\mid n} \mu(d) = 0$, to obtain the definition of $\mu(n)$.

8. Prove that

$$\tau^2(n) = \sum_{c\mid n} \sum_{b\mid c} \sum_{a\mid b} \mu^2(a)$$

This is E2235, *American Mathematical Monthly*, Vol. 77 (1970), p. 522.

9. Let $F(n) = \sum_{d\mid n} f(d)g(n/d)$. Prove that F is multiplicative if f and g are multiplicative. Compare Theorem 1.11.3.

10. Prove that $\sum_{d\mid n} \sigma(d)\phi(n/d) = n\tau(n)$.

11. A necessary and sufficient condition that n be a prime is that $\sigma(n) + \phi(n) = n\tau(n)$. This is E1674, *American Mathematical Monthly*, Vol. 71 (1964), p. 317.

2.6   ALTERNATE FORMS OF A REDUCED RESIDUE SYSTEM

The simplest choice of a reduced residue system mod 18 might seem to be 1, 5, 7, 11, 13, 17. It is possible that ± 1, ± 5, ± 7 might be considered more compact and more symmetrical. Two other possibilities will be considered in this section, both of which lead to very basic and important theorems. Is it possible to write a reduced residue system in which every element is a multiple of some integer? Is it possible to write a reduced residue system in which every element is a power of some integer?

EXAMPLE 2.6.1. $m = 18$. If r is an element of a reduced residue system, r cannot be a multiple of (or a power of) 2, 3, 4, 6, 8, 9, 10, 12, 14, 15, 16 because $(r, 18) = 1$. Consider the set 5, 25, 35, 55, 65, 85. Is this a reduced system mod 18? Certainly each of the numbers is relatively prime to 18, and there are six of them. Hence all we need to show is that no two belong to the same residue class mod 18. In a numerical case, the simplest way to check this is to reduce each to its least non-negative residue mod 18: 5, 7, 17, 1, 11, 13. A reduced residue system consisting of multiples of 5 has been found by multiplying each element of a given reduced residue system by 5. Try this technique to obtain multiples of 7, of 11, of 13, of 17.

THEOREM 2.6.1.   *A Reduced Residue System of Multiples.* Let $r_1, r_2, \ldots, r_{\phi(m)}$ be a reduced residue system mod m. Let a be an integer such that $(a, m) = 1$. Then $ar_1, ar_2, \ldots, ar_{\phi(m)}$ is a reduced residue system mod m.

Proof: Since $(a, m) = 1$ and $(r_i, m) = 1$, $(ar_i, m) = 1$ by the multiplicative property of relatively prime pairs. Also, since $(a, m) = 1$, if $ar_i \equiv ar_j \bmod m$, then $r_i \equiv r_j \bmod m$, by the arithmetic of congruence. Hence the set $ar_1, ar_2, \ldots, ar_{\phi(m)}$ consists of $\phi(m)$ integers from distinct residue classes, each of which is relatively prime to m. Thus the set is a residue system mod m.   ⬢

Can we construct a reduced residue system from powers of an integer? We have already ruled out any integer that is not relatively prime to m. There are others that can be eliminated. Certainly $1^k = 1$ for every

k, and we have seen that $(m - 1)^2 \equiv 1 \bmod m$, so that the powers of $(m - 1)$ are congruent alternately to $m - 1$ and 1. What about some other integers in the reduced residue system?

EXAMPLE 2.6.2. The following table gives the powers of a mod 18 and mod 7.

	mod 18								mod 7				
a	1	5	7	11	13	17		a	1	2	3	4	5	6
a^2	1	7	13	13	7	1		a^2	1	4	2	2	4	1
a^3	1	17	1	17	1	17		a^3	1	1	6	1	6	6
a^4	1	13	7	7	13	1		a^4	1	2	4	4	2	1
a^5	1	11	13	5	7	17		a^5	1	4	5	2	3	6
a^6	1	1	1	1	1	1		a^6	1	1	1	1	1	1

Note that, in every case, $a^6 \equiv 1 \bmod 18$ and $a^6 \equiv 1 \bmod 7$. Why 6? Is this number related in some special way to 18 and to 7? In some but not in *every* case, the powers of a certain a generate a reduced residue system mod m. Construct similar tables for other values of m, say $m = 5, 6, 9, 10, 11, 12$.

If you have studied modern algebra, you will not be surprised that, for each unit a, there is an n such that $a^n = 1$ in Z_m. This is a result of the fact that the set of units is a finite group under multiplication. If you could keep obtaining different elements by multiplying a by itself, then there would have to be an infinite number of elements in the group. So, someplace, the powers of a start repeating. If $a^{k+n} = a^k$ for some n, $a^n = 1$.

The results when the modulus is a prime were apparently known by Fermat and mentioned by him (without proof) in letters as early as 1640. (Fermat was a tremendous explorer but his map-keeping was rather poor.) The first published proof of Fermat's results appears to have been by Euler in 1736. Euler also studied the more general case of composite m. We consider the composite case first, and attach both names to the theorem. (It is a corollary to a general theorem on finite groups.)

THEOREM 2.6.2. *The Euler-Fermat Theorem.* Let a be an integer such that $(a, m) = 1$. Then $a^{\phi(m)} \equiv 1 \bmod m$.

$$\boxed{a^{\phi(m)} \equiv 1 \bmod m}$$

Proof: Let $r_1, r_2, \ldots, r_{\phi(m)}$ be a reduced residue system mod m. The set $ar_1, ar_2, \ldots, ar_{\phi(m)}$ is also a reduced residue system mod m, since $(a, m) = 1$. For each i, $i = 1, 2, \ldots, \phi(m)$, there exists a $j, j = 1, 2, \ldots, \phi(m)$ such that $ar_i \equiv r_j$ mod m. Therefore

$$\prod_{i=1}^{\phi(m)} (ar_i) \equiv \prod_{j=1}^{\phi(m)} r_j \text{ mod } m.$$

Each term in the left-hand product contains a factor of a. Combine these:

$$a^{\phi(m)} \prod_{i=1}^{\phi(m)} r_i \equiv \prod_{j=1}^{\phi(m)} r_j \text{ mod } m.$$

But for each i, $(r_i, m) = 1$. By the multiplicative property of relatively prime pairs, $(\Pi r_i, m) = 1$. The arithmetic of congruence allows us to conclude that $a^{\phi(m)} \equiv 1$ mod m.

COROLLARY 2.6.2a. Fermat's result. Let p be a prime and a any integer. Then $a^p \equiv a$ mod p.

Proof: If $p \nmid a$, $a^{p-1} \equiv 1$ mod p, which implies $a^p \equiv a$ mod p. If $p \mid a$, then $a \equiv 0$ mod p, $a^p \equiv 0$ mod p; hence $a^p \equiv a$ mod p. ●

COROLLARY 2.6.2b. If $p \nmid a$ and p is an odd prime, then $a^{(p-1)/2} \equiv \pm 1$ mod p.

Proof: Factor $a^{p-1} - 1 \equiv 0$ mod p. ●

EXAMPLE 2.6.3. Find the multiplicative inverse of 5 in the ring of residues mod 11. We have already done so by trial. Euler's theorem gives us a technique by which the multiplicative inverse can be calculated. Since 5 is a unit, $5^{10} \equiv 1$ mod 11. Hence $5(5^9) \equiv 1$ mod 11, and the residue class that is the inverse of 5 is represented by 5^9. Since $5^3 \equiv 4$ mod 11, $5^9 \equiv 4^3 \equiv 9$ mod 11.

EXAMPLE 2.6.4. Find x such that $5x \equiv 6$ mod 18. Since $5^6 \equiv 1$ mod 18 by Euler's theorem, $6 \cdot 5^6 \equiv 6$ mod 18; that is $5(6 \cdot 5^{6-1}) \equiv 6$ mod 18. An acceptable x must be $6 \cdot 5^5$. This number, reduced mod 18, gives 12. If we wish to check the calculation, we see that $5 \cdot 12 = 60 \equiv 6$ mod 18, and hence any x from the residue class congruent to 12 mod 18 is a solution of the congruence.

EXAMPLE 2.6.5. What is the last digit in 3^{145}? This is a good way to impress your friends with your ability to do lightning calculations. Remember that the last digit in any integer is its least non-negative residue mod 10. Since $(3, 10) = 1$ and $\phi(10) = 4$, we know that $3^4 \equiv 1$ mod 10. Now $145 = 4 \cdot 36 + 1$, so that $3^{145} = 3^{4 \cdot 36 + 1} = (3^4)^{36} \cdot 3 \equiv 3$ mod 10.

What about 2^{145}?

EXERCISES 2.6

Checks

1. Find the multiplicative inverse of 5 mod 16, using Euler's theorem.

2. Find x such that $3x \equiv 5$ mod 11, using Euler's theorem.

3. Find the remainder when 2^{463} is divided by 7.

4. Find the remainder when 3^{428} is divided by 20.

5. State a general method for finding the multiplicative inverse of a mod m.

6. State a general method for solving the conditional congruence $ax \equiv b$ mod m. [If $(a, m) = d \neq 1$, divide by d first.]

7. State a general method for finding the remainder when a^k is divided by m, when $(a, m) = 1$.

8. If $a^p \equiv b^p$ mod p, prove that $a \equiv b$ mod p, where p is prime.

Challenges

9. Find the remainder when 6^{385} is divided by 16.

10. Derive a method for finding the remainder when a^k is divided by m in the case $(a, m) = d \neq 1$.

11. If $a \equiv b$ mod p, prove that $a^p \equiv b^p$ mod p^2.

12. If p and q are distinct primes, prove

$$p^{q-1} + q^{p-1} \equiv 1 \text{ mod } pq,$$
$$p^q + q^p \equiv p + q \text{ mod } pq.$$

13. Is the converse of Fermat's theorem true? That is, if $a^{m-1} \equiv 1$ mod m, is m necessarily a prime? Prove that $2^{340} \equiv 1$ mod 341, but $341 = 11 \cdot 31$.

14. Prove that for $p > 3$ and prime, $ab^p - ba^p$ is divisible by $6p$. This is E1407, *American Mathematical Monthly*, Vol. 67 (1960) p. 290.

15. Prove that if p is a prime then $\binom{2p}{p} \equiv 2$ mod p. This is E1346, *American Mathematical Monthly*, Vol. 66 (1959), p. 61. *Hint*: Consider $(1 + 1)^{2p}$.

2.7   PRIMITIVE ROOTS

We return to the tables of Example 2.6.2, listing the powers of the units mod m. We know that $a^{\phi(m)} \equiv 1$ mod m, so that we can calculate a^k mod m for any k if we know the powers of a for exponents between 1 and $\phi(m)$. But we see that in some cases $a^n \equiv 1$ mod m for $n < \phi(m)$. Is there any way of predicting the smallest exponent for which $a^n \equiv 1$ mod m? For some units a, $\phi(m)$ is itself this smallest exponent. For how many a's will this be the case? In order to guess the answers to these and related questions, we need to look at more tables. First a definition or two will make the situation easier to talk about.

DEFINITION 2.7.1. *a Belongs to* t mod m. Let a be an integer such that $(a, m) = 1$. Then a belongs to the exponent t mod m if t is the smallest positive integer such that $a^t \equiv 1$ mod m.

EXAMPLE 2.7.1. From the table in Example 2.6.2, we see that 5 and 11 belong to the exponent 6 mod 18, whereas 7 and 13 belong to the exponent 3 and 17 belongs to the exponent 2 mod 18. The integers 3 and 5 belong to the exponent 6 mod 7, while 2 and 4 belong to 3 and 6 belongs to 2 mod 7.

DEFINITION 2.7.2. *Primitive Root.* An integer a such that $(a, m) = 1$ is a primitive root mod m if it belongs to the exponent $\phi(m)$ mod m.

EXAMPLE 2.7.2. 5 and 11 are primitive roots mod 18; 3 and 5 are primitive roots mod 7. 2 is not a primitive root mod 7, but 2 is a primitive root mod 11 and mod 13.

a^n mod 10

	1	3	7	9
a	1	3	7	9
a^2		9	9	1
a^3		7	3	
a^4		1	1	

a^n mod 11

	1	2	3	4	5	6	7	8	9	10
a	1	2	3	4	5	6	7	8	9	10
a^2		4	9	5	3	3	5	9	4	1
a^3		8	5	9	4	7	2	6	3	
a^4		5	4	3	9	9	3	4	5	
a^5		10	1	1	1	10	10	10	1	
a^6		9				5	4	3		
a^7		7				8	6	2		
a^8		3				4	9	5		
a^9		6				2	8	7		
a^{10}		1				1	1	1		

a^n mod 12

	1	5	7	11
a	1	5	7	11
a^2		1	1	1

a^n mod 13

	1	2	3	4	5	6	7	8	9	10	11	12
a	1	2	3	4	5	6	7	8	9	10	11	12
a^2		4	9	3	12	10	10	12	3	9	4	1
a^3		8	1	12	8	8	5	5	1	12	5	
a^4		3		9	1	9	9	1		3	3	
a^5		6		10		2	11			4	7	
a^6		12		1		12	12			1	12	
a^7		11				7	6				2	
a^8		9				3	3				9	
a^9		5				5	8				8	
a^{10}		10				4	4				10	
a^{11}		7				11	2				6	
a^{12}		1				1	1				1	

EXAMPLE 2.7.3. Study the four tables shown. Observe that there are no primitive roots mod 12. How many primitive roots are there mod 10, mod 11, mod 13? How many residues belong to 3 mod 10, mod 11, mod 13? How many residues belong to 5 mod 10, mod 11, mod 13? Think of your answers in relation to the numbers $\phi(m)$ for $m = 10, 11, 13$.

You have probably guessed, on the basis of the preceding example, what exponents a residue can belong to mod m. We derive this as a corollary to the following more general theorem.

THEOREM 2.7.1. *The Minimal Property of t.* If a belongs to the exponent t mod m, and $a^s \equiv 1$ mod m, then $t \mid s$.

Proof: By the division algorithm, $s = qt + r$, where $0 \leq r < t$. Thus

$$a^s = a^{qt+r} = a^{qt}a^r \equiv a^r \text{ mod } m$$

since $a^{qt} = (a^t)^q \equiv 1 \bmod m$. Hence $a^r \equiv 1 \bmod m$. But t is the smallest positive exponent such that $a^t \equiv 1 \bmod m$. This implies that $r = 0$ and $t \mid s$. ●

COROLLARY 2.7.1a. If a belongs to the exponent $t \bmod m$, then $t \mid \phi(m)$.

Proof: Use the Euler-Fermat theorem. ⬢

COROLLARY 2.7.1b. If g is a primitive root mod m, and $g^s \equiv 1 \bmod m$, then $\phi(m) \mid s$.

Proof: Since g is a primitive root, g belongs to $\phi(m)$. ●

We have seen in the tables of Examples 2.6.2 and 2.7.2 that the powers of a primitive root constitute a reduced residue system mod m. That this is the case for any m follows readily from the following theorem.

THEOREM 2.7.2. *Powers of a Primitive Root.* If g belongs to the exponent $t \bmod m$, then $g^a \equiv g^b \bmod m$ if and only if $a \equiv b \bmod t$. In particular, if g is a primitive root mod m, $g^a \equiv g^b \bmod m$ if and only if $a \equiv b \bmod \phi(m)$. The set $\{g, g^2, g^3, \ldots, g^{\phi(m)}\}$ is a reduced residue system mod m.

Proof: If $a \equiv b \bmod t$, $a = b + kt$, $g^a = g^{b+kt} = g^b g^{kt} \equiv g^b \bmod m$. Now assume that $g^a \equiv g^b \bmod m$. The conclusion is trivial if $a = b$. Assume that $a > b$. Since $g^t \equiv 1 \bmod m$, then $(g, m) = 1$ and $(g^b, m) = 1$. Thus $g^{a-b} \equiv 1 \bmod m$. By the minimal property of t, this implies $t \mid a - b$, which says that $a \equiv b \bmod t$. If g is a primitive root, $t = \phi(m)$ and $g^a \equiv g^b \bmod m$ implies $a \equiv b \bmod \phi(m)$. The set $\{g, g^2, g^3, \ldots, g^{\phi(m)}\}$ consists of $\phi(m)$ residues, each of which is relatively prime to m and no two of which are congruent mod m. This set is therefore a reduced residue system mod m. ●

To what exponent does a given residue belong mod m? If we know the exponent to which a belongs, it is not difficult to determine the exponent to which a^k belongs.

EXAMPLE 2.7.4. Refer to the table of a^n mod 13. We see that 10 belongs to 6 mod 13. To what exponent does 10^4 belong? If $10^{4k} \equiv 1$ mod 13, $6|4k$, since 10 belongs to 6. What is the least value of k such that $6|4k$? If $6|4k$, then $3|k$. Thus $k = 3$ is the least possible value of k, and 10^4 belongs to 3 mod 13.

If we know a primitive root mod m, the method of the preceding example can be applied to find the exponent to which any unit belongs, as well as to find how many integers belong to a given exponent.

EXAMPLE 2.7.5. 7 is a primitive root mod 13. To what exponent does 5 belong? We see from the table that $5 \equiv 7^3$ mod 13. Therefore $5^k \equiv 1$ implies $7^{3k} \equiv 1$, and we wish to determine the smallest suitable value of k. This means the smallest value of k for which $12|3k$; that is, $4|k$. Thus $k = 4$ and 5 belongs to 4 mod 13. How many integers belong to 4? Any integer will belong to 4 if it is congruent to 7^s mod 13, for a value of s which has the property that $12|sk$ implies $4|k$. This property means that $(12, s) = 3$. There are $\phi(4)$ such numbers s, $s = 3$ and $s = 9$. The integers that belong to 4 are, therefore, 7^3 and 7^9—that is, 5 and 8.

THEOREM 2.7.3. *The Exponent to Which a^s Belongs.* If a belongs to the exponent t mod m, then a^s belongs to the exponent t/d, where $d = (s, t)$.

Proof: Let a^s belong to k. We can write $s = ds'$ and $t = dt'$, where $(s', t') = 1$. We wish to show that $k = t'$. Since $(a^s)^k \equiv 1$ mod m, $t|sk$ by the minimal property of t. Rewritten, $t|sk$ is $dt'|ds'k$ or $t'|s'k$. But $(t', s') = 1$ and therefore $t'|k$. On the other hand, $a^t \equiv 1$ mod m implies that $a^{s't} \equiv 1$ mod m. Write $t = dt'$. We have $a^{s'dt'} \equiv 1$ mod m or $a^{st'} \equiv 1$ mod m. But a^s belongs to k, so $k|t'$. Therefore $k = t'$.   ●

COROLLARY 2.7.3a. If g is a primitive root mod m and $a \equiv g^s$ mod m, then a belongs to the exponent $\phi(m)/d$, where $d = (s, \phi(m))$.

COROLLARY 2.7.3b. If g is a primitive root mod m, g^s is a primitive root if and only if $(s, \phi(m)) = 1$.

COROLLARY 2.7.3c. If m has a primitive root, then m has $\phi[\phi(m)]$ primitive roots.

Proof: This result follows from the preceding corollary, since there are $\phi[\phi(m)]$ values of s such that $(s, \phi(m)) = 1$. ◆

Does a primitive root exist for every m? The answer is evidently *no*, since we saw that no primitive root exists for $m = 12$. Try $m = 8$, $m = 14$, $m = 16$. For what values of m do primitive roots exist? The study of this question involves counting the roots of the conditional congruence $x^n - 1 \equiv 0 \bmod m$. This step must be postponed until the next unit when we explore conditional congruences (see Section 3.3). To give us confidence that primitive roots do exist in a significant number of cases, we state without proof the following rather unusual theorem: A natural number $m > 1$ has a primitive root if and only if it is one of the numbers 2, 4, p^k, $2p^k$, where p is an odd prime and k is an integer greater than or equal to 1.

EXERCISES 2.7

Checks

1. Check the results of Theorem 2.7.3 and its corollaries with the numerical case $m = 13$.

2. Prove that if g is a primitive root mod p, where p is an odd prime, then $g^{(p-1)/2} \equiv -1 \bmod p$. Is the converse true?

3. Prove that if p is an odd prime greater than 3 of the form $2^n + 1$, then $p - a$ is a primitive root whenever a is a primitive root.

4. Prove that a necessary and sufficient condition for an integer g to be a primitive root mod p, where p is an odd prime, is that $(g, p) = 1$ and $g^{(p-1)/d} \not\equiv 1 \bmod p$ for every prime divisor d of $p - 1$.

Challenges

5. Use Theorem 2.7.2 to prove Wilson's theorem.

6. Refer to the table in Example 2.6.2. In the fifth row of the powers of the residues mod 18, a reduced residue system occurs. The same is true in the fifth row of the table mod 7. The first and fifth rows are the only rows that

are reduced residue systems. Prove that $\{r_1^s, r_2^s, \ldots, r_{\phi(m)}^s\}$ is a reduced residue system mod m if and only if $(s, \phi(m)) = 1$.

7. Let m be an odd integer and g be a primitive root mod m. If g is odd, prove that g is a primitive root mod $2m$. If g is even, prove that $g + m$ is a primitive root mod $2m$.

8. If g is a primitive root mod p^k, $k > 1$, prove that g is a primitive root mod p.

9. Use the fact that there are no primitive roots mod 8 to prove that there are no primitive roots mod 2^k, where $k \geq 3$.

10. Let m and n be positive integers such that $m \mid n$. If there are no primitive roots mod m, can there exist primitive roots mod n?

2.8 ARITHMETIC USING PRIMITIVE ROOTS

Let us refer to the table of a^n mod 13 (Example 2.7.2) and experiment with some arithmetic.

EXAMPLE 2.8.1. 2 is a primitive root mod 13. Every residue in the reduced residue system mod 13 can be written as a power of 2. Use this fact to calculate $5 \cdot 7$ mod 13. Since $5 \equiv 2^9$ and $7 \equiv 2^{11}$, $5 \cdot 7 \equiv 2^{9+11} = 2^{20}$ mod 13. But $2^{20} \equiv 2^8 \equiv 9$ mod 13; thus $5 \cdot 7 \equiv 9$ mod 13. This may seem like a complicated way to do a simple product, but it illustrates a useful idea.

Calculate 7^6 mod 13. Since $7 \equiv 2^{11}$ mod 13, $7^6 \equiv 2^{66} \equiv 2^6 \equiv 12$ mod 13.

The calculations in the preceding example are somewhat similar to calculations using logarithms. This similarity can be emphasized by introducing a little more notation. For simplicity, we shall assume in the remainder of this section that p is an odd prime and that g is a primitive root mod p.

DEFINITION 2.8.1. Ind$_g$ a. Let a be an integer, $(a, p) = 1$. The smallest positive integer k such that $g^k \equiv a$ mod p is called the *index of a with respect to the primitive root g for the prime p*. It is denoted by ind$_g$ a or, simply, ind a.

EXAMPLE 2.8.2. Let $p = 13$, $g = 2$.

$$a = \ 1 \quad 2 \quad 3 \quad 4 \quad 5 \quad 6 \quad 7 \quad 8 \quad 9 \quad 10 \quad 11 \quad 12$$
$$\text{ind}_2\, a = 12 \quad 1 \quad 4 \quad 2 \quad 9 \quad 5 \quad 11 \quad 3 \quad 8 \quad 10 \quad 7 \quad 6$$

Let $p = 11$, $g = 6$.

$$a = \ 1 \quad 2 \quad 3 \quad 4 \quad 5 \quad 6 \quad 7 \quad 8 \quad 9 \quad 10$$
$$\text{ind}_6\, a = 10 \quad 9 \quad 2 \quad 8 \quad 6 \quad 1 \quad 3 \quad 7 \quad 4 \quad 5$$

Theorem 2.7.2, phrased in terms of the idea of index, and using $m = p$, becomes $r \equiv s \bmod p$ if and only if $\text{ind}_g\, r = \text{ind}_g\, s$.

EXAMPLE 2.8.3. Refer to Examples 2.8.1 and 2.8.2. $\text{Ind}_2\, 5 = 9$, $\text{ind}_2\, 7 = 11$, $\text{ind}_2\, 5 \cdot 7 \equiv 20 \bmod 12$; $\text{ind}_2\, 7^6 \equiv 6 \cdot 11 = 66 \bmod 12$.

THEOREM 2.8.1. *Operations with Index.*

1. $a \equiv g^{\text{ind}_g\, a} \bmod p$
2. $\text{ind}_g\, (ab) \equiv \text{ind}_g\, a + \text{ind}_g\, b \bmod p - 1$
3. $\text{ind}_g\, a^k \equiv k\, \text{ind}_g\, a \bmod p - 1$
4. If h is a primitive root mod p, then $\text{ind}_g\, a \equiv \text{ind}_h\, a \cdot \text{ind}_g\, h \bmod p - 1$.

Proof:

1. Statement (1) is a direct consequence of Definition 2.8.1.
2. Since $a \equiv g^{\text{ind}\, a}$ and $b \equiv g^{\text{ind}\, b}$, $ab \equiv g^{\text{ind}\, a + \text{ind}\, b} \bmod p$. Therefore $g^{\text{ind}\, ab} \equiv g^{\text{ind}\, a + \text{ind}\, b} \bmod p$, or $\text{ind}\, ab \equiv \text{ind}\, a + \text{ind}\, b \bmod p - 1$.
3. The proof is similar to that of (2).
4. Since $a \equiv g^{\text{ind}_g\, a}$, we have $\text{ind}_h\, a = \text{ind}_h\, (g^{\text{ind}_g\, a})$. By (3), $\text{ind}_h\, (g^{\text{ind}_g\, a}) \equiv \text{ind}_g\, a \cdot \text{ind}_h\, g \bmod p - 1$. ●

EXAMPLE 2.8.4. Compute $3^{24} \cdot 5^{13} \bmod 11$. Use the table of Example 2.8.2, $p = 11$, $g = 6$.

$$\begin{aligned}
\text{ind}\, (3^{24} \cdot 5^{13}) &\equiv \text{ind}\, 3^{24} + \text{ind}\, 5^{13} \bmod 10 \\
&\equiv 24\, \text{ind}\, 3 + 13\, \text{ind}\, 5 \bmod 10 \\
&\equiv 24 \cdot 2 + 13 \cdot 6 \bmod 10 \\
&\equiv 6 \bmod 10.
\end{aligned}$$

But the integer whose index is 6 mod 10, from Example 2.8.2, is seen to be 5. Hence $3^{24} \cdot 5^{13} \equiv 5$ mod 11.

Operations with indices give us much valuable information about the solution of equations in the rings of residues. This question will be studied systematically in the next unit. An example will help you predict much that is to come.

EXAMPLE 2.8.5. For what x is $5x \equiv 9$ mod 13? We have ind 5 + ind $x \equiv$ ind 9 mod 12; that is, ind $x \equiv 8 - 9$ mod 12, ind $x = 11$ and $x \equiv 7$ mod 13. For what x is $x^7 \equiv 3$ mod 13? We have 7 ind $x \equiv$ ind 3 mod 12; that is, 7 ind $x \equiv 4$ mod 12. Since $(7, 12) = 1$, 7 has a multiplicative inverse mod 12, and ind $x = 4$; that is, $x \equiv 3$ mod 13.

EXERCISES 2.8

Checks

1. Let $p = 13$, $g = 7$. Construct a table of ind$_7$ as in Example 2.8.2. Calculate $5 \cdot 7$ mod 13 and 7^6 mod 13, using this table.

2. Check statement (4) of Theorem 2.8.1, using Example 2.8.2 and problem 1.

3. Use the table of Example 2.8.2 to calculate: $2 \cdot 4 \cdot 6 \cdot 8$ mod 11; 5^7 mod 11; 10! mod 11.

4. What is ind a if $a = 1$? What is ind a if $a = p - 1$?

5. Find x such that $3x \equiv 5$ mod 11, using the method of Example 2.8.5.

Challenges

6. Construct a table showing the elements of the reduced residue system mod 29 as powers of the primitive root 2. From this table, construct a table of ind$_2$ a for $p = 29$.

7. To what exponents can integers belong mod 29? Classify the elements of the reduced residue system according to the exponent to which they belong. What are the other primitive roots of 29?

8. Find the multiplicative inverse of 3, 14, 25 mod 29, using the table in problem 6.

9. Solve the conditional congruence $5x \equiv 27$ mod 29, using the table in problem 6.

10. If possible, find x such that $x^2 \equiv 3$ mod 29. If possible, find x such that $x^2 \equiv 7$ mod 29.

11. Prove that $x^n \equiv a$ mod 29 has a solution for every a provided that $(n, 28) = 1$.

12. Prove Wilson's theorem, using the idea of index.

CONDITIONAL CONGRUENCE— SOME INTERESTING FORMATIONS

3.1 LINEAR CONGRUENCES

We have seen many times that the relation "congruence modulo m" separates the integers into disjoint sets called residue classes. In exploring congruence we were led at several points to attempt to identify a residue class, the elements of which satisfy certain conditions. In Section 2.3, for example, we proved that there is an x such that $ax \equiv 1 \bmod m$ provided that $(a, m) = 1$. In Section 2.8 we saw that the complicated condition $x^n \equiv b \bmod p$ can be replaced by $n \operatorname{ind}_g x \equiv \operatorname{ind}_g b \bmod p - 1$, where g represents a primitive root mod p. It is time to explore a little more systematically the possibility of finding an x that satisfies certain conditions expressed in terms of congruence.

Note that the arithmetic of congruence implies that if $a \equiv b$ mod m, then $f(a) \equiv f(b)$ mod m for any polynomial f. When we ask for the truth set of an open sentence like $ax \equiv b$ mod m, we are asking for a set of *residue classes*, since if x belongs to the truth set, so does any integer y such that $y \equiv x$ mod m. The conditional congruence $ax \equiv b$ mod m *could* be written as an equation in the ring of residues, Z_m: $ax = b$ in Z_m. However, we choose here to retain the congruence notation.

It is natural to ask the number of solutions of a conditional congruence. By this question, we must mean the number of distinct residue classes mod m satisfying the congruence.

EXAMPLE 3.1.1. $3x \equiv 1$ mod 7 if and only if $x \equiv 5$ mod 7 (Example 2.2.4). This conditional congruence has *one* solution, the residue class $x \equiv 5$ mod 7. $x^2 \equiv 1$ mod 7 if and only if $x \equiv 1$ or $x \equiv 6$ mod 7. Two distinct residue classes satisfy this conditional congruence—that is, *two* solutions. $x^2 \equiv 1$ mod 8 if and only if $x \equiv 1, 3, 5, 7$ mod 8; *four* solutions.

Let us begin by considering the simplest form of conditional congruence, $ax \equiv b$ mod m. Will a solution exist? How many? How shall we find them?

EXAMPLE 3.1.2

$2x \equiv 1$ mod 18 has no solution, since $(2, 18) = 2$ (Theorem 2.3.1).

$5x \equiv 1$ mod 18 has a unique solution, since $(5, 18) = 1$.

$5x \equiv 4$ mod 18? Think about equations in algebra. We handle them by multiplying both sides of $5x = 4$ by the multiplicative inverse of 5, so that the coefficient of x will be 1. Try this process with congruences. The multiplicative inverse of 5 mod 18 is 11. $55x \equiv 44$ mod 18 becomes $x \equiv 8$ mod 18. Thus if there is a solution, it belongs to $M_{18} + \{8\}$. On the other hand, $x \equiv 8$ mod 18 implies $5x \equiv 40 \equiv 4$ mod 18. Thus the solution is the unique residue class $M_{18} + \{8\}$.

$2x \equiv 3$ mod 18? If there is a solution, then $18 | 2x - 3$. But $2x - 3$ is odd for every x and therefore not divisible by 18.

$8x \equiv 12$ mod 18? The arithmetic of congruence implies—because $(4, 18) = 2$—that $2x \equiv 3$ mod 9. Since $(2, 9) = 1$, 2 has a multiplicative inverse mod 9—namely, 5. Therefore $10x \equiv 15$ mod 9, and $x \equiv 6$ mod 9. This residue class is the union of two residue classes mod 18, $M_{18} + \{6\}$

and $M_{18} + \{15\}$. Since we are counting solutions in terms of residue classes mod 18, we say that $8x \equiv 12$ mod 18 has *two* solutions.

In Example 3.1.2 we see that if a is a unit, $ax \equiv b$ mod 18 certainly has a solution. If $(a, 18) = d$, whether it has a solution or not depends on the relation between d and b.

THEOREM 3.1.1. *The Linear Congruence.* Consider the conditional congruence $ax \equiv b$ mod m, where $(a, m) = d$.

1. If $d \nmid b$, the solution set is empty.

2. If $d \mid b$, the solution set consists of a single residue class mod m/d.

3. If $d \mid b$, the solution set consists of exactly d residue classes mod m. These are $M_m + \{r\}$, where $r = x_0 + im/d$, $i = 0, 1, 2, \ldots, d - 1$ and x_0 is a representative of the residue class in (2).

Proof: Let $a = da'$ and $m = dm'$. If $ax \equiv b$ mod m has a solution, $m \mid ax - b$; that is, $dm' \mid da'x - b$, which implies $d \mid da'x - b$. But this means that $d \mid b$, by the properties of divides (Theorem 1.1.1.) If $d \nmid b$, a solution cannot exist and the solution set is empty. This proves (1).

Suppose that $d \mid b$; that is, $b = db'$. By the arithmetic of congruence, $ax \equiv b$ mod m becomes $a'x \equiv b'$ mod m'. Since $(a, m) = d$, $(a', m') = 1$, by a characterization of the greatest common divisor. Thus a' is a unit. There exists an a_0 such that $a_0 a' \equiv 1$ mod m'. Multiply both sides of the congruence by a_0. We get

$$a_0 a'x \equiv a_0 b' \text{ mod } m' \quad \text{or} \quad x \equiv a_0 b' \text{ mod } m'.$$

If a solution exists, it must belong to $M_{m'} + \{a_0 b'\}$. Also, if $x \equiv a_0 b'$ mod m',

$$a'x \equiv a'a_0 b' \equiv b' \text{ mod } m'.$$

Therefore every x in $M_{m'} + \{a_0 b'\}$ is a solution. Let x_0 be an element of this residue class.

Statement (3) follows from statement (2) and the theorem on nested residue classes (Theorem 2.1.4). $M_{m'} + \{x_0\}$ is the union of the d residue classes $M_m + r_i$, where $r_i = x_0 + im'$, $i = 0, 1, 2, \ldots, d - 1$. ●

EXAMPLE 3.1.3. $7x \equiv 4$ mod 18. Find x.

Since $\phi(18) = 6$, the inverse of 7 is 7^5 (Euler-Fermat theorem). Then $7^6 x \equiv 4 \cdot 7^5$ mod 18; that is, $x \equiv 4 \cdot 7^5$ mod 18. This is certainly not the simplest representative of the residue class we are looking for: $4 \cdot 7^5 \equiv 28 \cdot 49 \cdot 49 \equiv 10 \cdot 13 \cdot 13 \equiv 16$ mod 18.

$4x \equiv 6$ mod 18. Find x.

Since $(4, 18) = 2$, this becomes $2x \equiv 3$ mod 9. The inverse of 2 is 2^5; hence $x \equiv 3 \cdot 2^5 \equiv 6$ mod 9. The solution set consists of $M_{18} + \{6\}$ and $M_{18} + \{15\}$.

The theorem on linear congruences describes exactly the conditions under which a solution exists and also tells how many solutions to expect. The proof is a constructive one and demonstrates a method of obtaining a solution. However, the technique of the theorem, demonstrated in Example 3.1.3, is frequently not the simplest and quickest one.

EXAMPLE 3.1.4. $7x \equiv 4$ mod 18. Find x.

Look for an element of $M_{18} + \{4\}$ that is also a multiple of 7. We have $4 \equiv 22 \equiv 40 \equiv 58 \equiv 76 \equiv 94 \equiv 112$ mod 18. At this point we have hit a multiple of 7. (Are we sure we will come to one eventually? How many numbers do we need to try at worst?) The conditional congruence $7x \equiv 4$ mod 18 is the same as $7x \equiv 112$ mod 18. This one is easily solved: $x \equiv 16$ mod 18.

$4x \equiv 6$ mod 18. Find x.

Look for an element of $M_{18} + \{6\}$ that is divisible by 4. Since $6 \equiv 24$, we have $4x \equiv 24$ mod 18 or $x \equiv 6$ mod 9.

$7x \equiv 4$ mod 18. Find x.

An approach that sometimes shortens the work is to alter the number 7 by multiplying not by the multiplicative inverse but by any other number that suggests itself. $7x \equiv 4$ mod 18 implies $35x \equiv 20$ mod 18, $-x \equiv 2$ mod 18, and $x \equiv 16$ mod 18.

$4x \equiv 6$ mod 18. Find x.

Here we might multiply by 5. We get $20x \equiv 30$ mod 18, or $2x \equiv 12$ mod 18, which implies $x \equiv 6$ mod 9.

Facility in calculating the solutions of linear congruences will be most helpful in further exploration in this area. Experiment with the techniques suggested in Example 3.1.4 and have fun devising some short-cuts of your own. Unless the processes you use are reversible, however, you might introduce some extraneous values into your set of candidates for solutions. It is always safest to check.

EXAMPLE 3.1.5. $4x \equiv 6$ mod 18. Multiply by 6. $24x \equiv 0$ mod 18, or $x \equiv 0$ mod 3, since $(24, 18) = 6$. This gives six solutions mod 18. Are $x \equiv 0, 3, 6, 9, 12, 15$ mod 18 all solutions of the original congruence?

EXERCISES 3.1

Checks

1. Find the solutions of the following conditional congruences:
   (a) $3x \equiv 12$ mod 24,   (b) $5x \equiv 4$ mod 24,
   (c) $10x \equiv 4$ mod 24,   (d) $3x \equiv 4$ mod 24.

2. Find the solutions of the following conditional congruences:
   (a) $5x \equiv 6$ mod 21,   (b) $47x \equiv -15$ mod 21,
   (c) $7x \equiv 3$ mod 50,   (d) $11x \equiv 21$ mod 59.

3. How many solutions exist for each of the conditional congruences $15x \equiv 50$ mod m:
   (a) $m = 66$,   (b) $m = 65$,   (c) $m = 67$,   (d) $m = 60$.

4. What went wrong in Example 3.1.5? How can the multiplying factor be chosen so as to avoid this problem?

Challenges

5. There are $23 \cdot 24$ possible congruences of the form $ax \equiv b$ mod 24, since there are 23 nontrivial choices for a and 24 choices for b. For how many of these choices do solutions exist? For how many is the solution unique?

6. Prove that if $(a, m) = 1$, the solution set of $ax \equiv b$ mod m is the same as the solution set of $kax \equiv kb$ mod m if and only if $(k, m) = 1$. Give an example in which $(a, m) \neq 1$ and $(k, m) \neq 1$ but the solution set of $ax \equiv b$ mod m is the same as the solution set of $kax \equiv kb$ mod m.

7. Find integers x and y such that $612x + 84y = 156$. Do so by considering the congruence $612x \equiv 156$ mod 84. Compare with Example 1.7.3.

8. Using Theorem 3.1.1, formulate a condition under which the Diophantine equation $ax + by = n$ has a solution.

3.2   SIMULTANEOUS LINEAR CONGRUENCES

Suppose that the solution set has been found for the congruence $a_1 x \equiv b_1$ mod m_1 and that the solution set has been found for the congruence $a_2 x \equiv b_2$ mod m_2. Do the residue classes that make up these solution sets have any integers in common? Is there a residue class that is the solution of the system of simultaneous congruences

$$a_1 x \equiv b_1 \text{ mod } m_1$$
$$a_2 x \equiv b_2 \text{ mod } m_2 ?$$

EXAMPLE 3.2.1.   Consider the system

$$3x \equiv 2 \text{ mod } 7$$
$$4x \equiv 5 \text{ mod } 9.$$

Since $3x \equiv 2$ mod 7 implies $x \equiv 3$ mod 7, and $4x \equiv 5$ mod 9 implies $x \equiv 8$ mod 9, we want elements common to $M_7 + \{3\}$ and $M_9 + \{8\}$. An element of the first class is of the form $7t + 3$. It will also belong to the second class if t is chosen so that $7t + 3 \equiv 8$ mod 9. This implies $7t \equiv 5$ mod 9 or $t \equiv 2$ mod 9. Thus x belongs to $[M_7 + \{3\}] \cap [M_9 + \{8\}]$ if it has the form $7(2 + 9k) + 3$—that is, $63k + 17$, for k an integer. The given system does have a common solution, the residue class $M_{63} + \{17\}$.

Consider the system

$$x \equiv 4 \text{ mod } 6$$
$$x \equiv 5 \text{ mod } 8.$$

In this case, we require $6t + 4 \equiv 5$ mod 8, or $6t \equiv 1$ mod 8. Since $(6, 8) = 2$, no such t exists.

Consider the system

$$x \equiv 4 \text{ mod } 6$$
$$x \equiv 2 \text{ mod } 8.$$

This system requires $6t + 4 \equiv 2$ mod 8 or $6t \equiv 6$ mod 8, which implies $t \equiv 1$ mod 4. The simultaneous system has the solution $x \equiv 10$ mod 24.

Clearly we cannot expect the solution set of the system of congruences to be nonempty unless the solution set of each individual congruence is nonempty. Each congruence can be replaced by $x \equiv a'$ mod m' for a suitable a' and m'. For simplicity, we might as well consider that the simultaneous congruences are already in this form.

THEOREM 3.2.1. *Simultaneous Congruences.* The system of congruences

$$x \equiv a_1 \text{ mod } m_1$$
$$x \equiv a_2 \text{ mod } m_2$$

has a solution if and only if $(m_1, m_2) | a_1 - a_2$. If a solution exists, it is a residue class mod $[m_1, m_2]$.

Proof: Let $(m_1, m_2) = d$. If there is a solution, x_0, of the system of congruences, then $m_1 | x_0 - a_1$ and $m_2 | x_0 - a_2$. Since $d | m_1$ and $d | m_2$, this implies $d | x_0 - a_1$ and $d | x_0 - a_2$; that is, $d | a_1 - a_2$.

Conversely, suppose that $d | a_1 - a_2$. A solution of the first congruence has the form $x = m_1 t + a_1$, for t an integer. This is a solution of the second congruence if and only if $m_1 t + a_1 \equiv a_2$ mod m_2; that is, $m_1 t \equiv a_2 - a_1$ mod m_2. Since $(m_1, m_2) = d$ and $d | a_2 - a_1$, the solution set of this congruence is a single residue class mod m_2/d. The elements of this class are $t = km_2/d + t_0$ for some t_0 and $k = 0, \pm 1, \pm 2, \ldots$. It follows that the integers satisfying the system of congruences are given by

$$x = m_1 \left(\frac{km_2}{d} + t_0 \right) + a_1 = k[m_1, m_2] + a_1 + m_1 t_0.$$

Thus if $d | a_2 - a_1$, the solution set of the simultaneous congruences is nonempty and consists of a uniquely determined residue class mod $[m_1, m_2]$. ●

COROLLARY 3.2.1. The system of congruences

$$x \equiv a_i \text{ mod } m_i, \quad i = 1, 2, \ldots, k$$

has a solution if and only if $(m_i, m_j) | a_i - a_j$ for each pair $i, j, i \neq j$. The solution is a residue class mod $[m_1, m_2, \ldots, m_k]$.

Proof: Left as an exercise. Use induction.   ●

A particular case that is important in the study of conditional congruences occurs when the moduli are relatively prime in pairs. This special case of the preceding theorem is called the Chinese Remainder Theorem. It is stated separately, and an alternate proof, which is constructive and rather elegant, is included.

THEOREM 3.2.2. *The Chinese Remainder Theorem.* The system of congruences $x \equiv a_i \bmod m_i$, $i = 1, 2, \ldots, k$, where $(m_i, m_j) = 1$ for each pair i, j such that $i \neq j$, has a solution and the solution is a residue class mod $m_1 m_2 \cdots m_k$.

Proof: Let $M = m_1 m_2 \cdots m_k$, and $M_i = M/m_i$. Since $(M_i, m_i) = 1$, there is an integer b_i such that $M_i b_i \equiv 1 \bmod m_i$.

Let $x \equiv \sum_{i=1}^{k} a_i b_i M_i \bmod M$. Since $M_j \equiv 0 \bmod m_i$ if $i \neq j$, then $x \equiv a_i b_i M_i$—that is, $x \equiv a_i \bmod m_i$ for each i.

Let x_0 and x_1 be two solutions of the system. Then $x_0 - x_1 \equiv 0 \bmod m_i$ for each i and therefore $x_0 - x_1 \equiv 0 \bmod M$.   ●

EXAMPLE 3.2.2. Consider the system of congruences

$$x = 3 \bmod 5$$
$$x \equiv 2 \bmod 7$$
$$x \equiv 1 \bmod 4.$$

$M = 140$, $M_1 = 28$, $M_2 = 20$, $M_3 = 35$. Since $28b_1 \equiv 1 \bmod 5$, $b_1 \equiv 2 \bmod 5$. Since $20b_2 \equiv 1 \bmod 7$, $b_2 \equiv 6 \bmod 7$. Since $35b_3 \equiv 1 \bmod 4$, $b_3 \equiv 3 \bmod 4$. Thus $x \equiv 3 \cdot 2 \cdot 28 + 2 \cdot 6 \cdot 20 + 1 \cdot 3 \cdot 35 \bmod 140$; that is $x \equiv 93 \bmod 140$.

In numerical cases, the methods of Example 3.2.1 are sometimes faster than the direct application of the technique of the theorem. Thus $x = 5t + 3$; $5t + 3 \equiv 2 \bmod 7$ implies $t \equiv -3 \bmod 7$ and $x = 3 + 5(-3 + 7k)$; that is, $x = -12 + 35k$; $-12 + 35k \equiv 1 \bmod 4$ implies $35k \equiv 13 \bmod 4$ or $k \equiv 3 \bmod 4$. Thus $x = -12 + 35(3 + 4s) = -12 + 105 + 140s$. We conclude that $x \equiv 93 \bmod 140$.

EXERCISES 3.2

Checks

1. Find the solution of the simultaneous congruences

$$x \equiv 5 \bmod 7$$
$$x \equiv 10 \bmod 12.$$

2. Find the solution of the simultaneous congruences

$$x \equiv 5 \bmod 21$$
$$x \equiv 11 \bmod 12.$$

3. Find five consecutive integers such that the first is divisible by 2, the second is divisible by 3, the third is divisible by 5, the fourth is divisible by 7, and the fifth is divisible by 11.

4. In the arithmetic progression $11x + 7$, $x = 1, 2, 3, \ldots$ find three consecutive terms divisible by 2, 3, 5 respectively.

Challenges

5. Find the solution of the simultaneous congruences

$$5x \equiv 2 \bmod 3$$
$$2x \equiv 4 \bmod 10$$
$$4x \equiv 7 \bmod 9.$$

6. Let $f(x) = ax + b$. Prove that, given any positive integer k, there exists an integer m such that $f(m)$ has k distinct prime divisors.

7. Given any set of k primes, prove that there exists a set of consecutive integers divisible by $p_1, p_2, \ldots, p_k$ respectively.

8. Given any set of k primes, does there always exist a set of k consecutive terms of the arithmetic progression $ax + b$ that are divisible by $p_1, p_2, \ldots,$ p_k respectively? If not, what additional conditions are needed to make such a set of terms exist?

9. For what values of c do the simultaneous congruences

$$5x \equiv 2 \bmod 12$$
$$7x \equiv c \bmod 15$$

have a solution?

10. The sequence $a + bk$, $k = 1, 2, 3, \ldots$ is called an arithmetic progression. Let $(a_1, b_1) = 1$ and $(a_2, b_2) = 1$. Prove that if the arithmetic progression $a_1 + b_1 k$ and the arithmetic progression $a_2 + b_2 k$ have any common terms, these terms form an arithmetic progression $a_3 + b_3 k$ such that $(a_3, b_3) = 1$.

3.3   CONGRUENCES OF HIGHER DEGREE

What type of congruence should interest us? Since our concern is centered on the ring of integers, it is reasonable to want to consider the most general type of congruence that involves the elements and the operations of this ring; that is, congruences of the form $f(x) \equiv 0 \bmod m$, where f is a polynomial with integers as coefficients:

$$a_0 x^n + a_1 x^{n-1} + a_2 x^{n-2} + \cdots + a_{n-1} x + a_n \equiv 0 \bmod m.$$

Recall that if $a \equiv b \bmod m$, then $f(a) \equiv f(b) \bmod m$; and thus we consider as "a solution" a residue class of integers mod m, and as the "number of solutions" the number of distinct residue classes whose union is the solution set of the congruence.

For a given m, there is a finite set of residue classes. It is theoretically possible (tedious though it may be) to find the solution set by the trial method. Replace x by each of the m elements of a complete residue system and determine whether the open sentence $f(x) \equiv 0 \bmod m$ is true or false for that particular element. Can we identify some shortcuts and predict some results regarding the number of solutions?

EXAMPLE 3.3.1

$$m = 5, \quad f(x) = x^2 + 3x + 2$$

x	0	1	2	3	4
$f(x) \bmod 5$	2	1	2	0	0

Solution set: $\{x: x \equiv 3 \text{ or } 4 \bmod 5\}$. There are two solutions.

EXAMPLE 3.3.2

$$m = 5, \quad f(x) = 11x^2 + 28x - 3.$$

Since $11 \equiv 1$, $28 \equiv 3$, $-3 \equiv 2 \bmod 5$, $11x^2 + 28x - 3 \equiv x^2 + 3x + 2 \bmod 5$ for every x. The solution set in Example 3.3.2 is the same as the solution set in Example 3.3.1.

A first rule of procedure suggested by this example is: Replace the coefficients in $f(x)$ by the simplest possible representatives of the residue classes involved—the least numerical residue or least non-negative residue, depending on your preference.

EXAMPLE 3.3.3

$$m = 5, \quad f(x) = x^4 - 1$$

x	0	1	2	3	4
$f(x)$ mod 5	-1	0	0	0	0

Solution set: $\{x: x \equiv 1, 2, 3, 4 \bmod 5\}$. Thus every a such that $(a, 5) = 1$ is in the solution set. Is this result surprising?

If we remember the Euler-Fermat theorem, we can predict the result in Example 3.3.3 and also in the following example.

EXAMPLE 3.3.4

$$m = 5, \quad f(x) = x^5 - x.$$

Then $f(x) \equiv 0$ for every x.

EXAMPLE 3.3.5

$$m = 5, \quad f(x) = x^{11} + 4x^{10} + 4x^9 + 3x^8 - 4x^7 + 2x^6.$$

Since $x^5 \equiv x \bmod 5$ for every x,

$$x^{11} = x^5 \cdot x^5 \cdot x \equiv x^3 \bmod 5.$$

Also, $x^{10} \equiv x^2$, $x^9 \equiv x$, $x^8 \equiv x^4$, $x^7 \equiv x^3$, $x^6 \equiv x^2$ mod 5, so that

$$f(x) \equiv 3x^4 + 2x^3 + x^2 + 4x \bmod 5.$$

Let $f_1(x)$ designate the polynomial $3x^4 + 2x^3 + x^2 + 4x$. Since $f(x) \equiv f_1(x)$ mod 5 for every x, the solution set of the congruence $f(x) \equiv 0$ mod 5 is the same as the solution set of $f_1(x) \equiv 0$ mod 5.

EXAMPLE 3.3.6

$$m = 5, \quad f_2(x) = 2x^3 + x^2 + 4x + 3.$$

Note that $f_2(x)$ is obtained from $f_1(x)$ in Example 3.3.5 by replacing x^4 by 1. Compare the solution set of $f_2(x) \equiv 0$ mod 5 with the solution set of $f_1(x) \equiv 0$ mod 5. Since $x^4 \equiv 1$ mod 5 for every x such that $(x, 5) = 1$, the solution set of $f_2(x) \equiv 0$ mod 5 is contained in the solution set of $f_1(x) \equiv 0$ mod 5.

x	0	1	2	3	4
$f_1(x)$ mod 5	0	0	1	3	3
$f_2(x)$ mod 5	3	0	1	3	3

Is it possible for $x = 0$ to appear in the solution set of $f_1(x) \equiv 0$ mod 5 without appearing in the solution set of $f_2(x) \equiv 0$ mod 5?

If the modulus is a prime, a second rule of procedure can now be established. Use the relation $x^p \equiv x$ mod p to reduce the degree of the polynomial to an integer less than or equal to $p - 1$. If desired, the terms of degree $p - 1$ can also be eliminated, but if this is done, it is necessary to investigate separately whether or not $x \equiv 0$ belongs in the solution set.

EXAMPLE 3.3.7

$$m = 6, \quad f(x) = x^2 + 3x + 2, \quad f_1(x) = 3x + 3.$$

Here f_1 is obtained from f by setting $x^2 = 1$. Since $\phi(6) = 2$, $x^2 \equiv 1$ mod 6 for the elements of a reduced residue system mod 6, $f(x) \equiv 0$ mod 6 has four solutions $\{x: x \equiv 1, 2, 4, 5 \text{ mod } 6\}$.

x	0	1	2	3	4	5
$f(x)$ mod 6	2	0	0	2	0	0
$f_1(x)$ mod 6	3	0	3	0	3	0

$f_1(x) \equiv 0$ mod 6 has three solutions $\{x: x \equiv 1, 3, 5 \text{ mod } 6\}$. 1 and 5 belong to the solution set in both cases. For the elements that are not in the reduced residue system mod 6, no prediction can be made.

The method by which the solution set for a composite modulus can be obtained from the solution set for prime moduli will be discussed in the next section.

How many solutions do we expect? Example 3.3.1 is a quadratic congruence with two solutions, Example 3.3.3 a fourth-degree congruence with four solutions, Example 3.3.5 a congruence of degree 11 that reduces to a congruence of degree 4 having two solutions. In each case, the modulus is prime and the number of solutions is less than or equal to the degree of the congruence. This is the result expected from analogy with the number of solutions of an nth degree equation. In Example 3.3.7 the modulus was composite and the quadratic congruence has four solutions.

EXAMPLE 3.3.8

$$m = 5, \quad f_2(x) = 2x^3 + x^2 + 4x + 3.$$

Here $x \equiv 1$ mod 5 is a solution. Can we write $f_2(x) \equiv (x - 1)g(x)$ mod 5? We know that

$$\begin{aligned}
f_2(x) - f_2(1) &= 2(x^3 - 1^3) + (x^2 - 1^2) + 4(x - 1) + 3 - 3 \\
&= 2(x - 1)(x^2 + x + 1) + (x - 1)(x + 1) + 4(x - 1) \\
&= (x - 1)(2x^2 + 3x + 7).
\end{aligned}$$

Since $f_2(1) \equiv 0$ mod 5, this implies that

$$f_2(x) \equiv (x - 1)(2x^2 + 3x + 2) \text{ mod } 5.$$

THEOREM 3.3.1. *The Factor Theorem.* Let f be a polynomial of degree n. Then $f(b) \equiv 0$ mod m if and only if there is a polynomial g of degree $n - 1$ such that $f(x) \equiv (x - b)g(x)$ mod m.

Proof: If $f(x) \equiv (x - b)g(x)$ mod m, then clearly $f(b) \equiv 0$ mod m. Suppose that $f(b) \equiv 0$ mod m. Let

$$f(x) = a_0 x^n + a_1 x^{n-1} + \cdots + a_{n-1} x + a_n.$$

Then

$$f(x) - f(b) = a_0(x^n - b^n) + a_1(x^{n-1} - b^{n-1}) + \cdots + a_{n-1}(x - b).$$

Since

$$x^k - b^k = (x - b)(x^{k-1} + x^{k-2}b + \cdots + b^{k-1}),$$

the expression for $f(x) - f(b)$ has a common factor $x - b$. The remaining factors combine to form a polynomial of degree $n - 1$, which we call $g(x)$. Thus

$$f(x) - f(b) = (x - b)g(x).$$

Since $f(b) \equiv 0$ mod m,

$$f(x) \equiv (x - b)g(x) \text{ mod } m. \quad \blacklozenge$$

COROLLARY 3.3.1. Suppose that $f(x) \equiv (x - b)g(x)$ mod p, where p is a prime. Then if $a \not\equiv b$ mod p, $f(a) \equiv 0$ mod p if and only if $g(a) \equiv 0$ mod p.

Proof: $(a - b)g(a) \equiv 0$ mod p implies $g(a) \equiv 0$ mod p, since $(p, a - b) = 1$. $\quad \blacklozenge$

The factor theorem and its corollary can be used to establish an upper limit on the number of solutions of $f(x) \equiv 0$ mod p. The basic theorem is named after Lagrange (1736–1813), who, along with a great deal of work in other areas of mathematics, established some very beautiful theorems of arithmetic.

THEOREM 3.3.2. *Lagrange's Theorem.* Let f be a polynomial of degree n in which the coefficient of x^n is not divisible by the prime p. The congruence $f(x) \equiv 0$ mod p has at most n solutions.

Proof: Use induction on the degree of the polynomial. The theorem is true for $n = 1$, since $ax \equiv b$ mod p has exactly one solution when $(a, p) = 1$. Suppose that the theorem is true for polynomials of degree k. Let f be a polynomial of degree $k + 1$. Assume that the congruence $f(x) \equiv 0$ mod p has $k + 2$ incongruent solutions $x_1, x_2, \ldots, x_{k+2}$. By the factor theorem, $f(x) \equiv (x - x_1)g(x)$ mod p, where the degree of g is k. By Corollary 3.3.1, $x_2, x_3, \ldots, x_{k+2}$ are all solutions of $g(x) \equiv 0$

mod p. Thus $g(x) \equiv 0$ mod p has at least $k + 1$ incongruent solutions. This result contradicts the induction hypothesis. Therefore $f(x) \equiv 0$ mod p has at most $k + 1$ incongruent solutions, and the theorem follows by induction. ●

One of the important applications of Lagrange's theorem is the proof of the theorem mentioned in Section 2.7, which states that every odd prime has a primitive root. In fact, it is just as easy to prove a more general result.

THEOREM 3.3.3. *Existence of a Primitive Root mod p.* If p is a prime and $d|p - 1$, there are $\phi(d)$ residues that belong to d mod p. In particular, there are $\phi(p - 1)$ primitive roots mod p.

Proof: Let $N(t)$ be the number of elements of a reduced residue system mod p that belong to the exponent t. Each such element is a solution of $x^t \equiv 1$ mod p. If $N(t) \neq 0$, there exists an element a that belongs to t. The t integers $a, a^2, a^3, \ldots, a^t$ are incongruent mod p and are solutions of $x^t \equiv 1$ mod p. By Lagrange's theorem, there are at most t solutions of this congruence; therefore the set $a, a^2, a^3, \ldots, a^t$ is exactly the solution set. Not all the residues in the solution set belong to t mod p. In fact, a^s belongs to t only if $(s, t) = 1$, since a^s belongs to $t/(s, t)$, Theorem 2.7.3. This shows that $\phi(t)$ of these solutions belong to t. Hence $N(t) = 0$ or $N(t) = \phi(t)$. The only possible values of t are the divisors of $p - 1$. Also, each element of the reduced residue system belongs to some exponent mod p. We can conclude:

$$p - 1 = \sum_{d|p-1} N(d) \leq \sum_{d|p-1} \phi(d) = p - 1.$$

If $N(d) = 0$ for any d, strict inequality would hold in the second inequality, which is impossible. Hence there is a residue that belongs to d for each divisor of $p - 1$. ●

EXERCISES 3.3

Checks

1. Find the solution set of $5x^3 - 2x + 1 \equiv 0$ mod 7.

2. Find the solution set of $5x^2 + 4x - 3 \equiv 0$ mod 6.

3. Find the solution set of $8x^{11} - 12x^9 + 347x^7 + 35 \equiv 0$ mod 7.

4. If 2 and 5 are the only solutions of $f(x) \equiv 0$ mod 7, prove that

$$f(x) \equiv (x - 2)(x - 5)g(x) \text{ mod } 7,$$

where $g(x) \equiv 0$ mod 7 has no solution not congruent to 2 or 5 mod 7.

5. Given that 2 is a solution of $x^3 + 2x^2 + 2x + 1 \equiv 0$ mod 7, find $g(x)$ such that

$$x^3 + 2x^2 + 2x + 1 \equiv (x - 2)g(x) \text{ mod } 7.$$

6. Find the number of solutions of $x^2 + 4x + 3 \equiv 0$ mod 12.

7. Is Lagrange's theorem true for a composite modulus? Where does the fact that p is prime play an important role in the proof?

Challenges

8. Find a and b so that $x^2 + ax + b \equiv 0$ mod 15 has more than two solutions. Find a and b so that this congruence has two solutions. Find a and b so that it has no solutions. Can you state a general condition on a and b?

9. Consider Lagrange's theorem for the congruence $f(x) \equiv 0$ mod p^k, where p is prime and $k > 1$.

3.4   COMPOSITE MODULI

Let us consider Example 3.3.7 from a slightly different point of view.

EXAMPLE 3.4.1

$$m = 6, \quad f(x) = x^2 + 3x + 2.$$

If there is an x_0 such that $f(x_0) \equiv 0$ mod 6, then $f(x_0) \equiv 0$ mod 2 and $f(x_0) \equiv 0$ mod 3. Conversely, since $(2, 3) = 1$, if $f(x_0) \equiv 0$ mod 2 and $f(x_0) \equiv 0$ mod 3, then $f(x_0) \equiv 0$ mod 6. The solutions of $f(x) \equiv 0$ mod 6 are seen to be exactly the solutions of the simultaneous congruences

$$f(x) \equiv 0 \text{ mod } 2$$
$$f(x) \equiv 0 \text{ mod } 3.$$

$x^2 + 3x + 2 \equiv x + x \equiv 2x \equiv 0$ mod 2. Solution set: $\{x: x \equiv 0, 1 \text{ mod } 2\}$.
$x^2 + 3x + 2 \equiv x^2 + 2 \equiv 0$ mod 3. Solution set: $\{x: x \equiv 1, 2 \text{ mod } 3\}$. The

integers x such that $f(x) \equiv 0$ mod 6 are the integers common to the residue classes in the set $\{x: x \equiv 0, 1 \text{ mod } 2\}$ and the set $\{x: x \equiv 1, 2 \text{ mod } 3\}$. These integers are determined by the four sets of simultaneous congruences

$$x \equiv 0 \text{ mod } 2 \quad x \equiv 0 \text{ mod } 2 \quad x \equiv 1 \text{ mod } 2 \quad x \equiv 1 \text{ mod } 2$$
$$x \equiv 1 \text{ mod } 3 \quad x \equiv 2 \text{ mod } 3 \quad x \equiv 1 \text{ mod } 3 \quad x \equiv 2 \text{ mod } 3$$

Each pair can be solved by the methods of Section 3.2. We get $x \equiv 4, 2, 1, 5$ mod 6. This result agrees with the solution set in Example 3.3.7.

EXAMPLE 3.4.2

$$m = 15, \quad f(x) = x^3 + 9x^2 + 3x + 10.$$

Since $15 = 3 \cdot 5$ and $(3, 5) = 1$, consider

$$f(x) \equiv 0 \text{ mod } 3 \quad \text{and} \quad f(x) \equiv 0 \text{ mod } 5.$$

Since $f(x) \equiv x + 1$ mod 3, the solution set of $f(x) \equiv 0$ mod 3 is $\{x: x \equiv 2 \text{ mod } 3\}$. Also, $f(x) \equiv x^3 + 4x^2 + 3x$ mod 5 has the solution set $\{x: x \equiv 0, 2, 4 \text{ mod } 5\}$. We obtain three sets of simultaneous congruences

$$x \equiv 2 \text{ mod } 3 \quad x \equiv 2 \text{ mod } 3 \quad x \equiv 2 \text{ mod } 3$$
$$x \equiv 0 \text{ mod } 5 \quad x \equiv 2 \text{ mod } 5 \quad x \equiv 4 \text{ mod } 5$$

from which the solution set of $f(x) \equiv 0$ mod 15 is found to be $\{x: x \equiv 5, 2, 14 \text{ mod } 15\}$.

The argument of the preceding examples could be applied to $f(x) \equiv 0$ mod m, where $m = m_1 m_2$ and $(m_1, m_2) = 1$. Our experience in Sections 1.10 and 1.11 suggests that the most advantageous method of factoring m is into powers of distinct primes.

THEOREM 3.4.1. *Resolution of Congruences.* Let

$$m = p_1^{a_1} p_2^{a_2} \cdots p_k^{a_k}.$$

The solution set of the congruence $f(x) \equiv 0$ mod m is equal to the solution set of the system of simultaneous congruences: $f(x) \equiv 0$ mod $p_1^{a_1}$, $f(x) \equiv 0$ mod $p_2^{a_2}$, ..., $f(x) \equiv 0$ mod $p_k^{a_k}$.

Proof: Let b be an integer such that $f(b) \equiv 0$ mod m. Then $m|f(b)$. Since $p_i^{a_i}|m$, $p_i^{a_i}|f(b)$ and $f(b) \equiv 0$ mod $p_i^{a_i}$ for every i. That is, b satisfies the simultaneous congruences

$$f(x) \equiv 0 \text{ mod } p_i^{a_i}, \quad i = 1, 2, \dots, k.$$

Conversely, suppose that $f(b) \equiv 0$ mod $p_i^{a_i}$ for each i. Then $p_i^{a_i}|f(b)$ for every i. Since $(p_i^{a_i}, p_j^{a_j}) = 1$ when $i \neq j$, this implies $m|f(b)$ and b belongs to a residue class in the solution set of $f(x) \equiv 0$ mod m. ●

EXAMPLE 3.4.3

$$m = 360, \quad f(x) = x^2 + 5x + 66.$$

Since $m = 2^3 \cdot 3^2 \cdot 5$, we consider the system of congruences

$$f(x) \equiv 0 \text{ mod } 2^3, \quad f(x) \equiv 0 \text{ mod } 3^2, \quad f(x) \equiv 0 \text{ mod } 5.$$

$x^2 + 5x + 66 \equiv x^2 - 3x + 2 \equiv 0$ mod 8; solution set: $\{x: x \equiv 1, 2 \text{ mod } 8\}$.
$x^2 + 5x + 66 \equiv x^2 - 4x + 3 \equiv 0$ mod 9; solution set: $\{x: x \equiv 1, 3 \text{ mod } 9\}$.
$x^2 + 5x + 66 \equiv x^2 + 1 \equiv 0$ mod 5; solution set: $\{x: x \equiv 2, 3 \text{ mod } 5\}$.
From these results we obtain the eight sets of simultaneous congruences.

$x \equiv 1$ mod 8	$x \equiv 1$ mod 8	$x \equiv 2$ mod 8	$x \equiv 2$ mod 8
$x \equiv 1$ mod 9	$x \equiv 3$ mod 9	$x \equiv 1$ mod 9	$x \equiv 3$ mod 9
$x \equiv 2$ mod 5	$x \equiv 2$ mod 5	$x \equiv 2$ mod 5	$x \equiv 2$ mod 5

$x \equiv 1$ mod 8	$x \equiv 1$ mod 8	$x \equiv 2$ mod 8	$x \equiv 2$ mod 8
$x \equiv 1$ mod 9	$x \equiv 3$ mod 9	$x \equiv 1$ mod 9	$x \equiv 3$ mod 9
$x \equiv 3$ mod 5	$x \equiv 3$ mod 5	$x \equiv 3$ mod 5	$x \equiv 3$ mod 5

The solution set is $\{x: x \equiv 57, 73, 82, 138, 217, 273, 282, 298 \text{ mod } 360\}$.

Theorem 3.4.1 and the Chinese remainder theorem show how the solution of a conditional congruence with composite modulus m can be accomplished by combining the solutions of the congruence for appropriate prime power moduli. Thus we can direct our attention to a modulus of the form p^k, where p is prime. Can we reduce the problem still further, from p^k to p^{k-1}, and eventually to the prime itself? Certainly, $f(b) \equiv 0$ mod p^k implies $f(b) \equiv 0$ mod p^{k-1}, so the set of integers such that $f(x) \equiv 0$ mod p^k is a subset of the set of integers such that $f(x) \equiv 0$ mod p^{k-1}. The theorem on nested residue classes tells us that any residue class mod p^{k-1} is the union of residue classes mod p^k. Given a residue class mod p^{k-1} in the solution set of $f(x) \equiv 0$ mod p^{k-1}, which subdivision, if any, is a solution of $f(x) \equiv 0$ mod p^k?

EXAMPLE 3.4.4

$$m = 5^3, \quad f(x) = x^3 + 3x + 1.$$

If $f(x) \equiv 0$ mod 5^3, then $f(x) \equiv 0$ mod 5. First find the solution set of this congruence. $x^3 + 3x + 1 \equiv 0$ mod 5 has solution set $\{x : x \equiv 1, 2 \text{ mod } 5\}$. Now

$$M_5 + \{1\} = \bigcup_{t=0}^{4} M_{5^2} + \{1 + 5t\}.$$

Which value of t, if any, gives a solution of $f(x) \equiv 0$ mod 5^2? To determine this value, consider $f(1 + 5t) \equiv 0$ mod 5^2.

$$(1 + 5t)^3 + 3(1 + 5t) + 1 \equiv 1 + 3 \cdot 5t + 3 \cdot 5^2 t^2 + 5^3 t^3 + 3 + 15t + 1$$
$$\equiv 5 + 30t \text{ mod } 5^2.$$

Thus $f(1 + 5t) \equiv 0$ mod 5^2 becomes $5 + 30t \equiv 0$ mod 5^2; that is, $6t \equiv -1$ mod 5 or $t \equiv 4$ mod 5. This implies $x \equiv 21$ mod 5^2.

Now apply the same procedure to see whether the residue class $M_5 + \{2\}$ contains a subclass that is a solution of $f(x) \equiv 0$ mod 5^2.

$$f(2 + 5t) \equiv 15 + 75t \text{ mod } 5^2$$
$$\equiv 15 \not\equiv 0 \text{ mod } 5^2.$$

In this case, there are no values of t for which $f(2 + 5t) \equiv 0$ mod 5^2, and the residue class $x \equiv 2$ mod 5 contains no solutions of $f(x) \equiv 0$ mod 5^2.

Note that at this point we have found the solution set of $f(x) \equiv 0$ mod 5^2; namely, $\{x : x \equiv 21 \text{ mod } 5^2\}$.

To locate the solutions of $f(x) \equiv 0$ mod 5^3, we must proceed one step further. We look for them in the residue class $x \equiv 21$ mod 5^2. Again $M_{5^2} + \{21\}$ is the union of five residue classes mod 5^3, $M_{5^3} + \{21 + 5^2 t\}$, $t = 0, 1, 2, 3, 4$. Consider $f(21 + 5^2 t)$ mod 5^3.

$$(21 + 5^2 t)^3 + 3(21 + 5^2 t) + 1 \equiv 21^3 + 3 \cdot 21^2 \cdot 5^2 t + 3 \cdot 21 + 3 \cdot 5^2 t + 1 \text{ mod } 5^3$$
$$\equiv (3 + 3 \cdot 21^2) 5^2 t + 75 \text{ mod } 5^3.$$

$(3 + 3 \cdot 21^2) 5^2 t + 75 \equiv 0$ mod 5^3 implies $(3 + 3 \cdot 21^2) t + 3 \equiv 0$ mod 5; that is, $(3 + 3)t + 3 \equiv 0$ mod 5 or $t \equiv 2$ mod 5. This gives $x \equiv 71$ mod 5^3.

This example gives us the pattern for a theorem regarding the relation of solutions mod p^k to solutions mod p^{k-1}. The key to the problem lies in calculating $f(a + p^{k-1}t)$ mod p^k. Since f is a polynomial, consider

$$(a + p^{k-1}t)^n = a^n + na^{n-1}p^{k-1}t + \frac{n(n-1)}{2}a^{n-2}p^{2k-2}t^2 + \cdots$$

by the binomial theorem. The terms omitted contain increasing powers of p. If $k \geq 2$, $2k - 2 \geq k$, and therefore the terms involving p^{2k-2} and higher powers are congruent to zero mod p^k. Thus

$$(a + p^{k-1}t)^n \equiv a^n + na^{n-1}p^{k-1}t \text{ mod } p^k.$$

Now

$$f(x) = c_0 x^n + c_1 x^{n-1} + c_2 x^{n-2} + \cdots + c_{n-1}x + c_n.$$

Apply the result of the preceding paragraph to each term, and add.

$$
\begin{aligned}
c_0(a + p^{k-1}t)^n &\equiv c_0 a^n + c_0 na^{n-1}p^{k-1}t && \text{mod } p^k \\
c_1(a + p^{k-1}t)^{n-1} &\equiv c_1 a^{n-1} + c_1(n-1)a^{n-2}p^{k-1}t && \text{mod } p^k \\
c_2(a + p^{k-1}t)^{n-2} &\equiv c_2 a^{n-2} + c_2(n-2)a^{n-3}p^{k-1}t && \text{mod } p^k \\
&\cdots && \\
c_{n-1}(a + p^{k-1}t) &\equiv c_{n-1}a + c_{n-1}p^{k-1}t && \text{mod } p^k \\
c_n &\equiv c_n && \text{mod } p^k \\
f(a + p^{k-1}t) &\equiv f(a) + f'(a)p^{k-1}t && \text{mod } p^k
\end{aligned}
$$

where

$$f'(x) = nc_0 x^{n-1} + (n-1)c_1 x^{n-2} + (n-2)c_2 x^{n-3} + \cdots + c_{n-1}.$$

The notation f' is chosen for this polynomial because it is, in fact, the derivative of f with respect to x. Those who remember calculus will see that the foregoing congruence could have been derived more quickly without using the binomial expansion but instead using Taylor's theorem, which says

$$f(a + p^{k-1}t) = f(a) + f'(a)p^{k-1}t + \frac{f''(a)}{2!}(p^{k-1}t)^2 + \cdots.$$

With this information about $f(a + p^{k-1}t)$ mod p^k at hand, and the messy part of the argument out of the way, it is easy to state and prove the general theorem.

THEOREM 3.4.2. *Prime Power Moduli.* Let $x \equiv a$ mod p^{k-1} be a solution of $f(x) \equiv 0$ mod p^{k-1}.

1. If $p \nmid f'(a)$, then the congruence $f(x) \equiv 0$ mod p^k has the solution $x \equiv a + t_0 p^{k-1}$, where t_0 is the unique integer $0 \le t_0 < p$ such that $f'(a)t_0 \equiv -f(a)/p^{k-1}$ mod p.

2. If $p \mid f'(a)$ and $p^k \nmid f(a)$, then the congruence $f(x) \equiv 0$ mod p^k has no solution in the residue class $x \equiv a$ mod p^{k-1}.

3. If $p \mid f'(a)$ and $p^k \mid f(a)$, then every element of the residue class $x \equiv a$ mod p^{k-1} is a solution of $f(x) \equiv 0$ mod p^k.

Proof: Since $f(a + p^{k-1}t) \equiv f(a) + f'(a)p^{k-1}t$ mod p^k, the congruence $f(a + p^{k-1}t) \equiv 0$ mod p^k leads to the congruence $f'(a)p^{k-1}t \equiv -f(a)$ mod p^k. Since, by hypothesis, $f(a) \equiv 0$ mod p^{k-1}, $p^{k-1} \mid f(a)$, and the congruence becomes $f'(a)t \equiv -f(a)/p^{k-1}$ mod p. Theorem 3.1.1, on the linear congruence, yields the three statements of the theorem. ●

EXAMPLE 3.4.5. Consider the calculations of Example 3.4.4, using Theorem 3.4.2. Here $p = 5$, $f(x) = x^3 + 3x + 1$, $f'(x) = 3x^2 + 3$. For $a \equiv 1$ mod 5, $f(a) = 5$, $f'(a) \equiv 6$; $6t \equiv -1$ mod 5, and $t \equiv 4$ mod 5. For $a \equiv 2$ mod 5, $f(a) = 15$, $f'(a) \equiv 15$. We have case (2) of the theorem and no solution to $f(x) \equiv 0$ mod 5^2 in this residue class mod 5. For $a \equiv 21$ mod 5^2, $f(a) \equiv 9325$, so that $-f(a)/5^2 = -373 \equiv 2$ mod 5. $f'(a) = 3(21)^2 + 3 \equiv 1$ mod 5. Thus $t \equiv 2$ mod 5, and $x \equiv 71$ mod 5^2 is a solution of $f(x) \equiv 0$ mod 5^3.

The primitive roots mod p^n are among the solutions of the congruence $x^{\phi(p^n)} \equiv 1$ mod p^n. The primitive roots mod p satisfy the congruence $x^{\phi(p^n)} \equiv 1$ mod p, since $\phi(p^n) = p^{n-1}\phi(p)$. In view of this fact, it is not surprising that the argument used in the preceding theorem can be applied to the problem of determining a primitive root mod p^n if a primitive root mod p is known.

Let g be a primitive root mod p. To what exponent can g belong mod p^n? Now $g^t \equiv 1$ mod p^n implies $g^t \equiv 1$ mod p. From the first statement, we can conclude that $t \mid p^{n-1}(p - 1)$ and from the second that $p - 1 \mid t$. Hence the possible values of t are $p^k(p - 1)$, where $k = 0, 1, \ldots, n - 1$. Note also that $a^{p^i(p-1)} = [a^{p^k(p-1)}]^{p^{i-k}}$, so that $a^{p^k(p-1)} \equiv 1$ mod p^n implies $a^{p^i(p-1)} \equiv 1$ mod p^n for every $i \ge k$. These remarks enable us to prove the following theorem.

THEOREM 3.4.3. *Primitive Root of p^n*. If g is a primitive root mod p, then either g or $g + p$ is a primitive root mod p^n.

Proof: $(g + p)^{p^{n-2}(p-1)} \equiv g^{p^{n-2}(p-1)} + p^{n-1}(p - 1)g^{p^{n-2}(p-1)-1}$ mod p^n by the binomial theorem. There are only two possibilities for $g^{p^{n-2}(p-1)}$ mod p^n. If $g^{p^{n-2}(p-1)} \not\equiv 1$ mod p^n, then g is a primitive root mod p^n. If $g^{p^{n-2}(p-1)} \equiv 1$ mod p^n, we shall show that $g + p$ is a primitive root mod p^n. In this case,

$$(g + p)^{p^{n-2}(p-1)} \equiv 1 + p^{n-1}(p - 1)g^{p^{n-2}(p-1)-1} \text{ mod } p^n.$$

Since $(p - 1, p) = 1$ and $(g, p) = 1$, p does not divide $(p - 1)g^{p^{n-2}(p-1)-1}$, so that $(g + p)^{p^{n-2}(p-1)} \not\equiv 1$ mod p^n. This implies that $g + p$ is a primitive root mod p^n. ●

Theorem 3.4.3, together with Theorem 3.3.3 and problems 7 and 9 of Exercises 2.7, completes the proof of the theorem on the existence of primitive roots, Section 2.7. The important role played by primitive roots in the study of conditional congruences will be seen in the following section.

EXERCISES 3.4

Checks

1. Find the solutions of $5x^6 - 4x - 22 \equiv 0$ mod 35.

2. Solve $3x^2 + 5x + 3 \equiv 0$ mod 11 and $3x^2 + 5x + 3 \equiv 0$ mod 11^2.

3. Solve $2x^3 - 6x^2 + 6x + 23 \equiv 0$ mod 3^k for $k = 1, 2, 3, 4$. Which of the three situations in Theorem 3.4.2 does this equation illustrate?

4. Find a primitive root of 25, given that 2 is a primitive root of 5. Find a primitive root of 125.

5. Solve

(a) $x^3 + x - 3 \equiv 0$ mod 7,
(b) $x^3 + x - 3 \equiv 0$ mod 7^2,
(c) $x^3 + x - 3 \equiv 0$ mod 7^3,
(d) $x^3 + x - 3 \equiv 0$ mod 13,
(e) $x^3 + x - 3 \equiv 0$ mod 4459.

Challenges

6. Solve the system of congruences

$$5x^2 + 4x - 3 \equiv 0 \bmod 6$$
$$3x^2 + 10 \equiv 0 \bmod 11$$

7. Suppose that $f(x) \equiv 0 \bmod p$ has the unique solution $x \equiv a \bmod p$. Prove that if $f(x) \equiv 0 \bmod p^2$ also has a unique solution, then $f(x) \equiv 0 \bmod p^k$ has a unique solution for every integer k.

8. Let f be a polynomial of degree n, with leading coefficient 1. What is the maximum possible number of solutions of the congruence $f(x) \equiv 0 \bmod m$, where $m = p_1^{a_1} p_2^{a_2} \cdots p_k^{a_k}$?

9. If g is a primitive root mod p^2, must g be also a primitive root mod p^3? *Hint:* Recall that g must be a primitive root mod p so that $g^{p-1} \equiv 1 \bmod p$.

3.5 nth POWER RESIDUES

We have discovered how to build up a solution to a congruence with a composite modulus by starting with a set of congruences involving prime moduli. However, we have still not discovered any general methods of finding the solutions of the congruence $f(x) \equiv 0 \bmod p$. It would be rather optimistic to expect a general formula for the solution of $f(x) \equiv 0 \bmod p$. After all, there is no such general formula for the solution of the equation $f(x) = 0$ where $f(x)$ is a polynomial of degree n. However, some rather specific results can be stated in special cases, in particular for congruences of the type $x^n \equiv b \bmod p$.

EXAMPLE 3.5.1. Consider the congruence $x^3 \equiv b \bmod 7$, where b is an integer as yet unspecified. Look at the cubes of elements in the complete residue system mod 7.

$a = 0$	1	2	3	4	5	6
$a^3 \bmod 7 = 0$	1	1	6	1	6	6

These calculations show that $x^3 \equiv 1 \bmod 7$ has three solutions, $x \equiv 1, 2, 4 \bmod 7$; $x^3 \equiv 6 \bmod 7$ has three solutions, 3, 5, 6 mod 7; $x^3 \equiv 0 \bmod 7$ has one solution, $x \equiv 0 \bmod 7$; but $x^3 \equiv 2$, $x^3 \equiv 3$, $x^3 \equiv 4$, $x^3 \equiv 5$ have no solutions.

EXAMPLE 3.5.2. Consider $x^3 \equiv b$ mod 11.

$a = 0$	1	2	3	4	5	6	7	8	9	10
a^3 mod 11 = 0	1	8	5	9	4	7	2	6	3	10

For each b in the complete residue system mod 11, $x^3 \equiv b$ mod 11 has exactly one solution.

EXAMPLE 3.5.3. $x^3 \equiv b$ mod 13.

$a = 0$	1	2	3	4	5	6	7	8	9	10	11	12
a^3 mod 13 = 0	1	8	1	12	8	8	5	5	1	12	5	12

For $b = 1$, 5, 8, or 12, there are three solutions; for $b = 0$, there is one. Otherwise there are no solutions.

So it appears that a congruence of the form $x^3 \equiv b$ mod p may have a solution for all b, or it may have a solution only for certain b's in the complete residue system mod p. The solution, if it exists, may or may not be unique. Is there a relation between the number of the solutions and the form of the prime? Or perhaps $\phi(p)$? Is the number of b for which a solution exists related to the number of solutions?

We can take care of $b = 0$ once and for all. $x^n \equiv 0$ mod p if and only if $x \equiv 0$ mod p for every p. (This is a simple divisibility property of primes.) Because of the trivial nature of this case, and because the key to the situation lies in the "cyclic" structure of the reduced residue system, we confine our attention to integers b in the reduced residue system mod p.

DEFINITION 3.5.1. nth *Power Residue*. An integer b, not congruent to zero mod p, is called an nth power residue mod p if there exists an x such that $x^n \equiv b$ mod p. If there is no such x, b is called an nth power nonresidue mod p. For $n = 3$, the terms *cubic residue* and *cubic nonresidue* are usual. For $n = 2$, the terms used are *quadratic residue* and *quadratic nonresidue*.

EXAMPLE 3.5.4. 1 and 6 are cubic residues mod 7. The cubic nonresidues mod 7 are 2, 3, 4, 5. Every element of the reduced residue system is a cubic residue mod 11. 1, 5, 8, 12 are cubic residues mod 13, and 2, 3, 4, 6, 7, 9, 10, 11 are cubic nonresidues mod 13.

The consideration of powers of integers mod p brings to mind the work we did in Section 2.7 when we considered the exponent to which a belongs mod p and, in particular, when we considered primitive roots. Let us look again at Example 3.5.3.

EXAMPLE 3.5.5. $p = 13$, 2 is a primitive root of 13, $2^m \equiv 2^n$ mod 13 if and only if $m \equiv n$ mod 12.

k	1	2	3	4	5	6	7	8	9	10	11	12
2^k mod 13	2	4	8	3	6	12	11	9	5	10	7	1

Let $x = 2^k$. $x^3 = 2^{3k} \equiv 3$ mod 13 implies $2^{3k} \equiv 2^4$ mod 13 or $3k \equiv 4$ mod 12. Since $(3, 12) = 3$ and $3 \nmid 4$, there is no k with $3k \equiv 4$ mod 12. Hence 3 is not a cubic residue.

$x^3 \equiv 12$ mod 13 with $x \equiv 2^k$ yields $2^{3k} \equiv 2^6$ mod 13 or $3k \equiv 6$ mod 12. This implies $k \equiv 2$ mod 4 or $k \equiv 2, 6, 10$ mod 12; that is, $x \equiv 4$, 12, 10 mod 13. Which powers of 2 yield cubic residues mod 13?

$x^5 \equiv 3$ mod 13 yields $5k \equiv 4$ mod 12, or $k \equiv 8$ mod 12 and $x \equiv 9$ mod 13. Why is there only one x such that $x^5 \equiv 3$ mod 13 but three values of x such that $x^3 \equiv 12$ mod 13? Consider $x^{10} \equiv 3$ mod 13. In this case, there are two solutions. Why?

THEOREM 3.5.1. *nth Power Residues.* Let p be prime and g a primitive root mod p.

1. An element b of reduced residue system mod p is an nth power residue if and only if $b \equiv g^s$ mod p and $(n, p - 1)|s$.

2. If b is an nth power residue, $x^n \equiv b$ mod p has $(n, p - 1)$ solutions.

3. There are $p - 1/(n, p - 1)$ nth power residues mod p.

Proof: Let $x \equiv g^k$ mod p. $x^n \equiv b$ mod p if and only if $g^{nk} \equiv g^s$ mod p; that is, if and only if $nk \equiv s$ mod $(p - 1)$.

$nk \equiv s$ mod $(p - 1)$ has a solution if and only if $(n, p - 1)|s$. This proves statement (1). If $nk \equiv s$ mod $(p - 1)$ has a solution, it has $(n, p - 1)$

solutions incongruent mod $(p - 1)$. These solutions yield values of x incongruent mod p. This proves statement (2). The number of s divisible by $(n, p - 1)$ is $(p - 1)/(n, p - 1)$, which proves statement (3).   ●

EXERCISES 3.5

Checks

1. Construct a table showing the elements of a reduced residue system mod 29 as powers of the primitive root 2. (See Exercises 2.8, problem 6.) Use this table to answer the following questions:
    (a) What residues are nth power residues mod 29, $n = 2, 3, 4, \ldots, 10$?
    (b) How many solutions does the equation $x^n \equiv a$ mod 29 have if $n = 2$, $3, 4, \ldots, 10$?

2. Using the table in 1, find the solutions, if any, of each congruence: $x^3 \equiv 15$ mod 29; $x^4 \equiv 20$ mod 29; $x^7 \equiv 19$ mod 29; $x^5 \equiv 12$ mod 29.

3. If n is odd, prove that the set of nth power residues is symmetric in the sense that a and $p - a$ are both nth power residues or nth power nonresidues. Show that this is not true, in general, if n is even.

4. Make a table similar to the table in Example 3.5.1, showing a^3 mod 9.

Challenges

5. Prove that the product of two cubic residues is always a cubic residue. Is this true for nth power residues? Prove that the product of an nth power residue and an nth power nonresidue is always a nonresidue.

6. Let p be an odd prime and $(n, p - 1) = d$. Prove that a is an nth power residue mod p if and only if $a^{(p-1)/d} \equiv 1$ mod p.

7. Find the solutions, if any, of $x^3 \equiv 6$ mod 77; $x^3 \equiv 5$ mod 77. What general statement can you make about the solutions of $x^3 \equiv a$ mod pq, where p and q are distinct primes?

8. Under what conditions will the equation $x^n \equiv a$ mod p^2 have a solution? Does the case $n = p$ present special problems? What can you say about the pth power residues mod p?

3.6   QUADRATIC RESIDUES

The nth power residue theorem is particularly simple in case $n = 2$. The reason is that $p - 1$ is divisible by 2 for any odd prime. In this case, the theorem states that half the elements of the reduced residue system are quadratic residues, those b such that $b \equiv g^{2s}$ mod p. We can obtain this result without the use of a primitive root.

EXAMPLE 3.6.1

$p = 7$	n	1	2	3	4	5	6					
	n^2 mod 7	1	4	2	2	4	1					
$p = 11$	n	1	2	3	4	5	6	7	8	9	10	
	n^2 mod 11	1	4	9	5	3	3	5	9	4	1	
$p = 13$	n	1	2	3	4	5	6	7	8	9	10	11	12
	n^2 mod 13	1	4	9	3	12	10	10	12	3	9	4	1

In each case, the squares of $1, 2, \ldots, (p-1)/2$ are distinct mod p but are repeated in the opposite order by the squares of $(p+1)/2, \ldots, (p-1)$.

The following theorem gives a general statement of the pattern observed in this example. Since the case $p = 2$ is somewhat trivial because there is a single element in the reduced residue system and this element is always an nth power residue, we will confine our attention to *odd primes p*. In this and the two following sections, p will represent an odd prime.

THEOREM 3.6.1. *Quadratic Residues.* There are exactly $(p-1)/2$ quadratic residues mod p. These are $1^2, 2^2, 3^2, \ldots, \left(\dfrac{p-1}{2}\right)^2$.

Proof: By definition, a quadratic residue b must be congruent to one of $1^2, 2^2, \ldots, (p-1)^2$. For any k, $(p-k)^2 \equiv k^2$ mod p. Now let i and j be such that $1 \le i, j \le (p-1)/2$. If $p \mid i^2 - j^2$, then $p \mid i - j$ or $p \mid i + j$. But $2 \le i + j \le p - 1$; hence $i^2 \equiv j^2$ mod p implies $i \equiv j$ mod p. The complete set of quadratic residues is, therefore, $1^2, 2^2, 3^2, \ldots, (p-1)/2^2$. ●

The following notational device is useful in studying quadratic residues.

$$(a/p)$$

DEFINITION 3.6.1. *Legendre's Symbol.* If $(a, p) = 1$ and p is an odd prime, the notation (a/p) is defined by

$(a/p) = 1$ if a is a quadratic residue mod p
$(a/p) = -1$ if a is a quadratic nonresidue mod p.

EXAMPLE 3.6.2. The results of the preceding example written in this notation are: $(1/7) = 1, (2/7) = 1, (3/7) = -1, (4/7) = 1, (5/7) = -1,$ $(6/7) = -1$. Similarly for $p = 11$ and $p = 13$. Can we see patterns emerging? $(2/13) = -1, (3/13) = 1, (6/13) = -1; (2/13) = -1, (5/13) = -1, (10/13) = 1$. Is $(a/13)(b/13) = (ab/13)$? Check some other cases.

Note that $(4/7) = 1, (4/11) = 1, (4/13) = 1$. Why is 4 always a quadratic residue?

Stated in words, the conjecture suggested by the preceding example is that the product of two quadratic residues is a quadratic residue, the product of two quadratic nonresidues is a quadratic residue, but the product of a quadratic residue and a quadratic nonresidue is a quadratic nonresidue. The truth of this conjecture can be established in several ways. On the basis of our previous work, the most direct method is an application of Theorem 3.5.1 on nth power residues.

THEOREM 3.6.2. *Properties of the Legendre Symbol.* Let p be an odd prime, $(a, p) = 1$ and $(b, p) = 1$:
1. If $a \equiv b$ mod p, then $(a/p) = (b/p)$.
2. If a is the square of an integer, then $(a/p) = 1$.
3. $(a/p)(b/p) = (ab/p)$.

Proof:
1. This is immediate from the definition, since $a \equiv b$ mod p implies that a and b represent the same residue class mod p.
2. Let $a = n^2$; the congruence $x \equiv n^2$ mod p has the obvious solutions $x \equiv n$ and $x \equiv -n$ mod p, so a is a quadratic residue and $(a/p) = 1$.
3. Let g be a primitive root mod p. Let $a \equiv g^s$ and $b \equiv g^t$ mod p. Then $ab \equiv g^{s+t}$ mod p. If a and b are both quadratic residues, s and t are even and so is $s + t$. Therefore ab is a quadratic residue. If a and b are quadratic nonresidues, s and t are odd and $s + t$ is again even; thus ab is a quadratic residue. If one is a residue, and one a nonresidue, then $s + t$ is an even number plus an odd number; hence $s + t$ is odd and ab is a nonresidue. Statement (3) expresses these facts, using the Legendre symbol. ●

Is there a way of evaluating the Legendre symbol without finding a primitive root and the corresponding table of powers or without simply squaring the elements of a reduced residue system? The quadratic character of an integer a is closely related to its representation in terms of a primitive root, and it was the Euler-Fermat theorem that led to the concept of primitive roots. Thus it is not surprising that the Euler-Fermat theorem should be the basis of an important characterization of the quadratic residues and nonresidues mod p.

EXAMPLE 3.6.3. The table shows the powers of the elements of the reduced residue system mod 13, from 1 to 6.

a	1	2	3	4	5	6	7	8	9	10	11	12
a^2	1	4	9	3	12	10	10	12	3	9	4	1
a^3	1	8	1	12	8	8	5	5	1	12	5	12
a^4	1	3	3	9	1	9	9	1	9	3	3	1
a^5	1	6	9	10	5	2	11	8	3	4	7	12
a^6	1	12	1	1	12	12	12	12	1	1	12	1

It is evident that $a^6 \equiv 1$ or 12 mod 13. This is not surprising, since $a^{12} \equiv 1$ mod 13 implies $13 \mid a^{12} - 1 = (a^6 - 1)(a^6 + 1)$. Hence 13 divides one of the factors. In Example 3.6.1 we found that 1, 3, 4, 9, 10, 12 are the quadratic residues. For each of these numbers, $a^6 \equiv 1$ mod 13. When a is a quadratic nonresidue, $a^6 \equiv -1$ mod 13. Incidentally, we see that the primitive roots are quadratic nonresidues but the quadratic nonresidues are not necessarily primitive roots (for example, 5 and 8). The behavior of a^6 can be predicted by using primitive roots. $3 \equiv 2^4$ mod 13, so $3^6 \equiv 2^{24} \equiv (2^{12})^2 \equiv 1$ mod 13; $6 \equiv 2^5$, hence $6^6 \equiv 2^{30} = (2^6)^5 \equiv (-1)^5 = -1$ mod 13.

THEOREM 3.6.3. *Euler's Criterion.* Let p be an odd prime and $(a, p) = 1$. Then $(a/p) \equiv a^{(p-1)/2}$ mod p; that is, a is a quadratic residue mod p if and only if $a^{(p-1)/2} \equiv 1$ mod p.

Proof: According to the Euler-Fermat theorem, $a^{p-1} \equiv 1$ mod p. This implies that

$$p \mid a^{p-1} - 1 = [a^{(p-1)/2} - 1][a^{(p-1)/2} + 1].$$

Since p is a prime, there are only two possibilities, either $p \mid a^{(p-1)/2} - 1$ or $p \mid a^{(p-1)/2} + 1$. Let g be a primitive root and $a \equiv g^s$. Then $a^{(p-1)/2} \equiv g^{s(p-1)/2}$. If a is a quadratic residue, s is even and $s(p-1)/2$ is a multiple of $(p-1)$, so that $a^{(p-1)/2} \equiv 1$ mod p. If a is a quadratic nonresidue, $s = 2k + 1$ and

$$a^{(p-1)/2} \equiv g^{s(p-1)/2} \equiv g^{k(p-1)+(p-1)/2} \equiv g^{(p-1)/2} \equiv -1 \text{ mod } p,$$

since g is a primitive root. Note that ± 1 are the only possibilities. ●

COROLLARY 3.6.3. If $p \equiv 1$ mod 4, -1 is a quadratic residue mod p; if $p \equiv 3$ mod 4, -1 is a quadratic nonresidue.

EXERCISES 3.6

Checks

1. Verify Euler's criterion for $p = 11$. Compare with Example 3.6.2.
2. Derive the properties of the Legendre symbol (Theorem 3.6.2) from Euler's criterion.
3. Determine whether 3 is a quadratic residue mod 17 from Euler's criterion.
4. Prove Corollary 3.6.3.
5. Using the properties of the Legendre symbol, construct the set of quadratic residues and nonresidues mod 23 with the least possible computation. You should be able to construct them from 1, 4, 9, 16, and -1; the quadratic character of each is easy to identify.

Challenges

6. If a is a quadratic residue mod p, prove that the multiplicative inverse of a is also a quadratic residue. If a is a quadratic nonresidue mod p, what is the quadratic character of the multiplicative inverse of a? What about the additive inverse of a?
7. Prove or disprove: The sum of the quadratic residues mod p is divisible by p if $p > 3$.
8. Prove or disprove: The product of the quadratic residues mod p is congruent to ± 1 mod p.
9. Prove that $\sum_{a=1}^{p-1} (a/p) = 0$.
10. Consider the quadratic congruence $x^2 + 2bx + c \equiv 0$ mod p. Apply the algebraic process called "completing the square" to prove that this congruence has a solution if and only if $b^2 - c$ is zero or a quadratic residue. Apply your analysis to $x^2 + 4x - 3 \equiv 0$ mod 7.

11. Consider $ax^2 + 2bx + c \equiv 0$ mod p, where $(a, p) = 1$. Under what conditions does this congruence have a solution? Consider the particular cases $3x^2 + 5x + 6 \equiv 0$ mod 7; $2x^2 + 7x + 11 \equiv 0$ mod 13. For what values of c does $3x^2 - 3x + c \equiv 0$ mod 11 have solutions?

3.7 GAUSS' LEMMA

In this section we come to a theorem that bears the name of Gauss. The introduction of his name is long overdue, for a great deal of the path we have followed was charted by this remarkable German mathematician. He had a rare genius for mental calculations, which perhaps explains his early interest in the area we now identify as number theory. His *Disquisitiones Arithmeticae*, completed when he was 21 years old, gives a thorough treatment of the theory of congruences, much of it concerned with the congruence $x^n \equiv a$ mod p. The theorem known as Gauss' lemma puts the information contained in Euler's criterion into a slightly different form. In order to do so, we use the technique employed in the proof of the Euler-Fermat theorem.

EXAMPLE 3.7.1. $p = 13$. Consider the residues 1, 2, 3, 4, 5, 6 and their multiples mod 13. Reduce the multiples to *least numerical residues* mod 13.

	a	$2a$	$3a$	$4a$	$5a$	$6a$
$a = 1$	1	2	3	4	5	6
2	2	4	6	−5	−3	−1
3	3	6	−4	−1	2	5
4	4	−5	−1	3	−6	−2
5	5	−3	2	−6	−1	4
6	6	−1	5	−2	4	−3
7	−6	1	−5	2	−4	3
8	−5	3	−2	6	1	−4
9	−4	5	1	−3	6	2
10	−3	−6	4	1	−2	−5
11	−2	−4	−6	5	3	1
12	−1	−2	−3	−4	−5	−6

You might notice several things about this table. For the moment we are particularly interested in the rows. In what way are the rows similar? In what way are they different? Let t be the number of negative signs in a row. Multiply the elements of a row: $a \cdot 2a \cdot 3a \cdot 4a \cdot 5a \cdot 6a \equiv (-1)^t \ 6!$ mod 13. That is, $a^6 \equiv (-1)^t$ mod 13. For $a = 3$, $t = 2$, and 3 is a quadratic residue. For $a = 8$, $t = 3$, and 8 is a nonresidue. Try others.

THEOREM 3.7.1. *Gauss' Lemma.* Let p be an odd prime and $(a, p) = 1$. Then $a^{(p-1)/2} \equiv (-1)^t$ mod p, where t is the number of negative elements in the sequence resulting from $a, 2a, 3a, \ldots, (p-1)a/2$ when each term is replaced by its least numerical residue mod p.

Proof: No one of the integers ai, where $1 \le i \le (p-1)/2$, is divisible by p, since $(a, p) = 1$. Hence $ai \equiv \pm j$ mod p, for some j with $1 \le j \le (p-1)/2$. The ai represent distinct residue classes (a Reduced Residue System of Multiples, Theorem 2.6.1) so neither $+j$ nor $-j$ can occur twice.

Suppose that $ai_1 \equiv j$ and $ai_2 \equiv -j$ mod p. Then $a(i_1 + i_2) \equiv 0$ mod p; and since $(a, p) = 1$, $i_1 + i_2 \equiv 0$ mod p. But the restrictions on i imply that $2 \le i_1 + i_2 \le p - 1$. Hence $i_1 + i_2 \not\equiv 0$ mod p, so it is impossible for $+j$ and $-j$ both to occur in the set of least numerical residues mod p. Thus the set $\{ai\}$ is congruent to a set of integers, the absolute values of which are $1, 2, 3, \ldots, (p-1)/2$. The product of the ai is thus congruent to $(-1)^t[(p-1)/2]!$. That is,

$$a^{(p-1)/2} \ [(p-1)/2]! \equiv (-1)^t \ [(p-1)/2]!$$

which implies $a^{(p-1)/2} \equiv (-1)^t$ mod p.   ●

The application of Gauss' lemma in a particular numerical case still involves considerable work. Thus we are led to consider further how to identify the terms to be counted. This identification can be expressed in terms of inequalities, since $ai \equiv -j$ if ai falls between $p/2$ and p, or between $3p/2$ and $2p$, or between $5p/2$ and $3p$, etc. This approach to counting yields especially simple results in the case $a = 2$, since in this case all the ai are less than p and only the interval from $p/2$ to p need be considered.

EXAMPLE 3.7.2. Let $a = 2$ and $p = 17$. Then ai lies between 2 and 16 for $i = 1, 2, 3, \ldots, 8$. The least numerical residue is negative if $17/2 \le 2i \le 17$. This inequality implies $17/4 < i < 17/2$, or $i = 5, 6, 7, 8$. In this case, $t = 4$ and 2 is a quadratic residue.

Let $p = 19$. Negative signs occur if $19/4 < i < 19/2$; that is, $i = 5, 6, 7, 8, 9$. Hence $t = 5$ and 2 is a nonresidue.

Let $p = 23$. Here $23/4 < i < 23/2$, which gives $11 - 5 = 6$ choices for i.

Let $p = 29$. $29/4 < i < 29/2$ gives $14 - 7$ or 7 choices.

Calculate 17 mod 8, 19 mod 8, 23 mod 8 and 29 mod 8.

THEOREM 3.7.2. *Evaluation of* $(2/p)$. Let p be an odd prime. 2 is a quadratic residue mod p if $p \equiv 1$ or 7 mod 8, and 2 is a quadratic non-residue mod p if $p \equiv 3$ or 5 mod 8.

Proof: Since $1 \leq i \leq (p - 1)/2$, $2 \leq 2i \leq p - 1$. The least numerical residue is negative if $p/2 < 2i < p$; that is, $p/4 < i < p/2$. Consider the four cases in the theorem.

1. $p = 8k + 1$. $p/4 < i < p/2$ implies $2k + \frac{1}{4} < i < 4k + \frac{1}{2}$. In this case, the number of possible values of i is $4k - 2k = 2k$.

2. $p = 8k + 3$. $2k + \frac{3}{4} < i < 4k + 1 + \frac{1}{2}$; that is, $2k + 1$ choices for i.

3. $p = 8k + 5$. $2k + 1 + \frac{1}{4} < i < 4k + 2 + \frac{1}{2}$, or $2k + 1$ values of i.

4. $p = 8k + 7$. $2k + 1 + \frac{3}{4} < i < 4k + 3 + \frac{1}{2}$, or $2k + 2$ values of i.

Thus, from Gauss' lemma, 2 is a quadratic residue in cases (1) and (4), since t is even, and is a nonresidue in cases (2) and (3) when t is odd. ●

COROLLARY 3.7.2. $(2/p) = (-1)^{(p^2 - 1)/8}$. From Gauss' lemma, $(2/p) = (-1)^t$. The corollary does not imply that $t = (p^2 - 1)/8$. All that is needed is an exponent that is odd when t is odd and even when t is even.

Once $(2/p)$ is calculated, $(2^k/p)$ can be found easily, since $(2^k/p) = 1$ if k is even by the square property, and $(2^k/p) = (2/p)$ if k is odd, by the product property. The only integers that must be considered further are the odd integers.

What do we really want to know? We need not t itself, but $(-1)^t$. Can we find an integer k that is even when t is even and odd when t is odd? If so, $(-1)^k = (-1)^t$. With this in mind, look again at Gauss' lemma.

When we replace ai by its least nonnegative residue, we obtain, from the division algorithm, $ai = [ai/p]p + r_i$, where $0 < r_i < p$. If $0 < r_i < p/2$, r_i is already the least numerical residue. If $p/2 < r_i < p$, we replace r_i by $p - s_i$, so that $ai = [ai/p]p + p - s_i$.

EXAMPLE 3.7.3. $p = 13$, $a = 7$. Carry out the process described above.

$$7 \cdot 1 = \left[\frac{7 \cdot 1}{13}\right] 13 + 7 = \left[\frac{7 \cdot 1}{13}\right] 13 + 13 - 6.$$

$$7 \cdot 2 = \left[\frac{7 \cdot 2}{13}\right] 13 + 1.$$

$$7 \cdot 3 = \left[\frac{7 \cdot 3}{13}\right] 13 + 8 = \left[\frac{7 \cdot 3}{13}\right] 13 + 13 - 5.$$

$$7 \cdot 4 = \left[\frac{7 \cdot 4}{13}\right] 13 + 2.$$

$$7 \cdot 5 = \left[\frac{7 \cdot 5}{13}\right] 13 + 9 = \left[\frac{7 \cdot 5}{13}\right] 13 + 13 - 4.$$

$$7 \cdot 6 = \left[\frac{7 \cdot 6}{13}\right] 13 + 3.$$

Note that the extra 13 occurs whenever the least numerical residue is negative—that is, t times. Now add these equations for $1 \leq i \leq 6$.

$$\sum_{i=1}^{6} 7i = \sum_{i=1}^{6} \left[\frac{7i}{13}\right] 13 + 13t + (1 + 2 + 3) - (4 + 5 + 6).$$

Since $\sum_{i=1}^{6} i$ occurs on the left-hand side, we are tempted to force it to occur on the right by adding and subtracting $(4 + 5 + 6)$. We get

$$7 \sum_{i=1}^{6} i = 13 \sum_{i=1}^{6} \left[\frac{7i}{13}\right] + 13t + \sum_{i=1}^{6} i - 2(4 + 5 + 6).$$

This is not as bad as it looks, since we want only the value of t mod 2. Replacing each integer with its value mod 2, we get

$$\sum_{i=1}^{6} i \equiv \sum_{i=1}^{6} \left[\frac{7i}{13}\right] + t + \sum_{i=1}^{6} i \bmod 2;$$

that is,

$$t \equiv \sum_{i=1}^{6} \left[\frac{7i}{13}\right] \bmod 2.$$

THEOREM 3.7.3. *Gauss' Second Lemma.* Let a be an odd integer and p an odd prime with $(a, p) = 1$. Then $(a/p) = (-1)^k$, where

$$k = \sum_{i=1}^{(p-1)/2} \left[\frac{ai}{p}\right].$$

Proof: Proceed as in Gauss' lemma, Theorem 3.7.1, and write ai as its least numerical residue mod p for $i = 1, 2, \ldots, (p-1)/2$. We have either

$$ai = \left[\frac{ai}{p}\right] p + r_i, \qquad 0 < r_i < p/2,$$

or

$$ai = \left[\frac{ai}{p}\right] p + p - s_i, \qquad 0 < s_i < p/2.$$

The number of s_i is the t of Gauss' lemma. Add these equations for $i = 1, 2, 3, \ldots, (p-1)/2$. Note that a and p are common factors in the sums.

$$a \sum i = p \sum \left[\frac{ai}{p}\right] + pt + \sum r_i - \sum s_i.$$

Since the integers r and s together make up the integers from 1 to $(p-1)/2$, we can add and subtract the sum of the s_i and obtain

$$a \sum_{i=1}^{(p-1)/2} i = p \sum_{i=1}^{(p-1)/2} \left[\frac{ai}{p}\right] + pt + \sum_{i=1}^{(p-1)/2} i - 2 \sum s_i.$$

Since $a \equiv 1$ mod 2, $p \equiv 1$ mod 2, and $2 \equiv 0$ mod 2, this equation can be replaced by the following congruence:

$$\sum_{i=1}^{(p-1)/2} i \equiv \sum_{i=1}^{(p-1)/2} \left[\frac{ai}{p}\right] + t + \sum_{i=1}^{(p-1)/2} i \bmod 2$$

or

$$t \equiv \sum_{i=1}^{(p-1)/2} \left[\frac{ai}{p}\right] \bmod 2.$$

Since $t \equiv k$ mod 2 and $(a/p) = (-1)^t$, we conclude that $(a/p) = (-1)^k$. ●

EXAMPLE 3.7.4. Let $a = 7$ and $p = 23$.

$$k = \left[\frac{7}{23}\right] + \left[\frac{14}{23}\right] + \left[\frac{21}{23}\right] + \left[\frac{28}{23}\right] + \left[\frac{35}{23}\right] + \left[\frac{42}{23}\right] + \left[\frac{49}{23}\right] + \left[\frac{56}{23}\right] + \left[\frac{63}{23}\right]$$

$$+ \left[\frac{70}{23}\right] + \left[\frac{77}{23}\right] = 0 + 0 + 0 + 1 + 1 + 1 + 2 + 2 + 2 + 3 + 3 = 15$$

and 7 is a nonresidue.
Let $a = 13$. Then

$$k = \sum_{i=1}^{11} \left[\frac{13i}{23}\right] = 0 + 1 + 1 + 2 + 2 + 3 + 3 + 4 + 5 + 5 + 6 = 32$$

and 13 is a residue.

EXERCISES 3.7

Checks

1. Carry through a calculation similar to Example 3.7.3, using $a = 5$ and $p = 17$.
2. Determine whether 7 is a quadratic residue mod 31 in at least three ways.
3. Prove Corollary 3.7.2.
4. Use Gauss' second lemma to evaluate each of the following pairs:
   (a) $(13/7)$ and $(7/13)$,
   (b) $(13/11)$ and $(11/13)$,
   (c) $(7/11)$ and $(11/7)$,
   (d) $(17/19)$ and $(19/17)$,
   (e) $(11/19)$ and $(19/11)$,
   (f) $(7/19)$ and $(19/7)$.
   Can you make a prediction about (p/q) and (q/p)?

Challenges

5. In the proof of Gauss' second lemma, no use was made of the fact that a is odd until the equality was evaluated mod 2. Start with $a = 2$ and obtain

$$2 \sum i = p \sum \left[\frac{2i}{p}\right] + pt + \sum i - 2 \sum s_i.$$

Use this result to prove Theorem 3.7.2. *Hint*: Use Example 1.3.3 to evaluate $\sum i$.

6. Use the method of Theorem 3.7.2 to evaluate $(3/p)$.
7. Evaluate $(5/p)$ by the method of Theorem 3.7.2. Would this be a good method in general for (q/p)?

8. If p is a prime ($p \neq 2, 3, 5, 11$ or 17), prove that there are three distinct quadratic nonresidues of p whose sum is divisible by p. This is E2173, *American Mathematical Monthly*, Vol. 76 (1969), p. 553.

3.8   QUADRATIC RECIPROCITY

Although Gauss' second lemma is somewhat more convenient for calculating (a/p) for odd a than Euler's criterion or Gauss' first lemma, its importance does not lie primarily in its use for calculation. This theorem, and the geometric interpretation of its result, points the way to a deep relationship between the quadratic character of the prime p relative to the prime q and the quadratic character of q relative to p.

EXAMPLE 3.8.1. What does the sum

$$\sum_{i=1}^{(p-1)/2} \left[\frac{ai}{p} \right]$$

mean geometrically? $a = 13$, $p = 23$. The numbers $13i/23$ are the ordinates of the points $(i, 13i/23)$ which lie on the line $y = 13x/23$. The value of y cannot be an integer for any value of x between 0 and 23, since 23 is prime. Points in the plane, both of whose coordinates are integers, are called *lattice points*. The line $y = 13x/23$ does not pass through any lattice points if $0 < x < 23$. Recall that $[13i/23]$ is the greatest integer less than or equal to $13i/23$—that is, the number of positive integers less than $13i/23$. To each of these positive integers corresponds a lattice point on the line $x = i$,

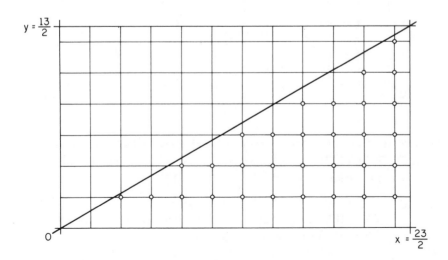

and below the line $y = 13x/23$. For example, $[13 \cdot 6/23] = 3$. Thus there are three lattice points on $x = 6$ and below the line $y = 13x/23$. These are $(6, 1)$, $(6, 2)$, $(6, 3)$. The integer k, then, is the number of lattice points in the triangle bounded by $y = 0$, $x = 23/2$, and $y = 13x/23$.

EXAMPLE 3.8.2. Reverse the roles of 13 and 23 in the preceding example and consider whether 23 is a quadratic residue mod 13. It would seem obvious to replace 23 by 10, since $(23/13) = (10/13)$, but we choose for the sake of symmetry not to do so. To evaluate $(23/13)$, we calculate

$$k' = \sum_{i=1}^{6} \left[\frac{23i}{13} \right] = 1 + 3 + 5 + 7 + 8 + 10.$$

This time think of the line $y = 13x/23$ in the form $x = 23y/13$. Now $[23i/13]$ is seen to be the number of lattice points on the line $y = i$ and lying above the line $x = 23y/13$. For $i = 3$, $[23 \cdot 3/13] = 5$, which means

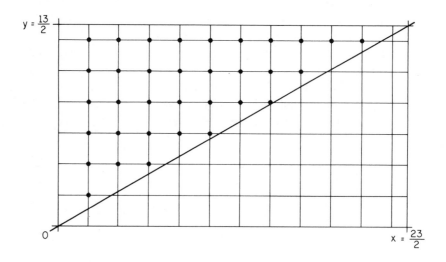

there are five lattice points on $y = 3$ and above $x = 23y/13$. These are $(1, 3)$, $(2, 3)$, $(3, 3)$, $(4, 3)$, $(5, 3)$. Altogether, k' is the number of lattice points in the triangle bounded by $x = 0$, $y = 13/2$ and $y = 13x/23$.

EXAMPLE 3.8.3. Put together the figures of Examples 3.8.1 and 3.8.2. We have the rectangle bounded by $x = 0$, $y = 0$, $x = 23/2$, $y = 13/2$.

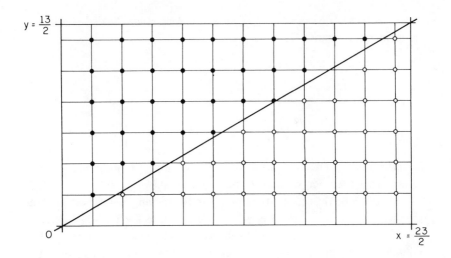

In this rectangle, there are $11 \cdot 6 = 66$ lattice points. These are divided by the line $y = 13x/23$ into two disjoint sets, one of which has k points and the other k' points. Thus $k + k' = 66$. But $(13/23) = (-1)^k$ and $(23/13) = (-1)^{k'}$. Hence

$$(13/23)(23/13) = (-1)^k(-1)^{k'} = (-1)^{k+k'} = (-1)^{66} = 1.$$

This tells us that if 13 is a quadratic residue mod 23, then 23 is a quadratic residue mod 13, or if 13 is a nonresidue mod 23, then 23 is a nonresidue mod 13. One of them must still be evaluated.

The general statement of the relationship pointed out in the previous examples is called the Quadratic Reciprocity Law. It was first noticed by Euler, who was unable to prove it. Legendre stated it in the form

$$(p/q)(q/p) = (-1)^{(p-1)/2 \cdot (q-1)/2}$$

but proved it only in special cases. Gauss worked very hard on this theorem and finally proved it in seven different ways. The geometric construction discussed above is due to Eisenstein, one of Gauss' students. As we have seen, it depends heavily on Gauss' lemma.

THEOREM 3.8.1. *The Quadratic Reciprocity Law.* Let p and q be distinct odd primes. Then

$$(p/q)(q/p) = (-1)^{(p-1)/2 \cdot (q-1)/2}.$$

Proof: Let

$$k = \sum_{i=1}^{(q-1)/2} \left[\frac{pi}{q} \right] \quad \text{and} \quad k' = \sum_{i=1}^{(p-1)/2} \left[\frac{qi}{p} \right].$$

Then $(p/q) = (-1)^k$ and $(q/p) = (-1)^{k'}$, so that $(p/q)(q/p) = (-1)^{k+k'}$. To prove the theorem, we need to prove that $k + k' = (p-1)/2 \cdot (q-1)/2$.

Let T be the set of lattice points in the rectangle in the (x, y) plane bounded by $x = 0$, $y = 0$, $x = q/2$, $y = p/2$. There are $(q-1)/2$ integral values of x and $(p-1)/2$ integral values of y and hence $(p-1)/2 \cdot (q-1)/2$ points in T. The line $px = qy$ is a diagonal of this rectangle and does not pass through any lattice points. For fixed i, the integer $[pi/q]$ is the number of lattice points on the line $x = i$ and below $px = qy$. Hence k is the number of points of T that lie below the diagonal. Similarly, $[qi/p]$ is the number of lattice points on the line $y = i$ and above $px = qy$. Thus k' is the number of lattice points in the rectangle and above the diagonal.

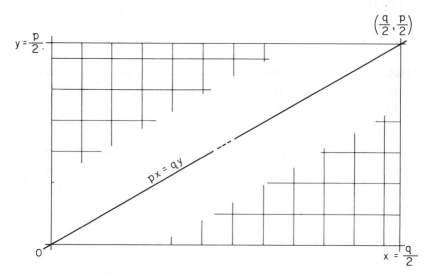

Consequently, the total number of points in T is $k + k'$, and we have

$$k + k' = \frac{p-1}{2} \cdot \frac{q-1}{2}. \quad \blacklozenge$$

COROLLARY 3.8.1.

$$(p/q) = (-1)^{(p-1)/2 \cdot (q-1)/2}(q/p).$$

Proof: Use the fact that $(q/p)^2 = 1.$ ●

The following examples give some rather straightforward applications of the quadratic reciprocity law. The full appreciation of the importance of this law can come only after much deeper study than is attempted here.

EXAMPLE 3.8.4. Determine the quadratic character of the prime 607 mod the prime 383.

$(607/383) = (224/383)$, since $607 \equiv 224$ mod 383,

    $= (16/383)(14/383)$, factoring the largest square factor,

    $= (2/383)(7/383).$

Since $383 \equiv 7$ mod 8, $(2/383) = 1.$

$(7/383)(383/7) = (-1)^{3 \cdot 191} = -1$, using the quadratic reciprocity law.
$(383/7) = (5/7)$, since $383 \equiv 5$ mod 7.

$(5/7)(7/5) = (-1)^{2 \cdot 3} = 1.$
$(7/5) = (2/5) = -1$, since the quadratic residues of 5 are 1 and 4, or
      because $5 \equiv 5$ mod 8.
Since 7 is a quadratic nonresidue mod 5, 5 must be a quadratic nonresidue mod 7. Therefore $(383/7) = -1$ and $(7/383) = 1$. Combine this with the fact that $(2/383) = 1$ to obtain $(607/383) = 1.$

EXAMPLE 3.8.5. For what odd primes p is 7 a quadratic residue? $(7/p)(p/7) = (-1)^{3 \cdot (p-1)/2}$; that is, $(7/p) = (p/7)(-1)^{3(p-1)/2}$. We want to determine the p for which $(7/p) = 1$. One possibility is for $(p/7) = 1$ and $(p - 1)/2$ to be even. Another possibility is $(p/7) = -1$ and $(p - 1)/2$ odd. This gives a set of conditions as follows:

$p \equiv 1$	2	4	mod 7	or	$p \equiv 3$	5	6	mod 7
$p \equiv 1$	1	1	mod 4	or	$p \equiv 3$	3	3	mod 4

We solve these by the Chinese remainder theorem:

$p = 1 + 4t \equiv 1, 2, 4$ mod 7   or $3 + 4t \equiv 3, 5, 6$ mod 7
   $4t \equiv 0, 1, 3$ mod 7   or $4t \equiv 0, 2, 3$ mod 7
    $t \equiv 0, 2, 6$ mod 7   or $t \equiv 0, 4, 6$ mod 7

These sets of simultaneous congruences have the solutions $p \equiv 1, 3, 9, 19,$ 25, 27 mod 28. Not all integers in these congruence classes are primes, but for any prime p in these classes, $(7/p) = 1$.

EXERCISES 3.8

Checks

1. Compare Theorem 3.8.1 with your results in Exercises 3.7, problem 4.

2. Calculate (85/223), (290/89), (1039/1553).

3. If p is a prime of the form $12k + 1$, prove that 3 is a quadratic residue mod p. Is -3 a quadratic residue or nonresidue?

4. Carry out the geometric argument used in the proof of the quadratic reciprocity theorem, for $p = 7$ and $q = 11$.

5. If q is an odd prime and $p = 4q + 1$, prove that $(q/p) = 1$.

Challenges

6. For what odd primes is 5 a quadratic residue?

7. For what odd primes is 6 a quadratic residue?

8. If p and q are distinct odd primes, each congruent to 3 mod 4, prove that p is a quadratic residue mod q if and only if q is a nonresidue mod p. State and prove an appropriate result for $p \equiv 1$ mod 4 and $q \equiv 3$ mod 4.

9. Define: a is a quadratic residue mod m if and only if $(a, m) = 1$ and the congruence $x^2 \equiv a$ mod m has a solution. Prove:
   (a) if $m = p_1 p_2 \cdots p_k$, where p_i is an odd prime, then a is a quadratic residue mod m if and only if a is a quadratic residue mod p_i for each i;
   (b) if a is a quadratic residue mod p, then a is a quadratic residue mod p^k for every k;
   (c) if $m = p_1^{a_1} p_2^{a_2} \cdots p_k^{a_k}$, then a is a quadratic residue mod m if and only if a is a quadratic residue mod p_i for each i.

10. Prove that 3 is a quadratic residue mod $11^2 \cdot 23$. How many solutions exist for the congruence $x^2 \equiv 3$ mod $11^2 \cdot 23$.

11. If p is a prime of the form $2^n + 1$, and $(a, p) = 1$, prove that a is a primitive root mod p if and only if it is a quadratic nonresidue mod p.

THE DISTRIBUTION OF THE PRIMES—SOME UNCLIMBED PEAKS

4.1 PRIMES REPRESENTED BY LINEAR FORMS

At the beginning of our exploration of the integers, we designated certain positive integers as prime. The set of primes plays a very basic role in the multiplicative structure of the integers, since every integer can be expressed as the product of a unit and a unique finite set of primes. Many of the basic properties of integers, and of collections of integers, have been related directly or indirectly to the prime factorization of these integers.

It is time to turn our attention to the set of primes itself, and to investigate some questions related to this set. The most obvious question might be how big is it? That the set is infinite was demonstrated in Section 1.3. But how is this infinite set of primes distributed among the infinite set of integers? For example, is every fifth integer starting with 3 a prime? This question is soon answered in the negative, since the set {3, 8, 13, 18, ...} contains many even integers, which, of course, cannot be prime. What about every fourth integer? The set {3, 7, 11, 15, ...} contains only odd integers, begins well, but we soon find a composite number in it. Sequences of positive integers in which the difference between successive terms is constant have the form $a + bk$ and are called *arithmetic sequences* or arithmetic progressions. Can there be an infinite arithmetic sequence consisting only of primes? Does every arithmetic sequence contain at least one prime? Do such sequences contain infinitely many primes? Primes that occur in arithmetic sequences can be thought of as primes represented by linear forms. The first section of this unit is concerned with the question: How many prime values are assumed by the function $f(k) = a + bk, k = 1, 2, 3, \ldots$?

After linear forms have been considered, it is natural to consider other types of functions that have integral values and might therefore have prime values. Some functions that are not linear but that are candidates for prime-producing functions are considered in Section 4.2.

We can look at the distribution of the primes from a different point of view. In Section 4.3 we ask how many primes occur in the first n integers; that is, what is the behavior of the function π? In Section 4.4 we consider briefly some recurring patterns in the set of primes. In general, the distribution of the primes is an area in which the questions are plentiful and easily grasped, but the answers are elusive.

EXAMPLE 4.1.1. Is there an infinite set of primes in the sequence

$$3, 7, 11, 15, \ldots, 3 + 4k, \ldots?$$

Recall Euclid's proof that the set of all primes is infinite (Section 1.3). Can we copy that method to show that a contradiction arises from the assumption that the set of primes of the form $3 + 4k$ is finite? In order to get a contradiction in this case, we must find a prime *of the correct form* that is not in the finite set. Consider $N_1 = 4(7 \cdot 11) + 3$; this is of the correct form and is not divisible by 3, 7, or 11; $N_1 = 311$ and is prime. Consider $N_2 = 4(7 \cdot 11 \cdot 19) + 3$; this is also of the correct form and is not divisible by 3, 7, 11, or 19; $N_2 = 5855 = 5 \cdot 1171$. Although N_2 is not prime, it has a prime divisor, $1171 = 4 \cdot 292 + 3$, which is a prime of the required form. Does a number of the form $3 + 4k$ always have a prime divisor of this form?

THEOREM 4.1.1. *Primes of the Form* $3 + 4k$. There is an infinite set of primes of the form $3 + 4k$.

Proof: Assume that the complete set of primes of the form $3 + 4k$ is the finite set $\{3, p_1, p_2, \ldots, p_r\}$. There are some primes other than 3 in this set, for example, 7. Let $N = 3 + 4(p_1 p_2 \cdots p_r)$.

By assumption, N cannot be prime, since it is of the form $3 + 4k$ and is not one of the primes listed. Therefore N is a product of primes. Since N is odd, every prime q that divides N is odd—that is, $q \equiv 1 \bmod 4$ or $q \equiv 3 \bmod 4$. If N is a product of primes each of which is congruent to 1 mod 4, then $N \equiv 1 \bmod 4$, which cannot be the case. Thus at least one of the prime divisors of N is congruent to 3 mod 4. Since 3 does not divide N, and since each p_i, $i = 1, 2, \ldots, r$, does not divide N, it follows that N must have a prime divisor of the form $3 + 4k$ that is not one of the given finite set. This is a contradiction. ●

Every prime except 2 falls into one of the congruence classes $n \equiv 1 \bmod 4$ or $n \equiv 3 \bmod 4$. Is there also an infinite set of primes in the congruence class $n \equiv 1 \bmod 4$? The technique of proof in the preceding theorem will not work here. (Why?) With a different form of N, our knowledge of quadratic residues can be used to settle this case.

THEOREM 4.1.2. *Primes of the Form* $1 + 4k$. There is an infinite set of primes of the form $1 + 4k$.

Proof: Let $N = (n!)^2 + 1$. If p is a prime less than or equal to n, then $p \mid n!$ and therefore $p \nmid N$. Let p be a prime such that $p \mid N$, so that $p > n$. Since $N \equiv 0 \bmod p$, $(n!)^2 \equiv -1 \bmod p$, which says that -1 is a quadratic residue mod p. According to Euler's criterion (Theorem 3.6.3), $p \equiv 1 \bmod 4$. Since $p > n$, this means that for every choice of n there is a prime p greater than n and such that $p \equiv 1 \bmod 4$. This implies that there is an infinite set of such primes. ●

The primes of the form $3 + 4k$ fall into two residue classes mod 8, those congruent to 3 and those congruent to 7. Are there primes in each of these classes, and, if so, how many?

THEOREM 4.1.3. *Primes of the Form* $3 + 8k$. The set of primes of the form $3 + 8k$ is infinite.

Proof: Consider $N = a^2 + 2$, where a is the product of the first n odd primes. Suppose that p is an odd prime such that $p|N$. Certainly $p > n$. Now $a^2 + 2 \equiv 0 \bmod p$ implies that -2 is a quadratic residue mod p. Thus either $p \equiv 1 \bmod 8$ or $p \equiv 3 \bmod 8$ (Section 3.7). If $p \equiv 1 \bmod 8$ for every prime p that divides N, then $N \equiv 1 \bmod 8$. But $a^2 \equiv 1 \bmod 8$, so that $N \equiv 3 \bmod 8$. Therefore at least one prime divisor of N must be congruent to 3 mod 8. Since $p > n$, this implies that the set of primes of this form is infinite. ●

The forms considered in the preceding three theorems are all special cases of primes that are congruent to a mod b—that is, primes given by linear expressions of the form $a + bk$. Many familiar names are associated with the problem of determining the number of primes in an arithmetic sequence of the form $a + bk$. Euler made the conjecture that the set of primes of the form $1 + bk$ is infinite for any integer b. Legendre claimed a proof of the existence of an infinite number of primes in the sequence $a + 2nb$, but his proof depended on a lemma that was later shown to be false. The credit for settling the problem goes to Dirichlet, a student of Gauss. He had mastered and enlarged on Gauss' *Disquisitiones Arithmeticae*. It is said that he slept with a copy under his pillow and worked through a paragraph or two in his spare moments. He had a good mastery of analysis as well. In 1837, with his innovative techniques of applying analysis to number theory, he proved the theorem that bears his name.

DIRICHLET'S THEOREM. If $(a, b) = 1$, there are infinitely many primes of the form $a + bk$.

The proof of this theorem in its general form is beyond the limits of our exploration here. It can be found, for example, in Le Veque [8]. We have already considered three special cases. Others can be proved in a similar manner and are suggested in the exercises. Special cases are still being investigated, not for the result but for the interest of the method of proof.

As a final special case, we shall consider primes of the form $1 + 2p^s k$. Here we prove that for each s there exists *one* prime of the form

$1 + 2p^s k$, making use of the exponent to which an integer belongs to do so. Then, from the fact that one prime exists for each s, we deduce the existence of an infinite set of primes for a fixed s.

EXAMPLE 4.1.2. Consider the case $p = 3$. Does there exist a prime of the form $1 + 2 \cdot 3^s k$ for each s? For a particular value of s, the most direct way to look for a prime of the required form is to write out the set of numbers $1 + 2 \cdot 3^s k$, $k = 1, 2, 3, \ldots$. Thus

$$s = 1. \quad 1 + 6k, \quad k = 1, 2, 3, \ldots = 7, 13, 19, 25, \ldots.$$
$$s = 2. \quad 1 + 18k, k = 1, 2, 3, \ldots = 19, 37, 55, 73, \ldots.$$
$$s = 3. \quad 1 + 54k, k = 1, 2, 3, \ldots = 55, 109, 163, 217, \ldots.$$

In this way we arrive rather quickly at prime numbers for a particular exponent s, but we get no clue as to how the form $1 + 2 \cdot 3^s k$ will help in the search for primes for a general value of s. Let us reconsider the case $s = 3$ and show that if $x = 2^{3^2}$, the expression $x^2 + x + 1$ must have a prime divisor of the form $1 + 2 \cdot 3^3 k$. We can argue as follows: Let q be a prime divisor of $x^2 + x + 1$. Then $q \mid x^3 - 1$; that is, $x^3 = 2^{3^3} \equiv 1 \bmod q$. We now wish to show that 2 belongs to the exponent $3^3 \bmod q$. If 2 belongs to $t \bmod q$, $t \mid 3^3$. Suppose that $2^{3^2} \equiv 1 \bmod q$. Then $x^2 + x + 1 \equiv 1 + 1 + 1 \equiv 3 \bmod q$. But q was assumed to be a divisor of $x^2 + x + 1$, so that $3 \equiv 0 \bmod q$. Since q is prime, this means that $3 = q$. But $2^3 \equiv 2 \bmod 3$, so that $2^{3^2} \equiv 2 \bmod 3$, which contradicts the assumption that $2^{3^2} \equiv 1 \bmod q$. Thus the assumption that $2^{3^2} \equiv 1 \bmod q$ leads to a contradiction. Since $2^3 \equiv 1 \bmod q$ implies that $2^{3^2} \equiv 1 \bmod q$, it also leads to a contradiction. Therefore 2 belongs to the exponent $3^3 \bmod q$. This implies that 3^3 divides $\phi(q)$. Since $x^2 + x + 1$ is odd, q must also be odd and $\phi(q)$ must be even. Therefore $2 \cdot 3^3$ divides $\phi(q)$; that is, $q = 1 + 2 \cdot 3^3 k$ for some k.

The argument in Example 4.1.2 establishes the existence of at least one prime of the form $1 + 2 \cdot 3^3 k$. Assume, for the moment, that a similar argument can be carried out to prove the existence of at least one prime of the form $1 + 2 \cdot 3^s k$, for any integer s. How can we use this to prove that there exists an infinite set of primes of the form $1 + 2 \cdot 3^3 k$?

EXAMPLE 4.1.3. Assume that for each $s > 0$ there exists a prime of the form $1 + 2 \cdot 3^3 k$. To prove that there exists an infinite set of primes of the form $1 + 2 \cdot 3^3 k$, we show that, given any integer n, there is a prime greater than n and of the form $1 + 2 \cdot 3^3 k$. Let n be a positive integer greater than 3. There exists a prime $q = 1 + 2 \cdot 3^n k_0$ for some integer k_0. Since $n > 3$, $3^n > n$, and $q > n$. Also, $q = 1 + 2 \cdot 3^3 (3^{n-3} k_0)$; that is, q is a prime of the form $1 + 2 \cdot 3^3 k$.

By combining the arguments illustrated in Examples 4.1.2 and 4.1.3, we can prove the following theorem.

THEOREM 4.1.4. *Primes of the Form* $1 + 2p^s k$. If p is a prime greater than 2 and s is a positive integer, there is an infinite set of primes of the form $1 + 2p^s k$.

Proof: Let $x = 2^{p^{s-1}}$ and let q be a prime divisor of $x^{p-1} + x^{p-2} + \cdots + x + 1$. Since x is even, q is odd. We first show that 2 belongs to $p^s \bmod q$.

Since $(x - 1)(x^{p-1} + x^{p-2} + \cdots + x + 1) = x^p - 1$, $q \mid x^p - 1$, or $x^p \equiv 1 \bmod q$. Also, $x^p = (2^{p^{s-1}})^p = 2^{p^s}$. We need to show that there is no power of 2 smaller than p^s that is congruent to 1 mod q. The only powers we need consider are the divisors of p^s. If we can show that $2^{p^{s-1}} \not\equiv 1 \bmod q$, then $2^{p^t} \not\equiv 1 \bmod q$ for $t < s - 1$.

Suppose that $2^{p^{s-1}} \equiv 1 \bmod q$. Then $x^{p-1} + \cdots + x + 1 \equiv p \bmod q$. But $x^{p-1} + \cdots + x + 1 \equiv 0 \bmod q$, so $p \equiv 0 \bmod q$ and $p = q$. To see that $p = q$ leads to a contradiction, consider $2^{p^{s-1}} \bmod p$. By the Euler–Fermat theorem, $2^p \equiv 2 \bmod p$, and (by induction) $2^{p^{s-1}} \equiv 2 \bmod p$. This contradicts the assumption that $2^{p^{s-1}} \equiv 1 \bmod q$, since this assumption implied that $q = p$. Therefore $2^{p^{s-1}} \not\equiv 1 \bmod q$, and 2 belongs to $p^s \bmod q$.

If 2 belongs to the exponent $t \bmod q$, t divides $\phi(q)$. Therefore $p^s \mid \phi(q)$. Since q is odd, $\phi(q) = q - 1$ is even, and 2 also divides $\phi(q)$. Since $(2, p) = 1$, this means that $2p^s \mid q - 1$; that is, $q = 1 + 2p^s k$ for some integer k.

To see that there is an infinite set of primes of this form, let n be any integer greater than s. There exists a prime q of the form $q = 1 + 2p^n k$. Since $p > 1$, $p^n > n$, and $q > n$. Now q can also be written

$$q = 1 + 2p^s(p^{n-s}k) = 1 + 2p^s k_1.$$

This result completes the proof of the theorem.   ●

EXERCISES 4.1

Checks

1. If $b = 24$, what values of a give arithmetic sequences containing an infinite set of primes?

2. If $(a, b) = d > 1$, prove that there is at most one prime of the form $a + bk$.

3. Prove that there is an infinite set of composite numbers of the form $a + bk$.

4. Calculate $N = x^2 + x + 1$, where $x = 2^{3^2}$. Discover whether N is prime. Remember that the prime divisors of N are of the form $1 + 2 \cdot 3^3 k$.

5. If $N = \prod_{i=1}^{n} q_i$ and $q_i \equiv 3 \bmod 4$ for each i, is $N \equiv 3 \bmod 4$?

6. Prove the equivalence of the statements: "There is an infinite set of primes" and "For every positive integer n, there is a prime greater than n."

Challenges

7. Prove that the set of primes of the form $5 + 6k$ is infinite.

8. Prove the same for primes of the form $7 + 8k$. *Hint:* Choose $N = 2(n!)^2 - 1$.

9. Prove the same for primes $5 + 8k$. *Hint:* Choose a as in Theorem 4.1.3 and $N = a^2 + 4$.

10. Prove the same for the primes $4 + 5k$. *Hint:* Choose $N = 5(n!)^2 - 1$.

11. Prove that Dirichlet's theorem is equivalent to the statement: If $(a, b) = 1$, there exists at least one prime in the sequence $a + bk$. This is E1218, *American Mathematical Monthly*, Vol. 63 (1956), p. 342.

12. If the integers are represented in base 2, every odd integer, and hence every odd prime, has units digit 1. Prove that given any m, there exists a prime whose representation in base 2 has exactly m zeros immediately preceding the units digit.

13. Prove that each prime divisor of $2^p - 1$ is greater than p, p a prime. As a corollary, derive that the set of primes is infinite. This is E1672, *American Mathematical Monthly*, Vol. 71 (1964), p. 317.

4.2   PRIMES REPRESENTED
IN OTHER WAYS

What about the number of primes in other special sets of integers? It would be natural to consider more general polynomial functions with integral coefficients, since these functions have integral values for integral values of x. Can we find a polynomial of degree higher than 1 that gives prime values? As in the case of linear expressions, certain polynomials can be eliminated immediately.

EXAMPLE 4.2.1

$f(x) = 4x^2 + 8x + 12; 4 \mid f(n)$ for every n.
$g(x) = x^2 + 4x + 3 = (x + 1)(x + 3); g(n)$ is composite for $n \neq 0$.
$h(x) = x^2 + 4x + 7; h(2)$ is prime; what about $h(7)$?
$m(x) = x^2 + 3x + 1; m(1) = 5$. Consider $m(y + 1)$. Since
$$m(y + 1) = (y + 1)^2 + 3(y + 1) + 1 = y(y + 5) + 5,$$
then $m(y + 1)$ is composite if $y = 5$. That is, $m(6)$ is composite.

We have a strong feeling, on the basis of the preceding example, that every polynomial takes on some composite values unless it is of degree zero—that is, a constant. Since we are interested in finding prime values, we are interested in polynomials for which $f(n)$ will be positive for an infinite set of n. This property will be ensured if the coefficient of the highest power of x is positive. Although there is reason to hope for an infinite set of primes, it is much easier to prove that the set $\{f(n), n = 1, 2, 3, \ldots\}$ contains an infinite set of composite integers.

THEOREM 4.2.1. *Composite Values of a Polynomial.* Let
$$f(x) = a_0 x^k + a_1 x^{k-1} + \cdots + a_{k-1}x + a_k,$$
where the coefficients a_i are integers, a_0 is positive, and $k \geq 1$. There is an infinite set of composite integers in the set $\{f(n), n = 1, 2, 3, \ldots\}$.

Proof: The condition $k \geq 1$ implies that f is not constant; $a_0 > 0$ implies that, for large enough $x, f(x)$ is positive and increases as x increases. By the argument used in the factor theorem (3.3.1), for any b,
$$f(x) = f(b) + (x - b)g(x).$$
Choose b such that $f(b) > 1$ and set $x = b + mf(b)$. Then
$$f(x) = f(b) + mf(b)g(x) = f(b)[1 + mg(x)].$$
As m increases, x increases; therefore $1 + mg(x)$ increases and will be greater than 1 for an infinite set of values of m. For each of these m, $f[b + mf(b)]$ is composite. ●

Even though, for a polynomial f, the set $\{f(n), n = 1, 2, 3, \ldots\}$ contains an infinite set of composite integers, the possibility still remains that the set contains an infinite set of prime integers as well.

EXAMPLE 4.2.2. The table shown helps us to consider some polynomials of the form $x^2 + x + a$. Since $x^2 + x = x(x + 1)$, we have the best chance of prime values if a is an odd prime.

x	$x^2 + x$	$x^2 + x + 5$	$x^2 + x + 7$	$x^2 + x + 11$	$x^2 + x + 17$
1	2	7	9	13	19
2	6	11	13	17	23
3	12	17	19	23	29
4	20	$25 = 5^2$	$27 = 3^3$	31	37
5	30	$35 = 5 \cdot 7$	37	41	47
6	42	47	$49 = 7^2$	53	59
7	56	61	$63 = 7 \cdot 9$	67	73
8	72	$77 = 7 \cdot 11$	79	83	89
9	90	$95 = 5 \cdot 19$	97	101	107
10	110	$115 = 5 \cdot 23$	$117 = 3^2 \cdot 13$	$121 = 11^2$	127
11	132	137	139	$143 = 11 \cdot 13$	149
12	156	$161 = 7 \cdot 23$	163	167	173
13	182	$187 = 11 \cdot 17$	$189 = 3^3 \cdot 7$	193	199
14	210	$215 = 5 \cdot 43$	$217 = 7 \cdot 31$	$221 = 13 \cdot 17$	227
15	240	$245 = 7^2 \cdot 5$	$247 = 13 \cdot 19$	251	257
16	272	277	279	283	$289 = 17^2$

For $a = 5, 11$, and $17, f(n)$ is prime for $n = 0, 1, \ldots, a - 2$. Such is not the case for $a = 7$. Check the factorization of the composite numbers that occur in each case. How does the case $a = 7$ differ from the cases $a = 5$ and $a = 11$?

The polynomial $x^2 + x + 17$, judged on its ability to yield distinct primes for consecutive values of n, certainly seems to be the best of the four polynomials checked, at least initially. The polynomial $x^2 + x + 41$, called the Euler trinomial, is even more impressive.

THEOREM 4.2.2. *Some Properties of the Euler Trinomial.* Let $f(x) = x^2 + x + 41$. The values of $f(x)$ are distinct primes, for $x = 0, 1, 2, \ldots, 39$. If $x > 40$, $f(x)$ has no divisor such that $1 < d < 40$. If $f(x)$ is composite for $x > 40$, it must have a prime divisor less than $x + 1$.

Proof: Calculation of $f(0)$, $f(1)$, $\ldots$, $f(39)$ verifies that these integers are distinct primes. Also, $f(40) = 41^2$. Suppose that $x > 40$ and $1 < d < 40$. By the division algorithm, $x = dq + r$, $0 \le r < d$. Then

$$f(x) = f(dq + r) = d(dq^2 + 2qr + q) + f(r).$$

If $d \mid f(x)$, then $d \mid f(r)$. But $f(r)$ is a prime greater than or equal to 41, which is a contradiction.

Suppose that x is an integer greater than 40. Then $2x > x + 40$, so that

$$x^2 + 2x + 1 > x^2 + x + 41.$$

This result implies that $(x + 1)^2 > f(x)$ for $x > 40$. If $f(x)$ is composite, it must have a prime divisor less than or equal to $\sqrt{f(x)}$—that is, less than or equal to $x + 1$. ●

In spite of the nice properties of this trinomial, no one has yet been able to establish whether or not the set $\{f(n), n = 1, 2, 3, \ldots\}$ contains an infinite set of primes. In fact, the answer to this question is not known even for the simplest type of quadratic polynomial, $x^2 + 1$.

There are other types of functions by means of which one might try to create primes. Mersenne searched for primes of the form $a^n - 1$. The rules of factorization lead quickly to the conclusion that if $a^n - 1$ is to be a prime, then a must be 2 and n must be a prime. (See Exercises 4.2, problem 6.)

EXAMPLE 4.2.3. Let $M(p) = 2^p - 1$. Then $M(2) = 3$, $M(3) = 7$, $M(5) = 31$, $M(7) = 127$. All of these integers are prime. But $M(11) = 2407 = 23 \cdot 89$.

Numbers of the form $M(n) = 2^n - 1$ are called *Mersenne numbers.* The question of whether or not Mersenne numbers are prime is especially interesting because of the relation of these numbers to even perfect numbers. An even number is perfect if and only if it is of the form $2^{s-1}(2^s - 1)$ and $2^s - 1$ is prime. (See Exercises 1.11, problems 6 and 8.) The study of Mersenne numbers has given rise to a great deal of research, particularly in relation to tests to determine whether or not an integer is prime. With the advent of the computer, it is now much less difficult to carry out these tests than it was at the time the numbers were first studied. Considerable work has been done during the 1960s in factoring numbers of the form $2^n - 1$ by using the computer. The machine time (and therefore the cost) required to study problems of this type is reduced considerably by making use of all the theoretical evidence available. For example, the prime factors of $2^p - 1$ must have the form $px + 1$ and, in addition, must be congruent to 1 or 7 mod 8. This information restricts considerably the divisors that must be tried to test whether $2^p - 1$ is prime. Bryant Tuckerman, using an IBM System/360 Model 91 computer, determined that $M(19,937)$ is a prime—a prime, in fact, that has 6002 digits in its decimal representation.

When we look at $a^n + 1$, we find that a must be even and n must be a power of 2 if such an integer can be prime (see Exercises 4.2, problem 7). Numbers of the form $F(n) = 2^{2^n} + 1$ are called *Fermat numbers*. These are prime for $n = 0, 1, 2, 3, 4$. Euler, however, was able to prove that 2^{2^5} has 641 as a divisor. Some of the interesting properties of Fermat numbers are suggested in the exercises. These numbers increase so rapidly with n that the task of testing is even more difficult for Fermat numbers than for Mersenne numbers. It was conjectured that the sequence

$$2 + 1, 2^2 + 1, 2^{2^2} + 1, 2^{2^{2^2}} + 1, \ldots$$

consists only of primes. By using a computer it was found that $2^{2^{16}} + 1$, the fifth number in this sequence, has a divisor $2^{19} \cdot 1575 + 1$ and is therefore composite. Factorizations have been found for many Fermat numbers for $n > 4$, but it has not yet been proved that there exist infinitely many composite Fermat numbers. On the other hand, it has not yet been proved that there exists a single prime Fermat number for $n > 4$. Here is an area where many questions are still unanswered.

Should we then abandon the search for a function having only prime values? It has been shown that such a function does exist, in fact, that infinitely many such functions exist. In 1947 W. H. Mills proved that there exists a real number θ such that $[\theta^{3^n}]$ is prime for all n. This result, which at first seems astonishing, is not exactly the type of result that was discussed when we considered representing primes by polynomials. The proof of the existence of the number θ involves the construction of a sequence of primes q_n such that

$$q_n^3 < q_{n+1} < (q_n + 1)^3 - 1.$$

In terms of these primes, the number θ is defined as $\lim_{n \to \infty} u_n$, where $u_n = q_n^{3^{-n}}$. The actual construction of θ thus is seen to involve the construction of large primes. A discussion of this result and some later work in this area can be found in the article "History of a Formula for Primes," by Underwood Dudley, *American Mathematical Monthly*, Vol. 76 (1969), pp. 23–28.

EXERCISES 4.2

Checks

1. What necessary conditions are suggested by Example 4.2.1 in order that a polynomial take on prime values?

2. Let $f(x) = x^2 - 3x + 1$. Let $b = 4$ and $x = b + mf(b)$ as in Theorem 4.2.1. Show how to find an infinite set of x such that $f(x)$ is composite.

3. Investigate the values of the polynomial $2x^2 + 29$.

4. Show that there exists at least one odd prime of the form $x^2 + 1$. What about $x^3 + 1$? $x^4 + 1$? What about primes of these forms that are greater than 1000?

5. If a is odd, $a > 1$, $a^n \pm 1$ cannot be prime. Why?

6. Prove:

$$a^n - 1 = (a - 1)(a^{n-1} + a^{n-2} + \cdots + a + 1),$$
$$a^{rs} - 1 = (a^r - 1)(a^{r(s-1)} + a^{r(s-2)} + \cdots + a^r + 1).$$

For what values of a is $a^n - 1$ necessarily composite? For what values of n is $a^n - 1$ necessarily composite?

Challenges

7. Prove that if $n = rs$, where s is odd, then

$$a^n + 1 = (a^r + 1)(a^{r(s-1)} - a^{r(s-2)} + \cdots - a^r + 1).$$

Deduce that $a^n + 1$ is composite unless a is even and n is a power of 2.

8. For F defined in this section, prove that $F(n) | F(n + k) - 2, k \geq 1$. Deduce $[F(n), F(n + k)] = 1$ for every k and use this result to prove that the set of primes is infinite.

9. Prove that $F(n) \equiv 5 \bmod 12, n \geq 1$.

10. Prove that if $p | F(n)$, then 2 belongs to the exponent $2^{n+1} \bmod p$. Show that this result implies that $p = 2^{n+1}k + 1$ for some integer k.

11. Prove that if q is a prime of the form $8k + 7$, then $q \mid M[(q - 1)/2]$. Use this result to find a prime divisor of $M(11)$. *Hint:* Use Euler's criterion and the fact that $(2/q) = 1$.

12. Prove that if $k > 0$, $m \mid M[k\phi(m)]$ if and only if m is odd.

4.3   THE FUNCTION π

Most of the number–theoretic functions we have studied have been multiplicative, and it has been possible to express $f(n)$ in terms of the prime factorization of n. Not so the function π. In Sections 4.1 and 4.2 we had trouble discovering whether a simple formula generates prime values. Naturally, then, we do not expect a simple formula for $\pi(n)$. However, the function $[x]$ can be used to describe the step-by-step process of calculating $\pi(n)$ for a particular n (Example 1.10.6). We state the result as a theorem without formal proof.

THEOREM 4.3.1. *A Formula for* $\pi(n)$. Let n be an integer and $\pi(\sqrt{n}) = r$. Then

$$\pi(n) = n - 1 + r - A_1 + A_2 - A_3 + \cdots + (-1)^r A_r,$$

where

$$A_k = \sum \left[\frac{n}{p_{i_1} p_{i_2} \cdots p_{i_k}} \right],$$

the summation consisting of $\binom{r}{k}$ terms, one for each of the possible products of k of the r smallest primes, $p_1, p_2, \ldots, p_r$.

This rather cumbersome formula does indeed represent accurately the function π, but in the process of calculating $\pi(n)$, we need to calculate the first r primes as well. Other formulas of this nature have been devised. In Exercises 4.3, problem 8, a suggestion is made, using Wilson's theorem, for an expression for $\pi(n)$ in terms of sums of cosines. Although this expression does not involve the first r primes explicitly, a careful look shows that the work needed to evaluate the cosine terms is exactly the work needed to derive the primes. The formula is nevertheless ingenious. Similar ones can be found, for example, in an article by C. P. Willans,

"On Formulae for the nth Prime Number," *Math. Gazette*, Vol. 48 (1964), pp. 413–415.

Formulas like the one in Theorem 4.3.1 tell us little about the growth and behavior of the function π. A logical place to start looking for information of this sort might be a study of the graph of the function based on tabulated values. We define $\pi(x)$ to be the number of primes less than or equal to x for any positive real number x. The function π is a non-decreasing step function. For large x, the rate of growth of π appears to decrease, which suggests that the ratio $\pi(x)/x$ might be of interest. This ratio gives the "density" of the primes in the first x integers. The accompanying figure shows that the behavior of $\pi(x)/x$, as x increases from 1 to 1,000,000, bears a marked resemblance to that of $1/\log x$.

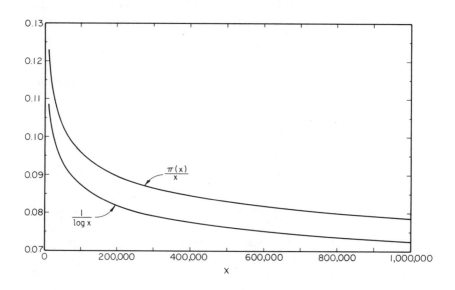

On the basis of rather limited tables of primes (up to 408,000), Legendre was able to recognize a relation between $\pi(x)/x$ and the function $1/\log x$. Here $\log x$ means the *natural* logarithm; that is, the base used is the transcendental number e. Legendre, in 1808, asserted that $\pi(x)/x$ is approximately $1/(\log x - 1.08366)$. Gauss asserted that $\pi(x)/x$ is approximately the function Li x defined by

$$\int_a^x \frac{du}{\log u}.$$

Neither mathematician made precise in what sense the word "approximately" was used. The statement that has come to be known as the Prime Number Theorem is as follows:

THEOREM 4.3.2. *The Prime Number Theorem.*

$$\lim_{x \to \infty} \frac{\pi(x)}{x/\log x} = 1.$$

The statement of this theorem does not imply that $\pi(x) = x/\log x$ for any x. It merely claims that the ratio

$$\frac{\pi(x)}{x/\log x}$$

is as close to 1 as we wish to make it for all x sufficiently large. The conjectures of Legendre and Gauss are quite consistent with this theorem, since

$$\lim_{x \to \infty} \frac{\pi(x)}{x/(\log x - 1.08366)} \quad \text{and} \quad \lim_{x \to \infty} \frac{\pi(x)}{\text{Li } x}$$

are also equal to 1.

For almost a century after this theorem was conjectured, it was not proved. Finally, in 1896, a proof was given by Hadamard and de la Vallée Poussin. This proof involved the use of analysis and hence cannot be considered "elementary." In particular, it is based on the properties of the complex-valued function

$$\zeta(s) = \sum_{n=1}^{\infty} \frac{1}{n^s}$$

which had been studied by Riemann and is called the Riemann zeta function.

For a long time it was thought that the prime number theorem was essentially a theorem of analysis, particularly since even its statement involved the concept of limit. But in 1949 A. Selberg and P. Erdös succeeded in giving an "elementary" proof of the prime number theorem based on an inequality involving logarithms that was proved by Selberg. The word "elementary" in this context does not mean "simple." It indicates only that the proof does not involve arguments from the theory of functions of a complex variable. Actually, the proof is quite lengthy and cannot be included here. See, for example, Levinson [9].

To get some feeling for the relation between the function $\pi(x)$ and the logarithm function, we will explore some elementary inequalities similar to the one proved by Tchebychef in 1850. The basic idea in deriving these inequalities is a study of the difference $\pi(2n) - \pi(n)$. For this purpose, we need an integer that is divisible by all the primes in the range $n < p < 2n$. The integer $(n + 1)(n + 2) \cdots (2n)$ would qualify, but a much smaller integer with the same property is the binomial coefficient

$$\binom{2n}{n} = \frac{(2n)!}{n!\,n!}.$$

EXAMPLE 4.3.1

$$\binom{12}{6} = \frac{12!}{6!\,6!} = \frac{7\cdot 8\cdot 9\cdot 10\cdot 11\cdot 12}{1\cdot 2\cdot 3\cdot 4\cdot 5\cdot 6}.$$

We can think of $\binom{12}{6}$ as the product of six rational numbers, each greater than or equal to 2. The result is

$$\binom{12}{6} > 2^6.$$

On the other hand, $\binom{12}{6}$ is one term in the expansion by the binomial theorem of $(1 + 1)^{12}$. This makes

$$\binom{12}{6} < 2^{12}.$$

In this numerical case, we could easily write $\binom{12}{6}$ in standard form by simple arithmetic. Instead, let us try to reason from our previous theorems so that we can see how to handle the general case. Only primes less than 12 can be involved. The largest power of 2 that divides the numerator is

$$\left[\frac{12}{2}\right] + \left[\frac{12}{2^2}\right] + \left[\frac{12}{2^3}\right] = 6 + 3 + 1.$$

The largest power of 2 that divides the denominator is

$$2\left\{\left[\frac{6}{2}\right] + \left[\frac{6}{2^2}\right]\right\} = 2\{3 + 1\} = 8.$$

Therefore the exponent of 2 in the prime factorization of $\binom{12}{6}$ is $10 - 8 = 2$. A similar analysis for the remaining primes is as follows:

The exponent of 3 is $[12/3] + [12/3^2] - 2[6/3] = 1$.
The exponent of 5 is $[12/5] - 2[6/5] = 2 - 2 = 0$.
The exponent of 7 is $[12/7] = 1$.
The exponent of 11 is $[12/11] = 1$.
Therefore $\binom{12}{6} = 2^2 \cdot 3 \cdot 7 \cdot 11$.

Note that for the primes between 6 and 12, the exponent must be at least 1, since these primes do not divide the denominator. Let k_p be the largest power of p that is less than or equal to 12. Then $k_2 = 3$, $k_3 = 2$, $k_5 = 1$, $k_7 = 1$, and $k_{11} = 1$. In each case, k_p is greater than or equal to the exponent of p in the standard form of $\binom{12}{6}$. Therefore we can conclude that

$$\binom{12}{6} \mid 2^3 \cdot 3^2 \cdot 5 \cdot 7 \cdot 11.$$

But $2^3 \cdot 3^2 \cdot 5 \cdot 7 \cdot 11 < 12^{\pi(12)}$, since each factor is less than 12 and there are $\pi(12) = 5$ factors. Thus

$$\binom{12}{6} < 12^{\pi(12)}.$$

The inequalities illustrated in this example, although they may seem rather crude in a numerical case, are good enough to yield some interesting bounds for $\pi(n)$. First, we derive the appropriate inequalities for the general case $\binom{2n}{n}$.

THEOREM 4.3.3. *Inequalities Involving* $\binom{2n}{n}$.

1. $$2^n < \binom{2n}{n} < 2^{2n}.$$

2. Let k_p be defined by $p^{k_p} \leq 2n < p^{k_p + 1}$. Then

$$\binom{2n}{n} = \prod_{p < 2n} p^\alpha$$

where $0 \leq \alpha \leq k_p$ for all p, and $\alpha > 0$ for $p > n$.

3. $$n^{\pi(2n) - \pi(n)} < \binom{2n}{n} < (2n)^{\pi(2n)}.$$

Proof: The inequality $2^n < \binom{2n}{n}$ is a consequence of the fact that $\binom{2n}{n}$ is a product of n rational numbers, each greater than or equal to 2. The inequality $\binom{2n}{n} < 2^{2n}$ is true, since $\binom{2n}{n}$ is one term of the expansion of $(1 + 1)^{2n}$ by the binomial theorem.

The largest power of p that divides $(2n)!$ is

$$\left[\frac{2n}{p}\right] + \left[\frac{2n}{p^2}\right] + \cdots + \left[\frac{2n}{p^{k_p}}\right].$$

The largest power of p that divides $n!\,n!$ is

$$2\left\{\left[\frac{n}{p}\right] + \left[\frac{n}{p^2}\right] + \cdots + \left[\frac{n}{p^{k_p}}\right]\right\}.$$

Some of the terms in this summation may be zero. The exponent of p in the prime factorization of $\binom{2n}{n}$ is

$$\sum_{i=1}^{k_p}\left\{\left[\frac{2n}{p^i}\right] - 2\left[\frac{n}{p^i}\right]\right\}.$$

Since $0 \le [x] - 2[x/2] \le 1$ for any x (see Example 1.4.6), each term of the sum is at most 1, and the sum itself is less than or equal to k_p. Therefore if α is the exponent of p, $\alpha \le k_p$ for each p. For those p greater than n, $[n/p^i] = 0$. If $p < 2n$, then $[2n/p] > 0$. Therefore, for the primes p such that $n < p < 2n$, we have $\alpha \ge 1$.

We prove statement (3) in two parts. Since $\alpha \le k_p$ for all $p < 2n$, then

$$\binom{2n}{n}\,\bigg|\,\prod_{p<2n} p^{k_p}.$$

But $p^{k_p} < 2n$ by definition, so

$$\prod_{p<2n} p^{k_p} < (2n)^{\pi(2n)}.$$

These two statements imply that

$$\binom{2n}{n} < (2n)^{\pi(2n)}.$$

There are $\pi(2n) - \pi(n)$ primes in the interval $n < p < 2n$. The product of these primes is greater than $n^{\pi(2n) - \pi(n)}$. But for each of these primes, the exponent α in the prime factorization of $\binom{2n}{n}$ is positive, so that

$$\prod_{n < p < 2n} p \,\Big|\, \binom{2n}{n}$$

and therefore

$$n^{\pi(2n) - \pi(n)} < \prod_{n < p < 2n} p < \binom{2n}{n}.$$

The inequality in statement (3) follows. ●

THEOREM 4.3.4. *An Upper Bound for* $\pi(2n) - \pi(n)$. For any positive integer $n > 1$,

$$\pi(2n) - \pi(n) < (2 \log 2) \frac{n}{\log n}.$$

Proof: The inequalities in statements (3) and (1) of the preceding theorem combine to give

$$n^{\pi(2n) - \pi(n)} < \binom{2n}{n} < 2^{2n};$$

that is, $n^{\pi(2n) - \pi(n)} < 2^{2n}$. Now take the logarithm of each side of this inequality. Since log is a function with the property $a < b$ implies $\log a < \log b$, we can write $\log n^{\pi(2n) - \pi(n)} < \log 2^{2n}$, which gives $\{\pi(2n) - \pi(n)\} \log n < 2n \log 2$. Since $\log n > 0$, this inequality is the same as

$$\pi(2n) - \pi(n) < 2 \log 2 \frac{n}{\log n}. \quad ●$$

Integers that are powers of 2 have the property that each is the double of the preceding element in the set. It is not surprising that the preceding theorem, which relates $\pi(n)$ and $\pi(2n)$, can be used to get an upper bound for $\pi(2^k)$.

THEOREM 4.3.5. *An Upper Bound for* $\pi(2^k)$.

$$\pi(2^k) < \frac{2^{k+1}}{k \log 2}.$$

Proof: The proof is by induction. If $k = 1$, then $\pi(2) = 1$, which is less than 4/log 2. Suppose that

$$\pi(2^k) < \frac{2^{k+1}}{k \log 2}.$$

The preceding theorem tells us that

$$\pi(2^{k+1}) < \pi(2^k) + 2 \log 2 \, \frac{2^k}{\log 2^k} = \pi(2^k) + 2^{k+1} \frac{\log 2}{k \log 2}.$$

Now use the induction hypothesis to eliminate $\pi(2^k)$.

$$\pi(2^{k+1}) < \frac{2^{k+1}}{k \log 2} + \frac{2^{k+1} \log 2}{k \log 2} = \frac{2^{k+1}(\log 2 + 1)}{k \log 2}.$$

We need some inequalities to bring this result into the required form. Since $\log 2 = 0.6931$, $\log 2 + 1 < 1.7$, and $(k + 1)1.7 < 2k$, if $0.3k > 1.7$; that is, $k > 5$. So, for $k > 5$,

$$\pi(2^{k+1}) < \frac{2^{k+1}}{k \log 2} \cdot \frac{2k}{(k + 1)} = \frac{2^{k+2}}{(k + 1) \log 2}.$$

The inequality is established provided that $k > 5$. The cases when $k \leq 5$ can be treated numerically:

$$\pi(4) < \frac{8}{2 \log 2}, \quad \pi(8) < \frac{16}{3 \log 2}, \quad \pi(16) < \frac{32}{4 \log 2}, \quad \pi(32) < \frac{64}{5 \log 2}. \quad \blacksquare$$

The upper bound for $\pi(2^k)$ can easily be extended to an upper bound for the function at any value of n. This step is done in the following theorem, and a lower bound for π is obtained as well.

THEOREM 4.3.6. *Bounds for* $\pi(n)$.

$$\frac{\log 2}{3} \cdot \frac{n}{\log n} < \pi(n) < 4 \frac{n}{\log n}.$$

Proof: Let k be the integer defined by the property $2^k \leq n < 2^{k+1}$. This inequality implies $(k + 1) \log 2 > \log n$. Now

$$\pi(n) \leq \pi(2^{k+1}) < \frac{2^{k+2}}{(k + 1) \log 2} < \frac{4n}{\log n}$$

which establishes the upper bound.

By property (1) of the binomial coefficient, $2^n < \binom{2n}{n}$ and by property (3), $\binom{2n}{n} < (2n)^{\pi(2n)}$. These results imply that $2^n < (2n)^{\pi(2n)}$. Again we take the logarithm of each side and obtain

$$n \log 2 < \pi(2n) \log 2n,$$

or

$$\pi(2n) > \frac{\log 2}{2} \frac{2n}{\log 2n},$$

which establishes a lower bound for even integers. Since $\pi(2n + 1) \geq \pi(2n)$, we have

$$\pi(2n + 1) > \log 2 \frac{n}{\log 2n}.$$

But $\log (2n + 1) > \log 2n$, which implies

$$\frac{n}{\log 2n} > \frac{n}{\log (2n + 1)}.$$

We wish to have $2n + 1$ in the numerator. Since $2n + 1 \leq 3n$ for all $n \geq 1$, we can write

$$\pi(2n + 1) > \frac{\log 2}{3} \frac{2n + 1}{\log (2n + 1)}.$$

Thus for every $n > 1$,

$$\pi(n) > \frac{\log 2}{3} \frac{n}{\log n},$$

and the lower bound for π is established. ◆

The inequalities stated in this theorem are only a very small step toward the proof of the prime number theorem. A good discussion of the

elementary proof of this theorem was published in 1969. This article by
Norman Levinson, "A Motivated Account of an Elementary Proof of the
Prime Number Theorem" *American Mathematical Monthly*, Vol. 76
(1969), pp. 225–245, was given a Lester R. Ford award as an expository
article in 1970.

EXERCISES 4.3

Checks

1. Use Theorem 4.3.1 to calculate $\pi(30)$.

2. How many primes would have to be considered in calculating $\pi(200)$?

3. Verify the properties of $\binom{2n}{n}$ for $n = 7$.

4. Verify the numerical cases $k \le 5$ in Theorem 4.3.5.

5. Use your knowledge of elementary calculus to prove that if A is a real
    number,

$$\lim_{x \to \infty} \frac{\pi(x)}{x/\log x} = 1 \quad \text{if and only if} \quad \lim_{x \to \infty} \frac{\pi(x)}{x/(\log x - A)} = 1.$$

6. Find an example of a pair of functions f and g such that

$$\lim_{n \to \infty} \frac{f(n)}{g(n)} = 1 \quad \text{but} \quad \lim_{n \to \infty} f(n) - g(n) = \infty.$$

Challenges

7. Prove that if n is prime,

$$\frac{\pi(n-1)}{n-1} < \frac{\pi(n)}{n}$$

and if n is composite,

$$\frac{\pi(n-1)}{n-1} > \frac{\pi(n)}{n}.$$

8. Define F as follows:

$$F(x) = \left[\cos^2 \pi \, \frac{(x-1)! + 1}{x} \right].$$

Show that $F(x) = 1$ if x is prime or if $x = 1$, and $F(x) = 0$ if x is composite. Hence show that

$$\pi(n) = -1 + \sum_{x=1}^{n} F(x).$$

9. Show that the formula of Theorem 4.3.1 can be written as

$$\pi(n) - \pi(\sqrt{n}) + 1 = \sum_{k \mid m} \mu(k) \left[\frac{n}{k}\right], \quad m = [\sqrt{n}]!. \qquad \text{(Jim Lawrence)}$$

10. Prove that the prime number theorem implies that

$$\lim_{n \to \infty} \frac{n \log p_n}{p_n} = 1$$

where p_n is the nth prime (according to magnitude). *Hint*: Consider $\pi(p_n)$.

11. Let $S(n) =$ the number of square-free integers less than or equal to n. Prove that $S(n) < 2^{\pi(n)}$. *Hint*: A square-free integer less than or equal to n must be a product of distinct primes less than or equal to n.

12. Use the preceding problem and the fact that every positive integer can be expressed uniquely as a product of the form $k^2 m$, where m is square free, to deduce the inequality $\pi(n) > \log n/(2 \log 2)$.

4.4 PRIME ARITHMETIC SEQUENCES

The prime number theorem describes the general behavior of the function π. When we look at this function in more detail, we see that its behavior is quite erratic. For example, arbitrarily long sequences of composite numbers exist. Consider the sequence $k! + 2, k! + 3, \ldots, k! + k$, or use Exercises 3.2, problem 7 or 8. We do not expect consecutive primes, except for 2 and 3, nor do we expect more than two consecutive odd primes, except for 3, 5, and 7 (see Exercises 1.2, problem 3). We know that we cannot find an arithmetic sequence in which every term is prime (Theorem 4.2.1). However, certain arithmetic sequences contain sets of consecutive terms that are prime. For example, 3, 7, 11 is a set of consecutive terms in an arithmetic sequence with common difference 4; also, 5, 11, 17, 23, 29 is a set of five terms from an arithmetic sequence with common difference 6. Such sets are referred to as *prime arithmetic sequences*. How many terms can a prime arithmetic sequence contain? Can such a sequence be found with any given common difference d?

EXAMPLE 4.4.1. What prime arithmetic sequences can be found beginning with 3?

3	3	3	3	3	3	3	3	3	3	3	3	3	3	...	a
5	7	9	11	13	15	17	19	21	23	25	27	29	31	...	$a+d$
7	11	15	19	23	27	31	35	39	43	47	51	55	59	...	$a+2d$
2	4	6	8	10	12	14	16	18	20	22	24	26	28	...	d

Here are some examples of arithmetic sequences in which the first term is 3. The common differences included are all even. (Why is it possible to discard odd values of d when searching for primes here?) Some even values of d could also be discarded. For example, d cannot be a multiple of 3. All three of the integers listed are prime if $d = 2, 4, 8, 10, 14, 20, 28$. Try to add a fourth term to the sequences that have three primes. We get 9, 15, 27, 33, 45, 63, 87. None is prime. What integer divides all of them? It is easy to see that if 3 is the first term, a prime arithmetic sequence cannot have more than three terms.

The restrictions on the number of terms and on the possible value of the common difference are given in the following theorems. You would enjoy proving some special cases before you read the general theorem.

THEOREM 4.4.1. *The Length of a Prime Arithmetic Sequence.* There does not exist a prime arithmetic sequence with $b + 1$ terms and with first term b.

Proof: Suppose that such a sequence exists; that is, suppose that b, $b + d$, $b + 2d$, ..., $b + bd$ are all prime. Since b is prime, $b > 1$, and therefore $b + bd$ is composite.   ●

THEOREM 4.4.2. *The Common Difference of a Prime Arithmetic Sequence.* If n and d are positive integers, and $n > 1$, and if b, $b + d$, $b + 2d$, ..., $b + (n - 1)d$ are odd primes, then d is divisible by every prime less than n.

Proof: By the preceding theorem, $b \geq n$. Suppose that p is prime and less than n. Consider $b + id$, for each $i = 0, 1, 2, ..., p - 1$. By the division algorithm, there exist m_i and r_i such that

$$b + id = m_i p + r_i, \quad 0 < r_i < p.$$

We know that $r_i \neq 0$, since $b + id$ is prime. There are p remainders that can take on the values 1, 2, 3, ..., $p - 1$. So two of them must be equal. Suppose that $r_k = r_j$. Then $b + kd - m_k p = b + jd - m_j p$ or $(k - j)d = (m_k - m_j)p$. Since $0 < |k - j| \leq p - 1$, we have $(p, k - j) = 1$, and therefore $p \,|\, d$. ◆

The two preceding theorems show that there is a limit on the length of a prime arithmetic sequence and a restriction on the common difference if a sequence of specified length exists. This still leaves unanswered the question of whether or not such prime arithmetic sequences do exist, of arbitrary length and given admissible starting point and common difference. The general answer to this question is not yet known.

Suppose that, for some n and d, we know that there exists a set of n primes that are consecutive terms of an arithmetic sequence with common difference d. How many such sets are possible? The simplest case of this question occurs when $n = 2$ and $d = 2$. Here we are looking for consecutive odd numbers, both prime. In Section 1.2 we noticed that many examples of such pairs of primes exist. They are called *twin primes*. One of the best known of the number–theoretic questions that are as yet unanswered is: Does there exist an infinite set of twin primes?

Twin primes occur when there is a k such that $6k - 1$ and $6k + 1$ are both prime. Obviously, the number of primes less than or equal to n that belong to a pair of twin primes is less than the total number of primes less than or equal to n. That the incidence of twin primes is significantly less is seen in the work of many who have considered the problem from various points of view. Let $\{p_i : i = 1, 2, 3, \ldots\}$ represent the set of primes arranged in order of magnitude. Let

$$P_n = \frac{1}{p_1} + \frac{1}{p_2} + \cdots + \frac{1}{p_n}.$$

It can be shown that, as n increases, P_n increases without bound, even though the difference between P_{n+1} and P_n becomes smaller as n becomes larger. In the terminology of calculus, the infinite series

$$\sum_{k=1}^{\infty} \frac{1}{p_i}$$

diverges.

Let $Q = \{q_1, q_2, q_3, \ldots, q_i, \ldots\}$ represent the set of primes that belong to pairs of twin primes, arranged in increasing order of magnitude. Of course, we do not know whether this set is finite or infinite. Let

$$Q_n = \frac{1}{q_1} + \frac{1}{q_2} + \cdots + \frac{1}{q_n}.$$

As n increases, Q_n also increases. If the set Q is finite, naturally there is a largest Q_n. Viggo Brun, by using a complicated process of sieving and some appropriate inequalities, was able to prove that even if the set Q is infinite, the values of Q_n do not increase without bound but are all less than some integer. In the terminology of calculus, this means that even if the set Q is infinite, $\sum (1/q_i)$ converges.

C. E. Fröberg has used the computer to obtain an estimate of the value of the series $\sum (1/q_i)$. Based on calculation of Q_n for high values of n, he arrived at the estimate $S = 1.70195$ for the value of this series. The error in the value of S, based on some conjectures of Hardy and Littlewood, was thought to be less than three units in the last place. Edgar Karst has computed Q_n for the first 2500 twin primes, from (3, 5) to (102761, 102763), and found its value to be 1.6733. This result indicates that the series converges quite slowly.

The fact that the series of inverses of twin primes converges certainly does not mean that the set of twin primes is finite. The geometric series

$$\frac{1}{2} + \frac{1}{4} + \cdots + \frac{1}{2^n} + \cdots$$

converges and has the value 1, even though the set of powers of $\frac{1}{2}$ is infinite. Viggo Brun's result simply shows that the twin primes are significantly more scarce than the set of primes, as n becomes large. On the other hand, there is still the possibility that the set of twin primes really is finite.

How will the question eventually be settled? Perhaps some refinement of the approach through inequalities will lead to a lower bound for the function representing the number of twin primes less than n. Another possibility is that some conjecture may be proved that can be shown to be equivalent to the conjecture that the set of twin primes is infinite. Perhaps the result about twin primes will be seen to be a special case of some general structure theorem as yet unrecognized. In any case, it remains a tantalizing question, so easy to formulate and yet so difficult to understand.

EXERCISES 4.4

Checks

1. Find examples of pairs of integers that are prime, in which d is not 2. For example, find p, $p + d$, both of which are prime for $d = 4$; $d = 6$; $d = 8$.

2. Find a prime arithmetic sequence of five terms with first term 5.

3. Find a prime arithmetic sequence of five terms with first term 13.

4. Can you find a prime arithmetic sequence of six terms with first term 5?

5. Sets of primes of the form p, $p + 2$, $p + 6$, $p + 8$ are called prime quadruplets. Find examples of prime quadruplets. Why is $p + 4$ omitted?

6. A prime arithmetic sequence of ten terms exists. What can you say about its common difference? Its first term is 199.

Challenges

7. Prove that if p, $p + 2$, $p + 6$, and $p + 8$ are prime and $p > 5$, the primes differ only in the units digit and these digits are 1, 3, 7, 9 respectively.

8. Prove that if p, $p + 2$, $p + 6$, and $p + 8$ are all prime, then $p \equiv 11$, 101, or 191 mod 210. *Hint:* Show that $p \equiv 5$ mod 6, $p \equiv 1$ mod 5, and $p \equiv 2$, 3, or 4 mod 7, and use the Chinese remainder theorem.

9. Twin primes occur when there is a k such that $6k - 1$ and $6k + 1$ are both prime. Show that it is impossible to obtain twin primes for three consecutive values of k, except for the case $k = 1, 2, 3$.

REFERENCES FOR UNIT 4

1. Bateman, P. T., and M. Low, "Prime Numbers in Arithmetic Progressions with Difference 24," *The American Mathematical Monthly*, Vol. 72 (1965), pp. 139–143.

2. Dudley, U., "History of a Formula for Primes," *American Mathematical Monthly*, Vol. 76 (1969), pp. 23–28.

3. Erdös, P., "On a New Method in Elementary Number Theory which Leads to an Elementary Proof of the Prime Number Theorem," *Proc. Nat. Acad. Sci.*, U.S.A., Vol. 35 (1949), pp. 374–384.

4. Fröberg, C. E., "On the Sum of Inverses of Primes and of Twin Primes," BIT., Vol. 1 (1961), pp. 15–20.

5. Fröberg, C. E., "On Some Number Theoretical Problems Treated with Computers," in R. F. Churchouse and J. C. Herz (Eds.), *Computers in Mathematical Research* (Amsterdam: North-Holland Publishing Company, 1968), pp. 84–88.

6. Karst, E., "The First 2500 Reciprocals and Their Partial Sums of All Twin Primes Between (3, 5) and (102761, 102763)," *Math. of Computation*, Vol. 23 (1969).
7. Lehmer, D. H., " Computer Technology Applied to the Theory of Numbers," *Studies in Number Theory*, W. J. LeVeque (Ed.), MAA Studies in Mathematics, Vol. 6 (1969), pp. 117–150.
8. LeVeque, W. J., *Topics in Number Theory*, Vol. II, (Reading, Mass.: Addison-Wesley Publishing Company, 1956).
9. Levinson, N., "A Motivated Account of an Elementary Proof of the Prime Number Theorem," *American Mathematical Monthly*, Vol. 76 (1969), pp. 225–245.
10. Selberg, A. "An Elementary Proof of Dirichlet's Theorem about Primes in Arithmetic Progressions," *Annals of Mathematics*, Vol. 50 (1949), pp. 297–313.
11. Willans, C. P., " On Formulae for the nth Prime Number," *Math. Gazette*, Vol. 48 (1964), pp. 413–415.

DIOPHANTINE EQUATIONS—
AN UNCHARTED SEA

5.1 LINEAR DIOPHANTINE
EQUATIONS

Most of us associate with the equation $ax + by + c = 0$ the geometric picture of a straight line in the plane. When we do so, we are thinking of the solution set of the equation as an infinite set of ordered pairs of real numbers (x, y). Each ordered pair (x, y) corresponds to a point in the plane, and the points corresponding to pairs in the solution set of the equation $ax + by + c = 0$ lie on a straight line. In this study, we are interested in the ring of integers rather than in the field of real numbers, and so we restrict our attention to a special type of linear equation—one in which the coefficients a, b, and c are integers—and to a special part of the solution set—those pairs (x, y) in which both x and y are integers—that is, those pairs corresponding to what are called *lattice points* in the plane. We might ask: (1) Do there exist any pairs (x, y) in the solution set in which both x and y are integers? (2) Is the set of such pairs infinite? (3) Does there exist a method of obtaining the complete set of solutions in integers for this equation?

EXAMPLE 5.1.1

$$8x + 5y - 3 = 0.$$

Since $(8, 5) = 1$, the characterization of the greatest common divisor, Theorem 1.6.1, tells us that integers x_0 and y_0 exist such that $8x_0 + 5y_0 = 1$. The integers $3x_0$ and $3y_0$ would have the property $8(3x_0) + 5(3y_0) = 3$. In this case, the answer to question (1) above is *yes*. Substitution shows that $x = 3x_0 - 5k$, $y = 3y_0 + 8k$ will also be in the solution set for every integer k, so the answer to the question (2) is also *yes*. The Euclidean algorithm gives us a technique for finding x_0 and y_0. We get $x_0 = 2$, $y_0 = -3$.

Let x_1 and y_1 be any pair of integers that is a solution of the equation. Then $8x_1 + 5y_1 = 3$ and $8(6) + 5(-9) = 3$, so that $8(x_1 - 6) + 5(y_1 + 9) = 0$. Since $(8, 5) = 1$, the unique factorization theorem implies that $8 | y_1 + 9$. In this way, we conclude that $y_1 = -9 + 8k$ and $x_1 = 6 - 5k$ for some integer k. Since every solution must have this form and every integer of this form is a solution, we have a method of obtaining every solution in integers, and the answer to question (3) is also *yes*.

Given any equation of the form $f(x_1, x_2, \ldots, x_n) = 0$, the same three questions can be asked. The answers, however, are not often as complete or as easy to obtain, even when the function f is a polynomial and the number of unknowns is only two. Questions of this nature, in which solutions in integers are studied, are called *Diophantine* problems. The name Diophantine comes from the Greek mathematician, Diophantus of Alexandria, who worked during the third century A.D. and considered many such problems. He was interested especially in situations that involved squares or cubes and that arose from geometric considerations. For a fascinating historical survey of the early history of Diophantine problems, see *Number Theory and Its History*, by Oystein Ore, Chapters 6, 7, and 8. A more recent brief elementary treatment is given by Martin

Gardner in the "Mathematical Games," *Scientific American*, July 1970, pp. 117–119. A much more technical account is contained in the article "A Brief Survey of Diophantine Equations" by W. J. LeVeque, in *Studies in Number Theory*, Volume 6 of MAA Studies in Mathematics. This article is well worth reading (even though some of the details may be beyond your range of complete comprehension), since it indicates the extent to which the study of Diophantine problems has stimulated the development of mathematical thought.

In this unit we consider a very few of the simpler Diophantine equations, and in Unit 6 we look at Diophantine problems from a slightly different point of view in discussing Waring's problem.

We begin by considering the general linear equation in two variables. Here, as we would expect from the example, a definitive theorem can be stated. (You have probably found this for yourself in Unit 1 or 3.) We can restrict our attention to an equation of the form $ax + by = n$ in which $(a, b) = 1$. If $(a, b) = d > 1$, then every integer of the form $ax + by$ is a multiple of d, so that the equation $ax + by = n$ cannot have a solution in integers if $d \nmid n$. If d does divide n, we have $a = a'd, b = b'd$, and $n = n'd$. The solution set of the equation $a'x + b'y = n'$ is the same as the solution set of the equation $ax + by = n$. Since $(a, b) = d$, $(a', b') = 1$. Thus there is no loss of generality in assuming at the outset that the common factor d has been removed, and the coefficients of x and y are relatively prime.

THEOREM 5.1.1. *Linear Equation in Two Variables.* Let a, b, and n be integers, $(a, b) = 1$. The integral solutions of $ax + by = n$ are given by $x = x_0 n - bk$, $y = y_0 n + ak$, where k is an integer and $ax_0 + by_0 = 1$.

Proof: Since $(a, b) = 1$, there exist integers x_0 and y_0 such that $ax_0 + by_0 = 1$. The integers $x_0 n$ and $y_0 n$, therefore, have the property $a(x_0 n) + b(y_0 n) = n$.

Let x_1, y_1 represent any pair of integers such that $ax_1 + by_1 = n$. We have $a(x_0 n - x_1) + b(y_0 n - y_1) = 0$. This result implies that $b \mid a(x_0 n - x_1)$. But since $(a, b) = 1$, the fundamental theorem of factorization (Theorem 1.8.1) tells us that $b \mid x_0 n - x_1$; that is, $x_1 = x_0 n - bk$ for some integer k. Also, $a(-kb) + b(y_0 n - y_1) = 0$, which shows that $y_1 = y_0 n + ak$. Substitution shows that any pair of integers x and y of the form $x = x_0 n - bk$, $y = y_0 n + ak$, is indeed a solution. The theorem follows. ●

The form of the solution obtained in this theorem is natural if we consider some of the basic theorems about the solution of equations that we have learned in high school algebra and again in linear algebra. In these courses we consider the solution of linear equations over a *field*—for example, the field of rational numbers. Let u represent the matrix $\binom{x}{y}$ and A the matrix (a, b). The equation $ax + by = n$ becomes $Au = n$. Let $u_0 = \binom{x_0}{y_0}$ be a particular solution of the equation so that $Au_0 = n$. Hence $A(u - u_0) = 0$, which implies that $u - u_0$ is a solution of the homogeneous equation $Au = 0$. The solutions of this equation have the form $k\binom{-b}{a}$, where k is any number from the field. We conclude that $u = u_0 + h$, where u_0 is a particular solution of the equation and h ranges over all the solutions of the corresponding homogeneous equation. That is,

$$u = \begin{pmatrix} x_0 - bk \\ y_0 + ak \end{pmatrix}.$$

In the Diophantine problem, we are choosing from the set of solutions over the field of rationals only those solutions in which x and y are integers. If the particular solution u_0 has both x_0 and y_0 integers, the complete set of integral solutions in the case of two variables will be obtained by choosing integral k.

EXAMPLE 5.1.2. Find integers x and y such that $12x + 20y = 44$. Since $(12, 20) = 4$ and $4|44$, integral solutions exist. The solution set of this equation is identical to the solution set of $3x + 5y = 11$. Since $3(2) + 5(-1) = 1$, $3(22) + 5(-11) = 11$ and a particular solution is $x_0 = 22$, $y_0 = -11$. The totality of integral solutions is obtained from $x = 22 - 5k$, $y = -11 + 3k$, where k is an integer. In matrix form, this result reads

$$u_0 = \begin{pmatrix} 22 \\ -11 \end{pmatrix}, \quad h = k\begin{pmatrix} -5 \\ 3 \end{pmatrix}.$$

Hence

$$u = \begin{pmatrix} 22 - 5k \\ -11 + 3k \end{pmatrix}.$$

The technique suggested in the theorem is not always the easiest method to use in the numerical calculation of the set of solutions. The following example shows a useful approach.

EXAMPLE 5.1.3. Consider again $3x + 5y = 11$. Then

$$x = \frac{11}{3} - \frac{5y}{3},$$

from which we can separate the part that is obviously an integer by writing

$$x = 3 - y + \frac{2 - 2y}{3}.$$

Now since x and y must be integers, $(2 - 2y)/3$ must be an integer. Call this integer t. Then $2 - 2y = 3t$, where y and t are integers. This is another Diophantine equation, but one in which the coefficients are smaller. Repeat the process.

$$y = 1 - t - \frac{t}{2}.$$

Set $t = 2s$. Then

$$y = 1 - 3s \quad \text{and} \quad x = 3 - (1 - 3s) + \frac{6s}{3} = 2 + 5s.$$

Thus the general solution is $x = 2 + 5s$, $y = 1 - 3s$. In matrix form, this reads

$$u_0 = \begin{pmatrix} 2 \\ 1 \end{pmatrix}, \quad h = s\begin{pmatrix} 5 \\ -3 \end{pmatrix}, \quad u = \begin{pmatrix} 2 + 5s \\ 1 - 3s \end{pmatrix}.$$

At first it may seem that the solution obtained in this way is different from the solution obtained in Example 5.1.2. The difference is only in form. To see this point, set $s = -k + 4$.

In many Diophantine problems, the nature of the problem suggests that we are interested not in all integral solutions but only in those for which both x and y are non-negative. Geometrically this means that we are interested in whether or not the line corresponding to the equation passes through lattice points in the first quadrant or on the axes.

EXAMPLE 5.1.4. Are there any solutions of $3x + 5y = 11$ such that both x and y are non-negative? Since $x = 22 - 5k \geq 0$ when $k \leq 4$, and $y = -11 + 3k \geq 0$ when $k \geq 4$, the only positive solution occurs when $k = 4$; that is, $x = 2$ and $y = 1$.

Consider $3x + 5y = 7$. Here $x = 14 - 5k$, $y = -7 + 3k$. The line $3x + 5y = 7$ passes through no lattice points in the first quadrant, since $x \geq 0$ requires $k < 3$, and $y \geq 0$ requires $k \geq 3$. The line representing $3x + 5y = 31$ passes through two lattice points $(2, 5)$ and $(7, 2)$.

For what values of n will the line $3x + 5y = n$ pass through lattice points in the first quadrant or on the coordinate axes? Since $x = 2n - 5k$ and $y = -n + 3k$, we are asking for what n is it possible to find k such that $2n - 5k$ and $-n + 3k$ are both non-negative. This condition requires that the integer k satisfy the inequality

$$\frac{2n}{5} \geq k \geq \frac{n}{3}.$$

For what values of n will such a k exist?

It is clear from algebra, as well as from geometry, that if n is large enough, there will be at least one lattice point in the first quadrant on $3x + 5y = n$. How far apart are the lattice points? How long is the portion of the line $3x + 5y = n$ that lies in the first quadrant? Further investigation along these lines is outlined in Exercises 5.1, problems 8, 9, 10.

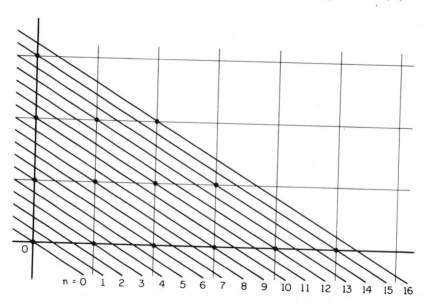

$$n = 0 \quad 1 \quad 2 \quad 3 \quad 4 \quad 5 \quad 6 \quad 7 \quad 8 \quad 9 \quad 10 \quad 11 \quad 12 \quad 13 \quad 14 \quad 15 \quad 16$$

EXERCISES 5.1

Checks

1. Which of the following equations have solutions in integers? In positive integers?
    (a) $3x + 7y = 17$.
    (b) $3x - 7y = 17$.
    (c) $15x + 12y = 17$.
    (d) $15x - 12y = 18$.

2. Write each of the equations in problem 1 as a conditional congruence.

3. Determine all the integral solutions of
    (a) $5x + 11y = 0$,
    (b) $5x + 11y = 1$,
    (c) $5x + 11y = n$.
    For what values of n does $5x + 11y = n$ have solutions with x and y non-negative?

4. Let a and b be positive integers. Express the solutions of $\pm ax \pm by = n$ in terms of the solutions of $ax + by = n$.

5. A student has $50, all of which he wants to spend on tapes and records. Tapes average $7 and records $4 in cost. How many of each can he buy?

6. Find all the solutions in non-negative integers of $71x + 97y = 384$.

Challenges

7. The first event during a contest at a trap and skeet club is 25 skeet targets at a cost of $1.50. The second is 100 trap targets at a cost of $5.00. The targets cost the club 2 cents each, and the club has an overhead of $100. How many entries in each event are necessary for the club to make a profit of $100 if the number of entry forms the club has is 80, and one entry form per event entered is needed?   (V. Koper)

8. Let a and b be positive integers, $(a, b) = 1$. Prove that $ax + by = n$ has a solution in positive integers for every $n > ab$, and no solution for $n = ab$.

9. Prove that the distance between consecutive lattice points on $ax + by = n$ is $\sqrt{a^2 + b^2}$. If a and b are positive integers, $(a, b) = 1$, and $n < ab$, prove that the line $ax + by = n$ passes through at most one lattice point with x and y both positive.

10. Prove that the equation in problem 8 has a solution in non-negative integers if $n > (a - 1)(b - 1)$. If $0 \leq n \leq (a - 1)(b - 1)$, for how many n does the equation have at least one solution in non-negative integers? Note that there exist positive solutions of $ax + by = n$ if and only if there exist non-negative solutions of $ax + by = n - a - b$.

11. Tickets for an Allied Arts event sold at $5.50, $4.50, and $3.00. Assume, for simplicity, that the seats are sold in blocks of a hundred. If a full house is required (9000 seats) and some tickets must be sold in each price range, how many blocks at each price level must be sold to pay the expenses ($44,000)? Is it possible to break even if 4000 tickets are reserved for students at $3.00 per ticket?

12. In the situation of the preceding problem, assume that 4000 student seats are sold at $3.00 and that the remaining seats are sold at two price levels— 2000 at a highest price and 3000 at a medium price. What prices should be charged for the performance to break even, assuming again that all tickets are sold?

5.2   MORE GENERAL LINEAR DIOPHANTINE PROBLEMS

Suppose that the Diophantine problem involves more than two unknowns or that a system of equations is required to describe it. Can a similar analysis be applied to this situation? Let us consider first a single equation with more than two unknowns:

$$a_1 x_1 + a_2 x_2 + \cdots + a_n x_n = b,$$

where b and a_i, $i = 1, 2, \ldots, n$, are integers, and integral values for the x_i are to be determined. It is clear that if $(a_1, a_2, \ldots, a_n) = d$, and $d > 1$, then

$$d \mid a_1 x_1 + a_2 x_2 + \cdots + a_n x_n$$

so that no solution in integers can exist unless $d \mid b$. If $d \mid b$, the solution set is unchanged if the common factor d is removed; thus we can again assume without loss of generality that $(a_1, a_2, \ldots, a_n) = 1$.

EXAMPLE 5.2.1. $6x - 4y + 8z = 6$. Here $(6, 4, 8) = 2$, and $2 \mid 6$, so that the equation is equivalent to $3x - 2y + 4z = 3$.

If integral solutions exist, there is some integer t such that $3x - 2y = t$ and $t + 4z = 3$. Conversely, to every t such that $t + 4z = 3$ and $3x - 2y = t$, there corresponds an integral solution of the equation $3x - 2y + 4z = 3$.

Because $t + 4z = 3$, z can be any integer and $t = 3 - 4z$.

Since $3x - 2y = t$, $x = t + 2k$ and $y = t + 3k$ for any integer k.

The solution set of the original equation is

$\{(x, y, z): x = 3 - 4z + 2k, y = 3 - 4z + 3k, z = z$, where z and k are
integers$\}$.

In matrix form, this is

$$\begin{pmatrix} x \\ y \\ z \end{pmatrix} = \begin{pmatrix} 3 \\ 3 \\ 0 \end{pmatrix} + z \begin{pmatrix} -4 \\ -4 \\ 1 \end{pmatrix} + k \begin{pmatrix} 2 \\ 3 \\ 0 \end{pmatrix}.$$

In the preceding example we could solve $3x - 2y = t$ because $(3, 2) = 1$. Even when $(a_1, a_2, \ldots, a_n) = 1$, it is possible that there will not be a pair of coefficients such that $(a_i, a_j) = 1$. For example, $(6, 10, 15)$ has this property. A slight modification of the procedure takes care of this case.

EXAMPLE 5.2.2. $6x + 10y + 15z = 4$. Set $6x + 10y = 2t$, since $(6, 10) = 2$ and $2t + 15z = 4$. (Was it just luck that the coefficients are relatively prime?) We obtain $t = -28 + 15s$, $z = 4 - 2s$, which gives the solution set $\{(x, y, z): x = -56 + 30s - 5k, y = 28 - 15s + 3k, z = 4 - 2s$, where k and s are integers$\}$.

The general theorem will be proved by induction, using the technique illustrated in the example.

THEOREM 5.2.1. *Linear Equation in n Variables.* Let b be an integer and let a_i, $i = 1, 2, \ldots, n$, be a set of integers such that $(a_1, a_2, \ldots, a_n) = 1$. The equation

$$a_1 x_1 + a_2 x_2 + \cdots + a_n x_n = b$$

has a solution in integers. The solution set is infinite and can be represented in terms of $n - 1$ arbitrarily assigned integers.

Proof: If $n = 2$, the theorem is exactly Theorem 5.1.1.

Assume that the statement of the theorem is true for $n = k$. Consider the statement for $n = k + 1$. Let $(a_1, a_2) = d$ and set $a_1 x_1 + a_2 x_2 = dt$. The set $x_1, x_2, \ldots, x_{k+1}$ is an integral solution of

$$a_1 x_1 + a_2 x_2 + \cdots + a_{k+1} x_{k+1} = b$$

if and only if the set $t, x_3, \ldots, x_{k+1}$ is an integral solution of

$$dt + a_3 x_3 + \cdots + a_{k+1}x_{k+1} = b,$$

and x_1, x_2 is an integral solution of $a_1 x_1 + a_2 x_2 = dt$. Since

$$(a_1, a_2, \ldots, a_{k+1}) = ((a_1, a_2), a_3, \ldots, a_{k+1}) = (d, a_3, \ldots, a_{k+1}),$$

the equation $dt + a_3 x_3 + \cdots + a_{k+1}x_{k+1} = b$ has the property $(d, a_3, \ldots, a_{k+1}) = 1$. This is an equation in k variables, so, by the induction hypothesis, its solution set is an infinite set of integers and can be represented in terms of $k - 1$ arbitrarily assigned integers. But x_1 and x_2 can be represented in terms of t and one arbitrarily assigned integer. Therefore the solution set of $a_1 x_1 + a_2 x_2 + \cdots + a_{k+1}x_{k+1} = b$ is infinite and can be expressed in terms of k arbitrarily assigned integers. The theorem follows by induction. ◆

The proof of the theorem outlines a method of constructing a solution set for a linear equation in n variables. However, for a particular situation, this method is not always the easiest. In the following example we approach an equation by several methods in order to illustrate some techniques and problems.

EXAMPLE 5.2.3

$$5x + 6y + 3z + 2w = 15.$$

Method 1. Copy the construction outlined in the theorem. Set $5x + 6y = t$; this implies

$$x = -t + 6u, \quad y = t - 5u.$$

The equation becomes $t + 3z + 2w = 15$. Now set $t + 3z = s$; this implies

$$t = 4s + 3v, \quad z = -s - v.$$

The equation becomes $s + 2w = 15$; from this result we obtain

$$s = -15 + 2k, \quad w = 15 - k.$$

We now substitute in turn for s and t and obtain

$$x = 6u - 3v - 8k + 60, \quad y = -5u + 3v + 8k - 60,$$

$$z = 15 - v - 2k, \quad w = 15 - k,$$

where u, v, and k are arbitrary integers. In matrix form

$$\begin{pmatrix} x \\ y \\ z \\ w \end{pmatrix} = \begin{pmatrix} 60 \\ -60 \\ 15 \\ 15 \end{pmatrix} + u \begin{pmatrix} 6 \\ -5 \\ 0 \\ 0 \end{pmatrix} + v \begin{pmatrix} -3 \\ 3 \\ -1 \\ 0 \end{pmatrix} + k \begin{pmatrix} -8 \\ 8 \\ -2 \\ -1 \end{pmatrix}$$

Method 2. Copy the method of Example 5.1.3. Since $2w = 15 - 5x - 6y - 3z$, we obtain

$$w = 7 - 2x - 3y - z + \frac{1 - x - z}{2}.$$

Set $1 - x - z = 2s$. The solution set of the equation is seen to be

$$\{(x, y, z, w): x = 1 - z - 2s, \quad y = y, \quad z = z, \quad w = 5 + z - 3y + 5s,$$

where z, y, and s are arbitrary integers}. In matrix form, this is

$$\begin{pmatrix} x \\ y \\ z \\ w \end{pmatrix} = \begin{pmatrix} 1 \\ 0 \\ 0 \\ 5 \end{pmatrix} + s \begin{pmatrix} -2 \\ 0 \\ 0 \\ 5 \end{pmatrix} + y \begin{pmatrix} 0 \\ 1 \\ 0 \\ -3 \end{pmatrix} + z \begin{pmatrix} -1 \\ 0 \\ 1 \\ 1 \end{pmatrix}.$$

In Example 5.1.3 it was easy to see that the two descriptions of the solution set are equivalent. In this case, it is more complicated. We need to show that the set of equations

$$1 - z - 2s = 60 + 6u - 3v - 8k,$$
$$y = -60 - 5u + 3v + 8k,$$
$$z = 15 - v - 2k,$$
$$5 + 5s - 3y + z = 15 - k,$$

can be solved for s, y, and z. That such is the case can be seen by solving the first three and substituting in the fourth equation.

Method 3. Obtain the integral solutions from the matrix solution.

A particular solution is $x = -14$, $y = 14$, $z = 0$, $w = 0$. The solutions of the homogeneous equation in the field of rationals are given by

$$t\begin{pmatrix} 6 \\ -5 \\ 0 \\ 0 \end{pmatrix} + u\begin{pmatrix} 0 \\ 0 \\ 3 \\ 2 \end{pmatrix} + v\begin{pmatrix} 0 \\ 3 \\ -6 \\ 0 \end{pmatrix},$$

where t, u, and v are rational numbers. In this case, we do not get the complete set of integral solutions by choosing t, u, and v to be integers. If $t = \frac{1}{2}$ and $v = \frac{1}{3}$, this representation gives the integral solutions $x = -11$, $y = 13$, $z = 3u - 3$, $w = -2u$, where u is an integer. A technique for isolating the integral solutions in the general case appears to be quite complicated.

We turn now to the question of solutions in integers for *systems* of linear equations. Certainly a necessary condition must be that each equation individually has a solution in integers. Also, the system of equations must be consistent; that is, it must possess a solution in the field of rational numbers before it can be expected to have a solution in integers.

EXAMPLE 5.2.4

$$3x - 2y + 4z = 3,$$
$$x + 2y - 6z = 1,$$

or in matrix form,

$$\begin{pmatrix} 3 & -2 & 4 \\ 1 & 2 & -6 \end{pmatrix}\begin{pmatrix} x \\ y \\ z \end{pmatrix} = \begin{pmatrix} 3 \\ 1 \end{pmatrix}.$$

Since $(3, -2, 4) = 1$ and $(1, 2, -6) = 1$, each equation has a solution in integers. To check the system for consistency, we compare the rank of the coefficient matrix

$$\begin{pmatrix} 3 & -2 & 4 \\ 1 & 2 & -6 \end{pmatrix}$$

and the rank of the augmented matrix

$$\begin{pmatrix} 3 & -2 & 4 & 3 \\ 1 & 2 & -6 & 1 \end{pmatrix}.$$

In this case, both matrices have rank 2. Since

$$\begin{vmatrix} 3 & -2 \\ 1 & 2 \end{vmatrix} = 8,$$

we can solve the equations

$$3x - 2y = 3 - 4z,$$
$$x + 2y = 1 + 6z,$$

for x and y and obtain

$$x = \frac{8 + 4z}{8}, \quad y = \frac{22z}{8}.$$

If the system of equations has a set of integral solutions, we must be able to choose z so that $8 + 4z \equiv 0 \bmod 8$ and $22z \equiv 0 \bmod 8$. The first requires $z \equiv 0 \bmod 2$, the second $z \equiv 0 \bmod 4$. Both are satisfied if $z \equiv 0 \bmod 4$. Thus the system of equations has a solution set given by

$$\{(x, y, z): x = 1 + 2t, y = 11t, z = 4t, \text{ where } t \text{ is an integer}\}.$$

In this case, the number of unknowns is one more than the rank of the coefficient matrix, and the solution is given in terms of one arbitrarily assigned integer. The preceding solution could be obtained by matrix methods without difficulty. A particular solution is given by

$$\begin{pmatrix} x \\ y \\ z \end{pmatrix} = \begin{pmatrix} 1 \\ 0 \\ 0 \end{pmatrix}.$$

The solution of the homogeneous equations is

$$k \begin{pmatrix} 4 \\ 22 \\ 8 \end{pmatrix}.$$

Thus all the integral solutions will be of the form

$$\begin{pmatrix} 1 \\ 0 \\ 0 \end{pmatrix} + k \begin{pmatrix} 2 \\ 11 \\ 4 \end{pmatrix},$$

where k is an integer.

EXAMPLE 5.2.5

$$x + 2y - 3z = -4,$$
$$4x - y + 6z = 8.$$

Again the equations are consistent and each has integral solutions. The method of the preceding example yields

$$x = \frac{-12 + 9z}{-9} \quad \text{and} \quad y = \frac{24 - 18z}{-9}.$$

But $-12 + 9z \equiv 0 \bmod 9$ has no solutions, since $(9, 9) = 9$ and $9 \nmid 12$. This system of equations has no simultaneous solution in integers.

EXAMPLE 5.2.6

$$x + 2y - 3z + 5t = -4,$$
$$4x - y + 6z + t = 8.$$

Solve for x and y:

$$x = \frac{-12 + 9z + 7t}{-9}, \quad y = \frac{24 - 18z + 19t}{-9}.$$

In this case, we need $7t \equiv 12 \bmod 9$ and $19t \equiv -24 \bmod 9$. Both congruences are satisfied if $t \equiv 3 \bmod 9$. The system of equations has the solution set

$$\{(x, y, z, t): x = -1 + 7u - z, y = -9 + 19u - 2z, z = z, t = 3 - 9u,$$
$$\text{where } u \text{ and } z \text{ are arbitrary integers}\}.$$

Note the similarity between the system of equations in Example 5.2.5 and that in Example 5.2.6. When $t = 0$, the equations in 5.2.6 reduce to those in 5.2.5, which have no solution in integers. Is $t = 0$ possible in the solution set described for Example 5.2.6?

The preceding examples indicate that the situation for a system of linear equations can be summarized as follows: Let the system consist of m linear equations in n unknowns. (1) It is necessary that the system be consistent. Let r be the rank of the coefficient matrix. Then the rank of the augmented matrix must also equal r. (2) It is necessary that each equation have a solution in integers. (3) By renumbering the variables and the equations, if necessary, we can let the coefficients of the variables $x_1, x_2, \ldots, x_r$ in the first r equations be an $r \times r$ matrix whose determinant is not zero. Call this determinant D. The remaining $m - r$ equations can be discarded, since they are linear combinations of the first r. (4) The first r variables can be expressed as functions of the remaining $n - r$ variables of the form $x_i = f_i(x_{r+1}, x_{r+2}, \ldots, x_n)/D$, where f_i is a linear expression. Then the system of equations has a solution in integers if and only if a solution exists for the system of congruences in $n - r$ variables:

$$f_i(x_{r+1}, x_{r+2}, \ldots, x_n) \equiv 0 \bmod D, \quad i = 1, 2, 3, \ldots, r.$$

EXERCISES 5.2

Checks

1. Carry out the details of Example 5.2.2.

2. Prove that $(a_1, a_2, a_3) = ((a_1, a_2), a_3)$.

3. In Example 5.2.1, set $-2y + 4z = 2w$ and $3x + 2w = 3$ in order to find the solutions. Is the solution set the same as that obtained in Example 5.2.1?

4. Give a geometric interpretation of $ax + by + cz = n$.

5. Which of the following systems of equations has solutions in integers?

$$\begin{array}{ccc} 9x + 12z = 15 & 9x + 12z = 15 & 9x + 12z = 16 \\ 9y - 4z = 2 & 9y - 4z = 1 & 9y - 4z = 1 \end{array}$$

6. Solve the system of equations in integers:

$$\begin{aligned} 3x - 2y + 4z &= 5, \\ x + 2y - 6z &= 1. \end{aligned}$$

Challenges

7. Find the solution in integers of the system of equations:

$$\begin{aligned} x + 2y - 3z + 5t &= -4, \\ 4x - y + 2z + t &= 8. \end{aligned}$$

8. A football team scores 3 for a field goal, 7 for a converted touchdown, and 8 for a touchdown with two-point conversion. (Assume, for simplicity, that conversions never fail and safeties do not occur.) What are the possible ways in which a team could score 35 points?

9. Three boys ate a meal together. Each had at least one coke (20¢), one order of fries (25¢), and one hamburger (45¢). When the bills came, they were for $2.00, $1.45, and $1.10 before tax. Which one was figured wrong? What were the possible meals of the other two boys?   (K. Dundas)

10. Let a_i, $i = 1, 2, \ldots, n$, be positive integers such that $(a_1, a_2, \ldots, a_n) = 1$. Assume, also, that $a_1 \leq a_2 \leq a_3 \leq \cdots \leq a_n$. Prove that the equation

$$a_1 x_1 + a_2 x_2 + \cdots + a_n x_n = b$$

has a solution in positive integers if

$$b > a_2 + a_3 + \cdots + a_{k-1} + a_1 a_k.$$

5.3   THE PYTHAGOREAN EQUATION

One of the most natural Diophantine equations of higher degree is suggested by the basic geometric property of every right triangle. The Pythagorean theorem says that the square of the length of the hypotenuse is the sum of the squares of the lengths of the other two sides. We are familiar with right triangles with sides of lengths 3, 4, and 5. How might we attempt, in a systematic way, to find all possible sets of integers that could be the lengths of sides of a right triangle—that is, have the property that $x^2 + y^2 = z^2$?

EXAMPLE 5.3.1. If $x^2 + y^2 = z^2$, then $y^2 = z^2 - x^2$; that is $y^2 = (z + x)(z - x)$. Suppose that $y = 6$. Then $y^2 = 36$. What factorizations of 36 into a product of two factors will yield integers z and x when $z + x$ is set equal to the larger of the two, and $z - x$ is equal to the smaller? Possible factorizations of 36 are $1 \cdot 36$, $2 \cdot 18$, $3 \cdot 12$, $4 \cdot 9$, $6 \cdot 6$.

$$z + x = 36,$$
$$z - x = 1$$

yields $2z = 37$, $2x = 35$ and z and x are not integers. The same would be true for $3 \cdot 12$ and $4 \cdot 9$. In order to obtain integers for z and x, we must have $z + x$ and $z - x$ both even or both odd—that is, of the same parity. Suppose that

$$z + x = 6,$$
$$z - x = 6.$$

Then $x = 0$ and the triangle is degenerate. We wish to reject this case. Thus we are led to

$$z + x = 18,$$
$$z - x = \ \ 2,$$

which yields $z = 10$, $x = 8$. This solution is acceptable although not of much interest, since it gives a triangle whose sides are proportional to 3, 4, 5.

Let us define as *acceptable* a factorization that yields two distinct integers of the same parity. The table shown indicates the sets of numbers that can be found by considering values of y from 3 to 12.

y	Acceptable Factorization of y^2		Resulting in		(x,y,z)
	$z + x$	$z - x$	z	x	
3	9	1	5	4	(4,3,5)
4	8	2	5	3	(3,4,5)
5	25	1	13	12	(12,5,13)
6	18	2	10	8	(8,6,10)
7	49	1	25	24	(24,7,25)
8	32	2	17	15	(15,8,17)
	16	4	10	6	(6,8,10)
9	81	1	41	40	(40,9,41)
10	50	2	26	24	(24,10,26)
11	121	1	61	60	(60,11,61)
12	72	2	37	35	(35,12,37)
	36	4	20	16	(16,12,20)
	24	6	15	9	(9,12,15)
	18	8	13	5	(5,12,13)

This table contains "duplications" in two ways. In the first place, the set (4, 3, 5) is not really distinct from the set (3, 4, 5), since the only difference lies in the choice of which side shall be called x and which side shall be called y. Moreover, (8, 6, 10) and (16, 12, 20) are both closely related to (4, 3, 5). Similarly, (12, 5, 13) and (24, 10, 26) are related to each other. If k is any integer not zero, then $x^2 + y^2 = z^2$ if and only if

$(kx)^2 + (ky)^2 = (kz)^2$. Thus we need search only for solutions such that $(x, y, z) = 1$, and with each of these solutions is associated a collection of solutions (kx, ky, kz), where k is any integer.

DEFINITION 5.3.1. *Primitive Pythagorean Triple.* A set of integers x, y, z such that $x^2 + y^2 = z^2$ and $(x, y, z) = 1$ is called a primitive Pythagorean triple.

EXAMPLE 5.3.2. In the table shown, the primitive Pythagorean triples are $(3, 4, 5)$, $(4, 3, 5)$, $(12, 5, 13)$, $(24, 7, 25)$, $(15, 8, 17)$, $(40, 9, 41)$, $(60, 11, 61)$, $(35, 12, 37)$, and $(5, 12, 13)$.

We observe that the third member of the primitive Pythagorean triples, the z value, is always odd and the first and second members are of opposite parity—that is, one is even and the other odd. (Why?)

THEOREM 5.3.1. *Properties of a Primitive Pythagorean Triple.* If x, y, z is a primitive Pythagorean triple, then
1. $(x, y) = 1$, $(z, x) = 1$, $(z, y) = 1$.
2. z is odd and x and y are of opposite parity.
3. If x represents the odd element, then $(z - x, z + x) = 2$.

Proof: Suppose that $(x, y) = d$. Then $d^2 | z^2$, which implies $d | z$, so that $d | (x, y, z)$; that is, $d = 1$. Similarly, $(z, x) = 1$ and $(y, z) = 1$. This result proves statement (1).

By statement (1), only one of x, y, and z can be even. If z is even, $z^2 \equiv 0 \bmod 4$. But z even implies x and y are both odd and $x^2 + y^2 \equiv 2 \bmod 4$. Thus z must be odd and either x or y even. Let y represent the even element.

Let $(z - x, z + x) = d$. Then $d | 2x$ and $d | 2z$; that is, $d | 2(x, z)$. But $(x, z) = 1$, so that $d | 2$. Since x and z are both odd, $z - x$ and $z + x$ are both even. Hence $d = 2$. ◆

We see now that we can avoid the duplication experienced in the table by restricting attention to primitive Pythagorean triples and by designating y to represent the even element of the triple. Can we, under these restrictions, find a means of generating exactly these triples?

THEOREM 5.3.2. *Solutions of the Pythagorean Equation.* x, y, z is a primitive Pythagorean triple in which y is even if and only if $x = u^2 - v^2$, $z = u^2 + v^2$, and $y = 2uv$, where $(u, v) = 1$ and u and v are integers of opposite parity.

Proof: Since $(z - x, z + x) = 2$, we can write $z + x = 2r$ and $z - x = 2s$, where $(r, s) = 1$. But $y^2 = (z - x)(z + x) = 4rs$; that is, $(y/2)^2 = rs$. This equation implies that integers u and v exist such that $r = u^2$ and $s = v^2$ (Example 1.10.5). Since $z + x = 2u^2$ and $z - x = 2v^2$, we obtain $z = u^2 + v^2$ and $x = u^2 - v^2$. Because $(r, s) = 1$, $(u, v) = 1$. But x and z are odd, so that u and v cannot be both odd; that is, u and v are of opposite parity.

Conversely, if $x = u^2 - v^2$, $y = 2uv$, and $z = u^2 + v^2$, a simple algebraic calculation shows that $x^2 + y^2 = z^2$. The fact that $(x, y, z) = 1$ follows from the fact that u and v are of opposite parity and $(u, v) = 1$. ●

The formulas of the preceding theorem can be used to construct primitive Pythagorean triples. The following example gives a brief sample.

EXAMPLE 5.3.3

u	v	x	y	z
2	1	3	4	5
3	2	5	12	13
4	1	15	8	17
4	3	7	24	25
5	2	21	20	29
5	4	9	40	41
6	1	35	12	37
6	5	11	60	61

Examine the table and see what conjectures you can make about Pythagorean triples. Compare yours with the ones suggested in the exercises.

The existence of triangles satisfying equations similar to the Pythagorean equation can also be studied. The following example suggests an investigation along this line.

EXAMPLE 5.3.4. What triangles, if any, have sides whose lengths are related by the equation $x^2 + 3y^2 = z^2$? Again, we look for sets such that $(x, y, z) = 1$. The following outline suggests how the solution set of this equation might be derived. You can supply the proof of each statement.

1. $(x, y, z) = 1$ implies that $(x, y) = 1$, $(y, z) = 1$, and $(x, z) = 1$.
2. x must be odd, and either y even and z odd or y odd and z even.
3. If y is even, then $(z + x, z - x) = 2$ and

$$z = 3u^2 + v^2, \quad x = |3u^2 - v^2|, \quad y = 2uv$$

where u and v are of opposite parity and $(u, v) = 1$.

4. If z is even, then $(z + x, z - x) = 1$ and

$$z = \frac{3u^2 + v^2}{2}, \quad x = \frac{|3u^2 - v^2|}{2}, \quad y = uv,$$

where u and v are both odd, $(u, v) = 1$, and $3 \nmid v$. For example, let $u = 2$ and $v = 1$ in the formulas in (3). Then $x = 11$, $y = 4$, $z = 13$. Let $u = 3$ and $v = 5$ in the formulas in (4). Then $x = 1$, $y = 15$, and $z = 26$. Each of these sets is a primitive solution of the equation $x^2 + 3y^2 = z^2$.

EXERCISES 5.3

Checks

1. In the table of Pythagorean triples on page 191, which of the triples are primitive?

2. Find all the primitive Pythagorean triples in which one side is 12.

3. Show that it is not possible for a Pythagorean triangle to have the hypotenuse equal to 7. For what triangles is one side equal to 7?

4. Prove that if three integers exist such that $x - y = a^2$, $y - z = b^2$, $x - z = c^2$, then a, b, c must be a Pythagorean triple. Find the numbers x, y, z corresponding to the triple 3, 4, 5.

5. Prove that if k is odd, there is no primitive Pythagorean triple containing the integer $2k$.

Challenges

6. Prove that in every primitive Pythagorean triple, one of the integers is divisible by 5, one is divisible by 4, and one is divisible by 3. (Two or more may be the same integer, as in 60, 11, 61.)

7. Complete the details of the development in Example 5.3.4.

8. What is the complete solution in integers of the equation $x^2 + 2y^2 = z^2$?

9. Extend this analysis to $x^2 + py^2 = z^2$. What can you say about $x^2 + ny^2 = z^2$, where n is composite?

10. Under what conditions does the hypotenuse of the triangle differ from one of the sides by 1?

11. Under what conditions do the two sides differ by 1?

12. Let n be an odd integer. How many primitive Pythagorean triples have $x = n$? Let $m = 4k$. How many primitive Pythagorean triples have $y = m$?

5.4 OTHER DIOPHANTINE PROBLEMS

Even if we restrict our attention to Diophantine problems involving polynomial functions, there are as many such problems as there are degrees, numbers of variables, and sets of integer coefficients. Many special cases have been studied rather deeply, but not all the results are as complete as those for the Pythagorean equation. A great deal of effort has been directed as well to a search for general methods and principles. Nevertheless, much of the material in this area remains in the realm of conjecture. The most we can attempt in this section is to look at a few problems that suggest some elementary approaches.

Linear Diophantine problems can be rewritten as conditional congruences and conditions for solvability can be determined from the study of congruences. It is natural to ask what our study of conditional congruences can contribute to our knowledge about the solutions of general Diophantine problems. Certainly we can set up the principle that integer solutions of the equation $f(x) = 0$ can exist *only if* the congruence $f(x) \equiv 0$ mod m has solutions for all integers m. In Unit 3 we saw that we need consider only m that are powers of primes. We can therefore expect to use congruences with some appropriate prime power modulus to get some information about the solutions of equations.

EXAMPLE 5.4.1. $x^2 + y^2 = 4z + 3$ has no solution in integers. Note that $x^2 \equiv 0$ or 1 mod 4 and $y^2 \equiv 0$ or 1 mod 4 so that $x^2 + y^2 \not\equiv 3$ mod 4.

EXAMPLE 5.4.2. $x^2 + 2y^2 = 8z + 5$ has no solution in integers. Here congruence mod 4 indicates that a solution might exist with $x \equiv 1$ and $y \equiv 0$ mod 4. But consider $x^2 \equiv 0$, 1 or 4 mod 8.

EXAMPLE 5.4.3. $ax^2 + by^2 = c$. If $a \equiv 1$ mod 3 and $b \equiv 1$ mod 3, for what c will a solution in integers be impossible? Since $x^2 \equiv 0$ or 1 mod 3 and $y^2 \equiv 0$ or 1 mod 3, $ax^2 + by^2 \equiv 0$, 1 or 2 mod 3. We have the case $ax^2 + by^2 \equiv 0$ mod 3 only if $x^2 \equiv 0$ and $y^2 \equiv 0$ mod 3. But this result implies that $ax^2 + by^2 \equiv 0$ mod 9. Therefore a solution in integers is impossible if $c \equiv 0$ mod 3 but $c \not\equiv 0$ mod 9; that is, $c = 3c_0$, where $(3, c_0) = 1$.

In the three preceding examples, the idea of congruence was used to obtain negative results. Can we get some positive results in this way? We would not expect to come up with a formula for the solution like the one obtained in the preceding section, but we might be able to prove that a solution in integers exists.

EXAMPLE 5.4.4. There exist integers x, y, z such that $x^2 + y^2 = 13z$. Since -1 is a quadratic residue mod 13, there exists an integer y such that $1 + y^2 \equiv 0$ mod 13. If y_0 is such an integer, then $1 + y_0^2 = 13k$ for some k, and 1, y_0, k is a set of integers that is a solution of $x^2 + y^2 = 13z$. Since it is possible to pick y_0 such that $1 \leq y_0 \leq 6$, it is possible to find a set 1, y_0, k such that k is less than 13. In this case, it is not difficult to find the actual values of y_0 and k. Suppose that the equation were replaced by $x^2 + y^2 = pz$. For what primes p could a similar argument be carried out?

EXAMPLE 5.4.5. Do there exist integers x, y such that $x^2 + y^2 = 13$? This is the same as asking if there is a solution of $x^2 + y^2 = 13k$ in which $k = 1$. Let us set up the problem for the more general case. Given a solution to the equation $x^2 + y^2 = zp$, $1 \leq z < p$, does there exist a solution in which $z = 1$?

Suppose that we have a solution in which z is even; that is, there exist integers a, b, c such that $a^2 + b^2 = 2cp$. This implies that a and b

must be of the same parity, so that $(a + b)/2$ and $(a - b)/2$ are integers. But

$$\left(\frac{a + b}{2}\right)^2 + \left(\frac{a - b}{2}\right)^2 = \frac{a^2 + b^2}{2} = cp.$$

Thus the integers $(a + b)/2$, $(a - b)/2$, c satisfy the equation $x^2 + y^2 = zp$. If c is still even, the same technique can be used to replace it by a smaller value. In this way we can eventually arrive at a set of integers a, b, c such that $a^2 + b^2 = cp$, and c is an odd number. If $c = 1$, we need go no further.

Suppose, then, that c is an odd integer greater than 1. The division algorithm says that we can find integers a_1, r, b_1, s such that $a = a_1 c + r$ and $b = b_1 c + s$. It is convenient in this case to pick r and s so that they are as small as possible in absolute value; that is,

$$-\frac{c}{2} < r < \frac{c}{2} \quad \text{and} \quad -\frac{c}{2} < s < \frac{c}{2}.$$

The equation $a^2 + b^2 = cp$ becomes

$$r^2 + s^2 + 2c(a_1 r + b_1 s) + (a_1^2 + b_1^2)c^2 = cp.$$

This shows that c divides $r^2 + s^2$, so that we can write $r^2 + s^2 = c_1 c$, where c_1 is, of course, positive unless c divides both a and b, in which case the problem of reducing c to 1 is trivial. We thus have

$$c_1 c + 2c(a_1 r + b_1 s) + (a_1^2 + b_1^2)c^2 = cp.$$

Divide by c and multiply by c_1.

$$c_1^2 + 2c_1(a_1 r + b_1 s) + (a_1^2 + b_1^2)c_1 c = c_1 p.$$

By definition, $c_1 c = r^2 + s^2$, and by simple algebraic manipulation

$$(a_1^2 + b_1^2)(r^2 + s^2) = (a_1 r + b_1 s)^2 + (a_1 s - b_1 r)^2.$$

This substitution for $(a_1^2 + b_1^2)c_1 c$ makes the equation take the form

$$(c + a_1 r + b_1 s)^2 + (a_1 s - b_1 r)^2 = c_1 p.$$

How large is c_1? Since

$$r^2 + s^2 < \frac{c^2}{2},$$

then $c_1 < c$. In this way a set of integers has been produced that satisfies $x^2 + y^2 = zp$ and in which the value of z is $c_1 < c$. If c_1 is still greater than 1, we can repeat this argument and produce a set of integers that satisfies $x^2 + y^2 = zp$, in which the value of z is $c_2 < c_1$. In a finite number of steps, we must reach a set of integers in which the value of z is 1. (Why?)

In Example 5.4.5 we demonstrated that for the Diophantine equation under consideration, given a solution in positive integers x_1, y_1, z_1, it is possible to find a solution in positive integers x_2, y_2, z_2 with the property that $z_2 < z_1$. This argument is like mathematical induction in reverse and is generally referred to as Fermat's *method of descent*. For certain equations, it can be applied effectively to prove the nonexistence of a solution. The following example illustrates the method in a different setting.

EXAMPLE 5.4.6. There is no Pythagorean triangle whose area is the square of an integer. Stated algebraically, this says that if x, y, z is a Pythagorean triple, then $xy/2$ cannot be a square. In order to prove this statement, we need consider only primitive triples. [If $(x, y, z) = d$, this condition introduces a factor of d^2 in $xy/2$, which does not affect the possibility of $xy/2$ being a square.] Suppose that $x^2 + y^2 = z^2$ and $xy = 2n^2$ for some integer n. These equations imply

$$z^2 + 4n^2 = (x + y)^2 \quad \text{and} \quad z^2 - 4n^2 = (x - y)^2.$$

The well-ordering principle tells us that if there is a set x, y, z, n that satisfies these equations, there must be one for which $(x + y)^2$ is a minimum. Suppose that we have such a set. We proceed to obtain a contradiction by constructing a set u, v, w, m such that

$$w^2 + 4m^2 = (u + v)^2, \quad w^2 - 4m^2 = (u - v)^2, \quad \text{and} \quad (u + v)^2 < (x + y)^2.$$

Since x, y, z is a Pythagorean triple, there exist integers a and b such that

$$x = a^2 - b^2, \quad y = 2ab, \quad z = a^2 + b^2,$$

where $(a, b) = 1$ and a and b are of opposite parity. Now $(a^2 - b^2)2ab = 2n^2$; that is,

$$n^2 = (a + b)(a - b)ab.$$

Since $(a, b) = 1$ and a and b are of opposite parity, $(a + b, a) = 1$, $(a + b, b)$ $= 1$, and $(a + b, a - b) = 1$. Therefore there exist integers r, k, s, and t such that

$$a + b = r^2, \quad a - b = k^2, \quad a = s^2, \quad b = t^2.$$

Here r and k are both odd and $(r, k) = 1$. This means that $(r + k)/2$ and $(r - k)/2$ are also relatively prime integers. Moreover,

$$\left(\frac{r + k}{2}\right)^2 + \left(\frac{r - k}{2}\right)^2 = \frac{r^2 + k^2}{2} = a$$

and a must be odd. Since b must be even, we can write $t^2 = 4m^2$. Then

$$\left(\frac{r + k}{2}\right)\left(\frac{r - k}{2}\right) = \frac{r^2 - k^2}{4} = \frac{b}{2} = 2m^2.$$

Let

$$u = \frac{r + k}{2}, \quad v = \frac{r - k}{2}, \quad w = s.$$

We have

$$u^2 + v^2 = w^2 \quad \text{and} \quad uv = 2m^2,$$

that is,

$$w^2 \pm 4m^2 = (u \pm v)^2.$$

We have already seen that $(u, v) = 1$. The fact that $(u, w) = 1$, also, follows from $u^2 + v^2 = w^2$. Thus u, v, w is a primitive Pythagorean triple. It remains to be shown that $(u + v)^2 < (x + y)^2$.

Since $u + v = r$,

$$(u + v)^2 = r^2 = a + b < 2a \le 2ab = y.$$

Also,

$$y < x + y < (x + y)^2.$$

Therefore

$$(u + v)^2 < (x + y)^2,$$

which contradicts the assumption that the triple x, y, z was the one for which $(x + y)^2$ was the least.

The method of descent has been applied to show the impossibility of finding solutions in integers for a great many Diophantine equations of the type $ax^4 + by^4 = cz^2$. For the case $a = b = c = 1$, see Exercises 5.4, problem 8.

The most famous of all Diophantine problems is the statement usually referred to as Fermat's Last Theorem or Fermat's Conjecture. In his copy of Bachet's edition of Diophantos' *Arithmetica*, Fermat wrote a note asserting that he had a remarkable proof, which the margin was too small to accommodate, of the statement: If n is an integer greater than 2, the equation $x^n + y^n = z^n$ has no solutions in integers except those given by $xyz = 0$.

What was this remarkable proof? Did Fermat overlook some important aspect of the problem? Was it a wry joke on his part to claim that he had a proof? Is the statement of the theorem really true? The answer to the first three questions we shall probably never know, nor does it really matter. What *is* important is that attempts to prove this statement have resulted in a great deal of work on the part of the best mathematical minds since Fermat's time, and this work has opened up many fruitful areas of mathematical research. The case $n = 4$ is readily established by the method of descent. The first complete proof for $n = 3$ is due to Legendre; Legendre and Dirichlet proved the statement for $n = 5$. It is not hard to see that we need to consider only odd prime values of n (see Exercises 5.4, problem 5). In the case of linear equations, we noted that a necessary condition for the existence of a solution in integers is the existence of a solution in rational numbers. For Fermat's conjecture, with $n = p$, it seems that a natural setting is not the ring of integers but the ring of algebraic integers in the algebraic number field generated by the pth root of unity. If $p \leq 19$, unique factorization holds in these fields and the proof of Fermat's conjecture is comparatively easy. Kummer (1810–1893) apparently believed at first that this technique would work for all p. Cauchy, however, showed that factorization is not always unique in such algebraic number rings, and, in fact, $p = 23$ is a case where unique factorization does not hold. Cauchy turned aside from the problem, but Kummer transformed his failure into fame. He undertook a detailed study of algebraic number fields, and of Fermat's conjecture, and laid the groundwork for many important advances in this area. He introduced a classification of primes into regular and irregular. (The conditions by which a regular prime can be identified are not simple.) He was able to get good results for regular primes and for some irregular primes. However, not all primes are regular; in fact, it has been shown that there exists an infinite set of irregular primes.

The computer has been applied extensively to this and other Diophantine problems. Through the work of D. H. Lehmer, E. Lehmer, H. Vandiver, J. L. Selfridge, and others, and by a combination of deep theoretical considerations and the computer, Fermat's theorem has been established for p less than 25,000.

During the past decade the computer has been used to study many Diophantine equations. As an example, consider the equation

$$x^5 + y^5 + z^5 + w^5 = s^5.$$

In connection with his study of Fermat's conjecture, Euler, about 1769, made the conjecture that it is impossible to find three fourth powers whose sum is a fourth power or four fifth powers whose sum is a fifth power. In an article, "Machines and Pure Mathematics," published in *Computers in Mathematical Research* (Amsterdam: North-Holland Publishing Co., 1968), D. H. Lehmer announced that

$$27^5 + 84^5 + 110^5 + 133^5 = 144^5.$$

This, says Dr. Lehmer, "not only demolishes Euler's assertion about fifth powers but even raises hopes that the equation

$$x^4 + y^4 + z^4 = w^4$$

may also be solvable."

EXERCISES 5.4

Checks

1. Prove that $x^2 + y^2 = (4k + 3)z^2$ has no solution in integers, where k is an integer.

2. Prove that $x^2 - 2y^2 = 8z + 3$ has no solution in integers.

3. Prove that

$$(a^2 + b^2)(r^2 + s^2) = (ar + bs)^2 + (as - br)^2.$$

4. Carry out the details of Examples 5.4.4 and 5.4.5 in the case $x^2 + y^2 = 13z$, and also $x^2 + y^2 = 29z$. Find a solution in which $z = 1$.

5. Show that if $p \mid m$, and if $x^m + y^m = z^m$ has a solution in integers, then the equation $x^p + y^p = z^p$ has a solution in integers.

Challenges

6. Prove that $x^3 + y^3 + z^3 = 9k + r$ has no solution in integers if $r \equiv 4, 5$ mod 9.

7. Discuss the solutions of $x^2 - y^2 = n$ for fixed integer n. In particular, how many solutions exist
   (a) if $n \equiv 1$ or $n \equiv 3$ mod 4,
   (b) if $n \equiv 2$ mod 4,
   (c) if $n \equiv 0$ mod 4?

8. Use the method of descent to prove that the equation $x^4 + y^4 = z^2$ has no solution in integers. The following steps are suggested.
   (a) Assume that the equation has a solution and attempt to find a solution with $z_1 < z$. The case $(x, y) = d > 1$ is easily handled.
   (b) If $(x, y) = 1$, then x^2, y^2, z is a primitive Pythagorean triple.
   (c) Write $y^2 = 2rs$, with $(r, 2s) = 1$, so that $r = r_1^2$ and $s = 2s_1^2$.
   (d) $x, 2s_1^2, r_1^2$ is a primitive Pythagorean triple. Express it in terms of u, v with $2s_1^2 = 2uv$, from which $u = u_1^2$ and $v = v_1^2$.
   (e) Show that $r_1^2 = u_1^4 + v_1^4$ and $r_1 < z$.

9. Deduce from the preceding problem that Fermat's last theorem is true for $n = 4$.

10. Use the method of descent to show that $x^4 + 4y^4 = z^2$ has no solution in positive integers.

REFERENCES FOR UNIT 5

1. Arpaia, P. J., "A Generating Property of Pythagorean Triples," *Mathematics Magazine*, January–February 1971, p. 26.
2. Gardner, M. "Diophantine Analysis and the Problem of Fermat's Legendary 'Last Theorem,'" *Scientific American*, July 1970, pp. 117–119.
3. Lehmer, D. H., "Machines and Pure Mathematics," in R. F. Churchouse and J. C. Herz (Eds.), *Computers in Mathematical Research* (Amsterdam: North-Holland Publishing Co. 1968), pp. 1–7.
4. LeVeque, W. J., "A Brief Survey of Diophantine Equations," in *Studies in Number Theory*, MAA Studies in Mathematics, Vol. 6 (1969), pp. 4–24.
5. Mordell, L. J., *Diophantine Equations* (London: Academic Press, 1969).
6. Sierpinski, W., *Elementary Theory of Numbers* (Warszawa, Poland: Panstwowe Wydawnictwo Naukowe, 1964).

WARING'S PROBLEM— A NARROWING GAP

6.1 INTRODUCTION TO WARING'S PROBLEM

In the study of linear Diophantine equations, we considered the geometric question: Under what conditions do there exist lattice points on the straight line $ax + by = c$, where the coefficients a, b, and c are integers? A simple generalization of this question would be: What circles $x^2 + y^2 = a$ contain lattice points? The circle $x^2 + y^2 = 3$ contains no lattice points; the circle $x^2 + y^2 = 4$ contains four lattice points $(2, 0)$, $(-2, 0), (0, 2), (0, -2)$; and the circle $x^2 + y^2 = 5$ contains the eight points $(1, 2)$, $(1, -2)$, $(-1, 2)$, $(-1, -2)$, $(2, 1)$, $(2, -1)$, $(-2, 1)$, $(-2, -1)$. Because of the symmetry of the circle, we need consider only non-negative values of x and y, since if (x, y) lies on the circle, so does $(\pm x, \pm y)$.

In algebraic terms, the question is: What integers, a, can be expressed as the sum of two squares? Naturally we will be concerned here only with non-negative values of a. Also, we will consider only non-negative values of x and y. For convenience, we allow zero as a value of x or y.

EXAMPLE 6.1.1. Let $5 = 2^2 + 1^2, 4 = 2^2 + 0^2$. 3 cannot be expressed as the sum of two squares.

It is clear that not all integers can be expressed as the sum of two squares. If $x^2 + y^2 = a$, then $a \equiv 0$, 1, or 2 mod 4. Is the converse of this statement true? Given $a \equiv 0$, 1, or 2 mod 4, can a be expressed as the sum of two squares?

EXAMPLE 6.1.2. Find x and y such that $x^2 + y^2 = 93$. Note that $93 \equiv 1$ mod 4. We can settle this question by a finite number of trials. Consider the set $\{93 - x^2, x = 1, 2, 3, \ldots, [\sqrt{93}]\}$. If $93 = x^2 + y^2$ for some pair x and y, then y^2 must be a member of this set. The integers in the set are $\{92, 89, 84, 77, 68, 57, 44, 29, 12\}$. None is a square. Thus 93 cannot be expressed as the sum of two squares.

If an integer cannot be expressed as the sum of two squares, it is natural to ask whether it can be expressed as the sum of more than two squares, say three squares. Not all integers can be expressed as the sum of three squares (see Exercises 6.1, problem 2).

EXAMPLE 6.1.3. Is it possible to express 93 as the sum of three squares? Experimentation shows that $93 = 8^2 + 5^2 + 2^2$. Note that $93 = 3 \cdot 31$. Neither 3 nor 31 can be expressed as the sum of two squares. In fact, 31 cannot be expressed as the sum of three squares. It can, however, be expressed as the sum of four squares, since $31 = 5^2 + 2^2 + 1^2 + 1^2$.

Which integers can be expressed as the sum of two squares? Can the answer to this question be reduced to a consideration of the primes involved in the factorization of the integer?

EXAMPLE 6.1.4. The integer $65 = 8^2 + 1^2$, the sum of two squares. Note that $65 = 5 \cdot 13$, and that $5 = 2^2 + 1^2$ and $13 = 3^2 + 2^2$. If a and b can be expressed as the sum of two squares, can ab be expressed in this way? (See Exercises 6.1, problem 1.)

The question of whether or not an integer can be expressed as a sum of squares, and, if so, how many squares are required to do so, is a particular case of a general class of Diophantine problems referred to as Waring's Problem. The English mathematician Edward Waring (1734–1798) had a book published in 1770 called *Meditationes Algebraicae*. In it he stated a number of propositions, two of which, when translated from the Latin, read: Every integer is a square, or the sum of 2, 3, or 4 squares. Every integer is either a cube or the sum of 2, 3, 4, 5, 6, 7, 8, or 9 cubes; every integer is also a fourth power or the sum of 2, 3, ..., up to 19 fourth powers. These are special cases of what has come to be known as Waring's Conjecture. This states that, for every positive integer k, there exists a smallest positive integer $g(k)$ such that any positive integer n can be expressed as the sum of at most $g(k)$ positive kth powers. Thus Waring's quoted statement is that if $k = 2$, $g(k) = 4$; if $k = 3$, then $g(k) = 9$; if $k = 4$, then $g(k) = 19$.

EXAMPLE 6.1.5. Let $n = 347$, $k = 2$; $347 = 18^2 + 23 = 18^2 + 4^2 + 2^2 + 1^2 + 1^2 + 1^2$. This representation of 347 uses six squares. However, $347 = 256 + 91$; that is, $347 = 16^2 + 9^2 + 3^2 + 1^2$. Can you find a representation requiring only three squares?
Let $n = 347$; $k = 3$.

$$347 = 7^3 + 1^3 + 1^3 + 1^3 + 1^3.$$

Let $n = 347$; $k = 4$.

$$347 = 4^4 + 3^4 + 1^4 + 1^4 + 1^4 + 1^4 + 1^4 + 1^4 + 1^4 + 1^4 + 1^4 + 1^4.$$

Let $n = 347$; $k = 5$.

$$347 = 3^5 + 2^5 + 2^5 + 2^5 + 1^5 + 1^5 + 1^5 + 1^5 + 1^5 + 1^5 + 1^5 + 1^5.$$

In a trivial sense, every integer n can be expressed as the sum of n kth powers, since $n = \sum 1^k$. The problem becomes meaningful when we look for an integer $g(k)$, independent of n and depending only on k, such that every integer can be expressed as a sum of $g(k)$ kth powers of non-negative integers. Waring did not give a proof of his assertions. It is assumed that he made his conjectures on the basis of examining tables of expansions of integers in sums of kth powers.

Research related to Waring's problem has been extensive. Does an integer $g(k)$ exist? It is possible to establish upper and lower bounds for $g(k)$? Is it possible to find the exact value of $g(k)$?

The proof of the existence of $g(k)$ for general k is surprisingly difficult and was not accomplished until 1909. David Hilbert (1862–1943), a German mathematician better known for his work in geometry and analysis than for his contributions to number theory, used the methods of integral calculus to give a quite complicated demonstration. Ten years later the British mathematician G. H. Hardy (1897–1947) and his collaborator John E. Littlewood (b. 1885) developed a new method for examining $g(k)$, this time using Cauchy's theorem on analytic functions of a complex variable. Their method also gives some information about the number of representations of a given integer as a sum of $g(k)$ kth powers. In 1942 the Russian number theorist Yu. V. Linnik gave an elementary existence proof for $g(k)$, based on the idea of density of sequences. Recall that the word "elementary" in this context means independent of the use of analysis. None of the proofs mentioned will be included here. On the basis of this work, however, we are justified in making the following definition.

$$\boxed{g(k)}$$

DEFINITION 6.1.1. $g(k)$. Let k be a positive integer; $g(k)$ is the smallest integer such that every positive integer n can be expressed as the sum of $g(k)$ non-negative kth powers.

In the case $k = 2$, the investigation of Waring's problem is most complete and accessible. In Section 6.2 the case $k = 2$ is discussed in detail. In Section 6.3 we will consider how the results of Section 6.2 can be used to obtain upper bounds for $g(4)$. Section 6.4 contains a similar investigation of $g(3)$. Sections 6.3 and 6.4 might be omitted on first reading if desired. Section 6.5 investigates a lower bound for $g(k)$ for all k and states some more general results.

EXERCISES 6.1

Checks

1. Verify the identity

$$(a^2 + b^2)(c^2 + d^2) = (ac + bd)^2 + (ad - bc)^2.$$

2. If $x^2 + y^2 + z^2 = a$, prove that $a \not\equiv 7 \bmod 8$.

3. Make a table representing each of the integers from 1 to 100 either as a square or as sums of as few squares as possible.

4. Make a table representing the integers from 1 to 100 either as a cube or as sums of as few cubes as possible.

5. On the basis of your tables, guess which of $g(2)$ and $g(3)$ is larger.

6. If $x^3 + y^3 = a$, prove that $a \equiv 0, 2,$ or $5 \bmod 7$. Check with your table.

Challenges

7. Prove that $g(k) \geq 2^k - 1$.

8. If p is a prime and $p \equiv 1 \bmod 4$, prove that there exist integers x, y, z such that $x^2 + y^2 = pz$. (See Example 5.4.4.)

9. Prove that every prime of the form $4k + 1$ can be expressed as the sum of two squares. (See Example 5.4.5.)

10. If $p \mid n$, and $p \equiv 3 \bmod 4$, prove that n cannot be written in the form

$$n = x^2 + y^2,$$

where $(x, y) = 1$.

11. If $n = p^k m$ and $(m, p) = 1$, p a prime such that $p \equiv 3 \bmod 4$, prove that n cannot be represented as the sum of two squares if k is odd.

12. Prove that n can be represented as the sum of two squares if and only if each prime factor of the form $4k + 3$ appears to an even power. *Hint*: Use problem 1 along with 9, 10, and 11.

6.2   THE CASE $k = 2$

Consideration of the table of the representation of the integers from 1 to 100 indicates that perhaps $g(2) = 4$; that is, every integer can be expressed as the sum of at most four positive squares. The situation will be easier to talk about if we consider squares of non-negative rather than positive integers. With 0^2 as a possible term in expressing integers as a sum of squares, $25 = 4^2 + 3^2 + 0^2 + 0^2$ and $25 = 5^2 + 0^2 + 0^2 + 0^2$ each represent 25 as a sum of four squares. Is it possible to represent every positive integer as a sum of four squares?

We note first an algebraic identity which shows that the product of two sums of four squares is again a sum of four squares. Since every integer is a product of primes, this identity makes it possible to restrict our attention to the problem of expressing primes as the sum of four squares.

THEOREM 6.2.1. *Euler's Identity.*

$$(x_1^2 + x_2^2 + x_3^2 + x_4^2)(y_1^2 + y_2^2 + y_3^2 + y_4^2) = z_1^2 + z_2^2 + z_3^2 + z_4^2,$$

where

$$z_1 = x_1 y_1 + x_2 y_2 + x_3 y_3 + x_4 y_4,$$
$$z_2 = x_1 y_2 - x_2 y_1 + x_3 y_4 - x_4 y_3,$$
$$z_3 = x_1 y_3 - x_3 y_1 + x_4 y_2 - x_2 y_4,$$
$$z_4 = x_1 y_4 - x_4 y_1 + x_2 y_3 - x_3 y_2.$$

Proof: See Exercises 6.2, problem 1. ●

The even prime 2 can be written as the sum of four squares, $2 = 1^2 + 1^2 + 0^2 + 0^2$. What about the odd primes?

EXAMPLE 6.2.1. Let $p = 31$. We can express it as the sum of four squares by trial and error without difficulty. Instead, let us examine a way of thinking that might suggest a method of handling every odd prime. Begin by considering a congruence instead of an equation. If $x_1^2 + x_2^2 + x_3^2 + x_4^2 = 31$, then it must be true that

$$x_1^2 + x_2^2 + x_3^2 + x_4^2 \equiv 0 \bmod 31.$$

The change to congruence enables us to use our knowledge of quadratic residues. There are 15 distinct quadratic residues mod 31, no one of which is congruent to zero, so the set $S_1 = \{0^2, 1^2, 2^2, \ldots, 15^2\}$ consists of 16 integers, no two congruent mod 31. Now form the set S_2 by adding -1 to the negative of each of the integers in S_1; $S_2 = \{-1, -1 - 1^2, -1 - 2^2, \ldots, -1 - 15^2\}$. This set also consists of 16 integers, no two of which are congruent mod 31. The set $S_1 \cup S_2$ contains 32 integers, and thus there must be at least one integer in S_1 that is congruent mod 31 to some integer in S_2. In this case, we find $4^2 \equiv -1 - 13^2 \bmod 31$; that is, $4^2 + 13^2 + 1 = 31m$. Calculation shows that $m = 6$. Can we now, by algebraic manipulation, find some way to reduce the integer m from 6 to 1? Note that 13 and 1 are both odd, so that $(13 - 1)$ and $(13 + 1)$ are even. Also,

$$(13 - 1)^2 + (13 + 1)^2 = 2(13^2 + 1^2).$$

If we make use of this fact, we can write $2(4^2 + 13^2 + 1) = 12 \cdot 31$ in the form

$$4^2 + 4^2 + (13 - 1)^2 + (13 + 1)^2 = 12 \cdot 31.$$

Both sides of this equation have a common factor 4, so that

$$2^2 + 2^2 + 6^2 + 7^2 = 3 \cdot 31.$$

The integer m has been reduced from 6 to 3, and $3 \cdot 31$ is expressed as the sum of four squares. But 3 is odd, so that the technique used to reduce 6 is no longer useful. Consider the effect of examining the equation mod 3. Replace each of the integers 2, 2, 6, 7 by its least numerical residue mod 3. We get

$$(-1)^2 + (-1)^2 + 0^2 + 1^2 \equiv 0 \bmod 3.$$

Calculation shows that $(-1)^2 + (-1)^2 + 0^2 + 1^2 = 3$. We now have two expressions involving the sum of four squares:

$$2^2 + 2^2 + 7^2 + 6^2 = 3 \cdot 31 \quad \text{and} \quad (-1)^2 + (-1)^2 + 1^2 + 0^2 = 3.$$

By Euler's identity, we can write

$$(-2 - 2 + 7)^2 + (-2 + 2 - 6)^2 + (2 + 7 - 6)^2 + (0 + 6 + 2 + 7)^2 = 9 \cdot 31,$$

that is,

$$3^2 + 6^2 + 3^2 + 15^2 = 9 \cdot 31.$$

Each of the integers in this sum is divisible by 3. Was this a lucky accident? Divide 9 from both sides of the equation: $1^2 + 2^2 + 1^2 + 5^2 = 31$.

We now set out to prove that every integer can be represented as a sum of four squares. The proof is accomplished in much the same way as

Example 6.2.1, by a succession of steps beginning with the use of quadratic residues and then effecting successive reductions of a certain quantity until the integer 1 is reached. In order to make the proof easier to understand, the steps are proved separately and labeled as lemmas.

THEOREM 6.2.2. *Sums of Squares.* Every positive integer can be expressed as a sum of four non-negative squares—that is, $g(2) \leq 4$.

Proof:

LEMMA 1. If p is an odd prime, there is an integer m, with $1 \leq m < p$, such that $mp = x_1^2 + x_2^2 + x_3^2 + x_4^2$ for some integers x_1, x_2, x_3, x_4.

Let S_1 be the set of integers $\{0^2, 1^2, \ldots, [(p-1)/2]^2\}$ and let S_2 be the set $\{-1, -1 - 1^2, -1 - 2^2, \ldots, -1 - [(p-1)/2]^2\}$. The set S_1 consists of 0 and the $(p-1)/2$ distinct quadratic residues mod p. There are $(p+1)/2$ integers in the set and no two are congruent mod p. The set S_2 also consists of $(p+1)/2$ integers, no two of which are congruent to each other mod p. The set $S_1 \cup S_2$ contains $p + 1$ integers, so that there must be two which are congruent to each other mod p—that is, at least one integer in S_1 is congruent to some integer in S_2. Let x^2 and $-1 - y^2$ be these integers. Then $x^2 \equiv -1 - y^2 \bmod p$, $0 < x < p/2$ and $0 < y < p/2$. Therefore there exists an integer m such that $x^2 + y^2 + 1 + 0 = mp$ and since

$$x^2 < \frac{p^2}{4}, \quad y^2 < \frac{p^2}{4}, \quad p > 2,$$

we have

$$mp < \frac{p^2}{4} + \frac{p^2}{4} + 1 < p^2.$$

This result implies that $m < p$ as the lemma requires.

LEMMA 2. If p is an odd prime and m is the *least* positive integer such that $mp = x_1^2 + x_2^2 + x_3^2 + x_4^2$, then m is odd.

Lemma 1 assures us that at least one such m exists. Suppose that m is even. Then either the x_i are all even or they are all odd, or two are even and two are odd. In any of these cases, they can be grouped in pairs such that the sum and difference of each pair is even. Assume that $x_1 + x_2 \equiv 0 \bmod 2$ and $x_3 + x_4 \equiv 0 \bmod 2$. Then

$$\frac{x_1 + x_2}{2}, \quad \frac{x_1 - x_2}{2}, \quad \frac{x_3 + x_4}{2}, \quad \frac{x_3 - x_4}{2}$$

are all integers, and the sum of the squares of these four integers is

$$\frac{x_1^2 + x_2^2 + x_3^2 + x_4^2}{2};$$

that is, $(m/2)p$. Thus there is an integer, $m/2$, smaller than m, such that $(m/2)p$ is the sum of four squares. This contradicts the assumption that m is the least such integer. Hence m cannot be even.

LEMMA 3. If p is an odd prime and m is an odd integer, $1 < m < p$, such that $mp = x_1^2 + x_2^2 + x_3^2 + x_4^2$, then there exists a positive integer n such that $n < m$, and $nm = y_1^2 + y_2^2 + y_3^2 + y_4^2$, for integers y_i such that $y_i \equiv x_i \bmod m$, $i = 1, 2, 3, 4$.

Since m is odd and greater than 1, $m \geq 3$. Choose y_i such that $y_i \equiv x_i \bmod m$ and $|y_i| < m/2$, $i = 1, 2, 3, 4$. Then

$$y_1^2 + y_2^2 + y_3^2 + y_4^2 \equiv x_1^2 + x_2^2 + x_3^2 + x_4^2 \equiv 0 \bmod m,$$

so there exists an integer n such that

$$y_1^2 + y_2^2 + y_3^2 + y_4^2 = nm.$$

The integer n is not zero, since this would imply $y_i = 0$ for $i = 1, 2, 3, 4$, which implies $x_i \equiv 0 \bmod m$, $i = 1, 2, 3, 4$. That is, $m \mid p$, which is not possible. Also, since $y_i^2 < m^2/4$,

$$y_1^2 + y_2^2 + y_3^2 + y_4^2 < m^2,$$

that is,

$$nm < m^2 \quad \text{and} \quad n < m.$$

LEMMA 4. If p is an odd prime and m is the *least* positive integer such that $mp = x_1^2 + x_2^2 + x_3^2 + x_4^2$, then $m = 1$.

From Lemma 1, m exists and is less than p. From Lemma 2, m

cannot be even. Suppose that m is greater than 1. Then, from Lemma 3, there exists an n such that $n < m$ and

$$nm = y_1^2 + y_2^2 + y_3^2 + y_4^2.$$

Now, from Euler's identity, we obtain

$$m^2 np = A_1^2 + A_2^2 + A_3^2 + A_4^2,$$

where

$$
\begin{aligned}
A_1 &= x_1 y_1 + x_2 y_2 + x_3 y_3 + x_4 y_4, \\
A_2 &= x_1 y_2 - x_2 y_1 + x_3 y_4 - x_4 y_3, \\
A_3 &= x_1 y_3 - x_3 y_1 + x_4 y_2 - x_2 y_4, \\
A_4 &= x_1 y_4 - x_4 y_1 + x_2 y_3 - x_3 y_2.
\end{aligned}
$$

Since $y_i \equiv x_i \bmod m$, we have

$$A_1 = x_1^2 + x_2^2 + x_3^2 + x_4^2 = mp \equiv 0 \bmod m.$$

Also,

$$A_2 \equiv x_1 x_2 - x_2 x_1 + x_3 x_4 - x_4 x_3 \equiv 0 \bmod m.$$

Similarly, $A_3 \equiv 0 \bmod m$ and $A_4 \equiv 0 \bmod m$, which implies that $m^2 \mid A_i^2$ and therefore that

$$np = \left(\frac{A_1}{m}\right)^2 + \left(\frac{A_2}{m}\right)^2 + \left(\frac{A_3}{m}\right)^2 + \left(\frac{A_4}{m}\right)^2.$$

But $n < m$, so the fact that np can be expressed as a sum of four squares contradicts the assumption that m is the least positive integer such that mp is a sum of four squares. The assumption that $m > 1$ leads to a contradiction, and therefore $m = 1$.

The proof of the theorem can now be completed. Lemma 4 says that an odd prime p can be expressed as a sum of four squares. We know that the even prime, 2, can be written as a sum of four squares. Every integer can be written as a product of primes, therefore Euler's identity and mathematical induction establish the fact that every integer is the sum of four squares. ⬣

We have seen that some integers can be expressed as a sum of two or a sum of three squares. In the following sections, it is useful to know exactly which integers can be expressed with fewer than four squares.

THEOREM 6.2.3. *Sums of Three Squares.* The integer $n > 0$ can be written as a sum of three squares if and only if n is not of the form $4^r(8k + 7)$, where r and k are non-negative integers.

Proof: The proof that $4^r(8k + 7)$ cannot be written as a sum of three squares is outlined in problems 6 and 7. For the proof of the other half of the theorem, see, for example, [9]. ●

It follows from Theorem 6.2.3 that $g(2) \geq 4$ and therefore, since $g(2) \leq 4$, $g(2) = 4$.

EXERCISES 6.2

Checks

1. Verify Euler's identity by expanding both sides of the equation. Show that the identity in Exercise 6.1, problem 1, is a particular case of Euler's identity.

2. Prove that $A_3 \equiv 0 \bmod m$ and $A_4 \equiv 0 \bmod m$.

3. Compare the steps in Example 6.2.1 with the lemmas in Theorem 6.2.2.

4. Derive a representation of 43 as a sum of squares, using the steps of the lemmas. Can you find other representations of 43? Can you find one requiring fewer than four squares? Answer the same questions for 47.

5. Find as many different representations as possible of the integer 112 as a sum of four squares.

Challenges

6. Prove that a positive integer of the form $8k + 7$ cannot be represented as the sum of three squares.

7. Prove that a positive integer of the form $4^r(8k + 7)$ cannot be represented as a sum of three squares. *Hint*: Let r be the least positive integer such that $4^r(8k + 7)$ can be represented as a sum of three squares. Show that if $r > 1$, then $4^{r-1}(8k + 7)$ can be represented as a sum of three squares.

8. Find as many representations as possible of 207 as a sum of four squares. Same for 658. *Hint*: It is easiest to do this directly without the use of the existence theorems.

9. Let $S(n) = $ the number of representations of n as a sum of four squares. Here representations are considered as different if $-x_i$ is used in place of x_i. Also, the same digits in a different order is considered a different representation, so that if x_1, x_2, x_3, x_4 are all distinct and nonzero, there are $16 \cdot 24$ representations associated with this set of digits. Jacobi has proved that $S(n) = 8\sigma(n)$ if n is odd and $S(n) = 24\sigma(m)$ if $n = 2^k m$, where m is odd and $k \geq 1$. Calculate $S(207)$ and $S(658)$ and compare with problem 8.

6.3   UPPER BOUNDS FOR $g(4)$

The facts about the expression of an integer as the sum of squares appear to have been known as early as Diophantus, although the first published proof was given by Lagrange in 1770. Much work was done on the estimation of $g(k)$ for various values of k, but the next published proof connected with this number appears to be the proof of Liouville in 1859 that $g(4) \leq 53$. How can we arrive at the upper bound 53? What are some techniques that can be used to reduce this upper bound? As in the case of squares, we begin with an algebraic identity. This identity, established by Lucas in 1876, gives a connection between the sum of squares and the sum of fourth powers.

THEOREM 6.3.1. *Lucas' Identity.*
$$6(x_1^2 + x_2^2 + x_3^2 + x_4^2)^2 = (x_1 + x_2)^4 + (x_1 - x_2)^4 + (x_1 + x_3)^4$$
$$+ (x_1 - x_3)^4 + (x_1 + x_4)^4 + (x_1 - x_4)^4$$
$$+ (x_2 + x_3)^4 + (x_2 - x_3)^4 + (x_2 + x_4)^4$$
$$+ (x_2 - x_4)^4 + (x_3 + x_4)^4 + (x_3 - x_4)^4.$$

Proof: See Exercises 6.3, problem 2.   ●

EXAMPLE 6.3.1. Express 191 as the sum of fourth powers. We can proceed directly and write

$$191 = 3^4 + 110 = 3^4 + 2^4 + 2^4 + 2^4 + 2^4 + 2^4 + 2^4 + 14$$
$$= 3^4 + 6 \cdot 2^4 + 14 \cdot 1^4$$

which is a sum of 21 fourth powers.

Let us look at 191, using Lucas' identity.

$$191 = 6 \cdot 31 + 5.$$

Now $31 = 5^2 + 2^2 + 1^2 + 1^2$, so that

$$191 = 6 \cdot 5^2 + 6 \cdot 2^2 + 6 \cdot 1^2 + 6 \cdot 1^2 + 5.$$

Lucas' identity gives $6a^2$ as a sum of fourth powers, where a is represented as a sum of squares. In this case, we obtain

$$6 \cdot 5^2 = 6(2^2 + 1^2 + 0^2 + 0^2)^2$$
$$= 3^4 + 1^4 + 2^4 + 2^4 + 2^4 + 2^4 + 1^4 + 1^4 + 1^4 + 1^4 + 0^4 + 0^4.$$
$$6 \cdot 2^2 = 6(1^2 + 1^2 + 0^2 + 0^2)^2$$
$$= 2^4 + 0^4 + 1^4 + 1^4 + 1^4 + 1^4 + 1^4 + 1^4 + 1^4 + 1^4 + 0^4 + 0^4.$$
$$6 \cdot 1^2 = 6(1^2 + 0^2 + 0^2 + 0^2)^2 = 6 \cdot 1^4.$$

Combine these equations, plus the equation $5 = 5 \cdot 1^4$. We obtain

$$191 = 3^4 + 5 \cdot 2^4 + 30 \cdot 1^4$$

which expresses 191 as a sum of 36 fourth powers.

Although the number of fourth powers when 191 was expanded by using Lucas' identity is much larger than the number found by using the direct approach, use of the former can be applied to a general integer and results in the inequality $g(4) \leq 53$.

THEOREM 6.3.2. $g(4) \leq 53$. Every integer is the sum of not more than 53 fourth powers.

Proof: By the division algorithm, every integer can be written in the form $6m + r$, where $0 \leq r \leq 5$. Since $g(2) = 4$, every non-negative integer m can be written in the form

$$m = a_1^2 + a_2^2 + a_3^2 + a_4^2.$$

Also, every non-negative integer a_i can be written in the form

$$a_i = x_1^2 + x_2^2 + x_3^2 + x_4^2.$$

From Lucas' identity, $6a_i^2$ is the sum of 12 fourth powers. Thus $6m$ is the sum of at most 48 fourth powers, and the integer $6m + r$ is the sum of at most 53 fourth powers, since $r \leq 5$. ◆

The integer 53 is far from being the least upper bound for $g(4)$. What methods can we devise to find a smaller upper bound? We have seen that not all integers require four squares in their representation as a sum of squares. If the integer m in the preceding theorem could be expressed as the sum of three squares, then the integer $6m$ could be expressed, by Lucas' identity, as a sum of not more than 36 fourth powers.

EXAMPLE 6.3.2.

$$228 = 6 \cdot 38 = 6(5^2 + 3^2 + 2^2).$$

Since $6 \cdot 5^2 = 3^4 + 4 \cdot 2^4 + 5 \cdot 1^4 + 2 \cdot 0^4$, $6 \cdot 3^2 = 3 \cdot 2^4 + 6 \cdot 1^4 + 3 \cdot 0^4$, and $6 \cdot 2^2 = 2^4 + 8 \cdot 1^4 + 3 \cdot 0^4$, direct application of Lucas' identity yields

$$228 = 3^4 + 8 \cdot 2^4 + 19 \cdot 1^4 + 8 \cdot 0^4.$$

That is, 228 is the sum of 36 fourth powers, of which 28 are positive. Now consider 234. This integer is $6 \cdot 39$, and 39 cannot be expressed as the sum of three squares, since $39 \equiv 7 \mod 8$. However, we can write $39 = 25 + 14$. Now $6 \cdot 14 = 84 = 3^4 + 3$ and is the sum of 4 fourth powers. Since 25 is a square, $6 \cdot 25$ is the sum of 12 fourth powers. We can deduce that 234 can be expressed as a sum of not more than 18 positive fourth powers. In this example we were unusually lucky in the fact that 25 was a square. If $m \equiv 7 \mod 8$, $m - 14 \equiv 1 \mod 8$, and since $m - 14$ can be expressed as the sum of at most three squares, we would obtain in this way the conclusion that $6m$ can be expressed using at most $36 + 4 = 40$ fourth powers. (See Exercises 6.3, problem 3.)

THEOREM 6.3.3. $g(4) \leq 45$. Every positive integer is the sum of not more than 45 fourth powers.

Proof: Let $n = 6m + r$, where $m \geq 0$ and $0 \leq r \leq 5$. If $m \equiv 1, 2, 3, 5, 6 \mod 8$, then m is the sum of not more than three squares. Hence $6m$ is the sum of not more than 36 fourth powers and n is the sum of not more than 41 fourth powers.

Suppose that $m \equiv 0$ or $4 \mod 8$. Then $m - 27 \equiv 5$ or $1 \mod 8$.

$$n = 6(m - 27) + 6 \cdot 27 + r.$$

Since $m - 27$, if it is positive, can be expressed as the sum of three squares, $6(m - 27)$ can be expressed as the sum of at most 36 fourth powers. Also, $6 \cdot 27 = 2 \cdot 3^4$ and $r \leq 5$, so that n is the sum of at most 43 fourth powers. The integers less than 162 will have to be considered separately, since $m < 27$ for these integers.

Suppose that $m \equiv 7 \mod 8$. In this case, if $m > 14$, we can consider $m - 14 \equiv 1 \mod 8$ and write

$$n = 6(m - 14) + 6 \cdot 14 + r.$$

Again, $6(m - 14)$ is the sum of 36 fourth powers, and $6 \cdot 14 = 3^4 + 3$ is the sum of 4 fourth powers. The greatest number of fourth powers required for n is thus $36 + 4 + 5 = 45$.

We have had to assume that $m > 27$ and $m > 14$ in the preceding argument. All the integers for which m is smaller than 27 are less than 162. By looking at tables, it is easy to determine that each of these integers can be expressed as the sum of not more than 19 fourth powers. As early as 1853, tables had been prepared showing that all integers less than or equal to 4100 can be written as the sum of not more than 19 fourth powers.   ●

In Theorem 6.3.2 we arrived at a rather large upper bound for the integer $g(4)$. Although 53 is a long way from the conjectured 19, the fact that the bound of 53 can be established proves the existence of $g(4)$. In Theorem 6.3.3 we saw how algebraic manipulation and the use of knowledge about sums of squares could be used to reduce the bound from 53 to 45. By similar and more ingenious techniques, this bound can be reduced further. The goal of 19 has, however, not yet been reached, nor has a counterexample been found to the conjecture that every integer can be expressed as the sum of at most 19 fourth powers.

EXERCISES 6.3

Checks

1. Prepare a table representing the integers from 1 to 100 as sums of 19 fourth powers.

2. Prove Theorem 6.3.1 by expanding both sides of the equation.

3. Carry out the method of Example 6.3.2 using $m = 6 \cdot 71 = 426$.

4. Find, by numerical experimentation, the representation of 426 as a sum of fourth powers that requires the smallest number of terms.

5. In Example 6.3.1, 191 is written as a sum of 21 fourth powers. Is this a counterexample to show that $g(4) \neq 19$?

Challenges

6. Prove that if $n \equiv j \bmod 48$, $j = 6, 12, 18, 30, 36$, then n is a sum of not more than 36 fourth powers.

7. Show that if $n \equiv 42 \bmod 48$, then n can be expressed as a sum of not more than 40 fourth powers. *Hint*: Write $48h + 42 = 48(h - 1) + 6 + 3^4 + 3 \cdot 1^4$.

8. Show that if $n \equiv 44 \bmod 48$, n can be expressed as a sum of not more than 38 fourth powers. *Hint*: Write $n = 48h + 12 + 2 \cdot 2^4$.

9. Using problems 6, 7, and 8, show that if $n \equiv 36, 37, 38, 39, 40, 41, 42, 43, 44, 45, 46, 47, \bmod 48$, then n can be expressed as a sum of not more than 41 fourth powers.

10. Prove $g(4) \leq 41$. *Hint*: Use the techniques of problems 7, 8, and 9 to handle special cases not already covered.

6.4   THE CASE $k = 3$

The investigation of $g(3)$ was brought to a successful conclusion by Arthur Wieferich, who published in 1909 a proof that $g(3) = 9$. An oversight in this proof was corrected by Kempner in 1912. The first attempts to establish a value for $g(3)$ were in the form of tables. By 1903 tables had been constructed for all integers up to 40,000, showing their representation as sums of cubes. Maillet was the first to find an inequality for $g(3)$. In 1895 he showed that $g(3) \leq 21$. This result was reduced to $g(3) \leq 17$ and later to $g(3) \leq 13$.

The proof that $g(3) = 9$ is not included here. We will discuss the inequality $g(3) \leq 21$. Like the investigation of $g(4)$, the proof of this inequality begins with an algebraic identity. It then uses an ingenious method of obtaining a proof for all integers by considering only the integers in a certain interval.

THEOREM 6.4.1. *Maillet's Identity.*

$$\sum_{i=1}^{4} (a + x_i)^3 + (a - x_i)^3 = 2a\{4a^2 + 3(x_1^2 + x_2^2 + x_3^2 + x_4^2)\}.$$

Proof: A simple calculation shows that

$$(a + x)^3 + (a - x)^3 = 2a^3 + 6ax^2.$$

In this expression, let $x = x_1, x_2, x_3, x_4$ and add the resulting equations. The cubes are non-negative if $a \geq x_i$.   ●

EXAMPLE 6.4.1. Let $a = 3, x = 2$.

$$2(3^3) + 18(2^2) = 5^3 + 1^3.$$

Consider the integer $8a^3 + 6am$, where $a = 3$ and $m = 7$; that is, 342. Since $7 = 2^2 + 1^2 + 1^2 + 1^2$, we obtain from Maillet's identity

$$8 \cdot 27 + 6 \cdot 3 \cdot 7 = 5^3 + 1^3 + 4^3 + 2^3 + 4^3 + 2^3 + 4^3 + 2^3.$$

This gives a representation of 342 as the sum of eight cubes. Since $a^2 > m$, each of the cubes is non-negative.

If we combine expressions of the form $8a^3 + 6am$ and $8a'^3 + 6a'm'$, we can derive information about integers that differ from $8a^3 + 8a'^3$ by multiples of 6.

EXAMPLE 6.4.2. Let $a' = 5$ and $m' = 11$. Since $11 = 3^2 + 1^2 + 1^2$, we obtain

$$8 \cdot 5^3 + 6 \cdot 5 \cdot 11 = 8^3 + 2^3 + 6^3 + 4^3 + 6^3 + 4^3 + 5^3 + 5^3.$$

Combine this result with the expansion obtained in Example 6.4.1.

$$8 \cdot 3^3 + 8 \cdot 5^3 + 6(3 \cdot 7 + 5 \cdot 11) = 8^3 + 2 \cdot 6^3 + 3 \cdot 5^3 + 5 \cdot 4^3 + 4 \cdot 2^3 + 1^3.$$

This is the sum of 16 cubes.

Let $B = 8 \cdot 3^3 + 8 \cdot 5^3 + 6 \cdot 42$. We can write 42 in the form $3m + 5m' = 42$. Since $(3, 5) = 1$, we can find integers m and m' to satisfy this equation. We wish the integers m and m' to be positive and also to be less than a^2 and a'^2, respectively, so that the cubes obtained from Maillet's identity will be non-negative. Write $m' = (42 - 3m)/5$. For some m, $0 \le m \le 4$, $42 - 3m \equiv 0 \bmod 5$. Here $m = 4$. The corresponding value of m' is 6.

$$8 \cdot 3^3 + 6 \cdot 3 \cdot 4 = 5^3 + 1^3 + 3^3 + 3^3 + 3^3 + 3^3 + 3^3 + 3^3.$$
$$8 \cdot 5^3 + 6 \cdot 5 \cdot 6 = 8 \cdot 5^3 + 6 \cdot 5(2^2 + 1^2 + 1^2)$$
$$= 7^3 + 3^3 + 6^3 + 4^3 + 6^3 + 4^3 + 5^3 + 5^3.$$

Add these: $8 \cdot 3^3 + 8 \cdot 5^3 + 6 \cdot 42 = 7^3 + 2 \cdot 6^3 + 3 \cdot 5^3 + 2 \cdot 4^3 + 7 \cdot 3^3 + 1^3$, a sum of 16 cubes.

A similar expansion could be performed for any integer of the form $8 \cdot 3^3 + 8 \cdot 5^3 + 6K$, where $15 \le K \le 125$. Why consider only these values of K? We must be able to find non-negative solutions m and m' for the equation $3m + 5m' = K$. These are guaranteed for $K \ge 15$. We must also have $m < 9$ and $m' < 25$. One can be chosen arbitrarily. In this case, we picked $m < 5$. To ensure that $m' < 25$, we make $5m' < 125$; that is, $K \leq 125$.

In Example 6.4.2 we found that every integer of the form $8 \cdot 3^3 + 8 \cdot 5^3 + 6K$, and lying between $8 \cdot 3^3 + 8 \cdot 5^3 + 90$ and $8 \cdot 3^3 + 8 \cdot 5^3 + 750$, can be expressed as a sum of 16 cubes. What about the other integers in this interval? Although not of the form treated in Maillet's identity, they differ from an integer of the desired form by at most 5.

EXAMPLE 6.4.3. In Example 6.4.2 we considered $B = 1468 = 8 \cdot 3^3 + 8 \cdot 5^3 + 6 \cdot 42$ and expressed B as the sum of 16 cubes. Each of the integers 1469, 1470, 1471, 1472, 1473 differs from B by at most 5. Each of these integers can be expressed as a sum of cubes by adding 1^3 an appropriate number of times to the representation of 1468. Thus each can be represented as a sum of not more than 21 cubes. What about 1474? Since $1474 = 1216 + 258 = 8 \cdot 3^3 + 8 \cdot 5^3 + 6 \cdot 43$, it can be expressed as a sum of 16 cubes by finding m and m' to satisfy the equation $3m + 5m' = 43$, and then applying Maillet's identity. In a similar manner, all the integers between $1216 + 90$ and $1216 + 750$ can be represented as a sum of not more than 21 non-negative cubes.

We are now in a position to discuss the first step in proving that $g(3) \leq 21$. The proof of the theorem follows the pattern of Examples 6.4.2 and 6.4.3.

THEOREM 6.4.2. *Some Integers That Are Sums of 21 Cubes.* Let a and a' be positive integers such that $a < a' < a^2$ and $(a, a') = 1$. Any integer B such that

$$8(a^3 + a'^3) + 6aa' \leq B \leq 8(a^3 + a'^3) + 6a'^3$$

is the sum of at most 21 cubes.

Proof: Let $2A = 2a(4a^2 + 3m)$, where $0 \leq m \leq a^2$. Since m can be expressed in the form

$$m = x_1^2 + x_2^2 + x_3^2 + x_4^2, \qquad x_i \leq a,$$

it follows from Maillet's identity that $2A$ can be written as the sum of 8 cubes. Similarly,

$$2A' = 2a'(4a'^2 + 3m'), \qquad 0 \le m' \le a'^2$$

can be written as the sum of 8 cubes. Consider the integer $C = 2A + 2A'$. This integer can be written as the sum of 16 cubes, using Maillet's identity. Note that C is of the form $8a^3 + 8a'^3 + 6K$, where $K = am + am'$.

Now consider any integer K such that $aa' \le K \le a'^2$. Is it possible to write K in the form $K = am + a'm'$, where a and a' are given integers such that $(a, a') = 1$ and m and m' are to be found to satisfy the inequalities $0 \le m \le a^2$ and $0 \le m' \le a'^2$? Since $am + a'm' = K$ is a linear Diophantine equation with relatively prime coefficients, integral solutions exist. Consider

$$m' = \frac{K - am}{a'}.$$

Since $K - ai, i = 0, 1, 2, \ldots, a' - 1$ is a complete residue system mod a', there is a value of i such that $K - ai \equiv 0 \bmod a'$. The corresponding value of m' is an integer. Now, since $K \ge aa' > ai$, we obtain $m' > 0$. Also, $m' \le K/a' \le a'^2$. Since m has been chosen less than a', we have also $m < a^2$. It follows that any integer C of the form $8a^3 + 8a'^3 + 6K$, where $aa' \le K \le a'^3$, can be written as the sum of no more than 16 cubes.

Now consider any integer B in the interval described by the theorem. The integer B differs from some C in the interval by at most 5. Hence B can be expressed as the sum of not more than 21 cubes.   ●

In order to prove that $g(3) \le 21$, we need to show that every non-negative integer can be included in some interval such as that described in the theorem, for some suitable choice of a and a'. As long as $a \ge 3$, it is clear that $a - 1$ and a will be a suitable choice of a and a', since $(a - 1, a) = 1$ and $a < a^2 - 2a + 1$. Also, a and $a + 1$ are a suitable choice for a and a'. If we form the interval corresponding to the choice $a - 1$ and a, will it overlap with the interval corresponding to the choice a and $a + 1$?

EXAMPLE 6.4.4. Let $a = 4$. Corresponding to $a - 1$ and a, we have the interval from

$$8 \cdot 3^3 + 8 \cdot 4^3 + 6 \cdot 3 \cdot 4 \quad \text{to} \quad 8 \cdot 3^3 + 8 \cdot 4^3 + 6 \cdot 4^3;$$

that is the interval from 800 to 1112. The interval corresponding to a and $a + 1$ is the interval from

$$8 \cdot 4^3 + 8 \cdot 5^3 + 6 \cdot 4 \cdot 5 \quad \text{to} \quad 8 \cdot 4^3 + 8 \cdot 5^3 + 6 \cdot 5^3;$$

that is, from 1632 to 2262. These intervals do not overlap. Notice that the second interval is longer than the first. Will the intervals be more likely to overlap if a larger value of a is chosen? Let $a = 11$. The right-hand endpoint of the interval corresponding to $a - 1$ and a is

$$8 \cdot 10^3 + 8 \cdot 11^3 + 6 \cdot 11^3 = 26634.$$

The left-hand endpoint of the interval corresponding to a and $a + 1$ is

$$8 \cdot 11^3 + 8 \cdot 12^3 + 6 \cdot 132 = 25064.$$

These intervals do overlap. Will this always be the case for large enough a?

THEOREM 6.4.3. $g(3) \leq 21$. Every positive integer is the sum of at most 21 positive cubes.

Proof: Choose a and a' to be $k - 1$ and k. Then every integer in the interval

$$8(k - 1)^3 + 8k^3 + 6k(k - 1) \leq B \leq 8(k - 1)^3 + 8k^3 + 6k^3$$

can be expressed as the sum of not more than 21 cubes, since $k - 1$ and k satisfy the conditions of Theorem 6.4.2. The same is true of the integers such that

$$8(k + 1)^3 + 8k^3 + 6k(k + 1) \leq B \leq 8(k + 1)^3 + 8k^3 + 6(k + 1)^3.$$

These intervals overlap if

$$8(k - 1)^3 + 8k^3 + 6k^3 \geq 8(k + 1)^3 + 8k^3 + 6(k + 1)k;$$

that is, if $3k^3 - 27k^2 - 3k - 8 \geq 0$. This inequality is satisfied if $k \geq 10$. (See Exercises 6.4, problem 2.) For $k = 10$, the lower bound of the intervals is 14,372. Since, by making k large enough, the right-hand endpoint can be made as large as desired, any interval greater than 14,372 can be included in one of the sets of overlapping intervals defined by successive choices of k.

From numerical tables, it is known that the integers less than 14,372 can be expressed as the sum of not more than nine cubes. ●

EXERCISES 6.4

Checks

1. Express 342 as the sum of cubes without using Maillet's identity. What is the least number of cubes required?

2. Draw a graph of the cubic equation $y = 3x^3 - 27x^2 - 3x - 8$. Show that $y > 0$ when $x \geq 10$.

Challenges

3. Prove that every integer of the form $6a^3 + 6am$ is the sum of six or fewer cubes if $0 \leq m \leq a^2$ and $m \neq 4^h(8n + 7)$.

4. If a and a' are odd, $a < a' < a^2/8$ and $(a, a') = 1$, prove that the equation $ma + m'a' = K'$ has eight solutions such that $0 \leq m \leq a^2$ and $0 \leq m' \leq a'^2$, provided that $8aa' \leq K' \leq a'^3$.

5. Consider the system of integers $m = m_1 + ja'$, $j = 0, 1, 2, 3, 4, 5, 6, 7$ and a' odd. Prove that at most three of these integers are of the form $4^h(8n + 7)$, where h and n are non-negative integers.

6. From the eight solutions m, m' in problem 4, prove that there are at least two pairs m, m' such that neither m nor m' is of the form $4^h(8n + 7)$.

7. Use problems 3, 4, 5, and 6 to prove that every integer $6A$ with

    $$6(a^3 + a'^3) + 48aa' \leq 6A \leq 6(a^3 + a'^3) + 6a'^3,$$

    where $a < a' < a^2/8$ and a, a' are odd, $(a, a') = 1$, is the sum of 12 or fewer positive cubes.

8. Use the argument of Theorem 6.4.3 to complete the proof that every integer above a certain limit is the sum of 13 or fewer cubes. *Hint*: If $6 \nmid a$, $a^3 \equiv a$ mod 6.

6.5   A LOWER BOUND FOR
$g(k)$

We have seen that arguments leading to an upper bound for $g(3)$ and $g(4)$ frequently yield an expression of an integer in terms of cubes or fourth powers which involves many more terms than is really necessary. In any particular case, we can usually find a much shorter representation as a sum of powers. Let us look now at ways of deciding the minimum number of terms needed—that is, at a lower bound for $g(k)$.

EXAMPLE 6.5.1. Let n be an integer less than 8. Since $2^3 = 8$, the only way to express n as a sum of cubes is to express n as the sum of n terms, each equal to 1^3. Thus

$$7 = 1^3 + 1^3 + 1^3 + 1^3 + 1^3 + 1^3 + 1^3$$

and therefore $g(3) \geq 7$. Similarly, since $2^4 = 16$, the integers from 1 to 15 can be expressed only as sums of 1^4, and so $g(4) \geq 15$.

THEOREM 6.5.1. *A Lower Bound for $g(k)$.*

$$g(k) \geq 2^k - 1$$

Proof: The integer $2^k - 1$ can be expressed as a sum of positive kth powers only by writing it in the form $2^k - 1 = 1^k + 1^k + \cdots + 1^k$. $2^k - 1$ such terms are required and $g(k) \geq 2^k - 1$.   ●

The lower bound obtained so easily in the preceding theorem is not the greatest lower bound for $g(k)$. We can see this fact by constructing for any k an integer that requires more than $2^k - 1$ terms.

EXAMPLE 6.5.2. Consider $2^3 + 7 \cdot 1^3 = 15$. This number requires 8 cubes. Can we find an integer requiring more than 8 cubes? The integer 16 requires only 2 cubes. However, $16 + 7 = 23$ requires 9 cubes. Does $24 + 7 = 31$ require 10 cubes? Why not? Similarly,

$31 = 2^4 + 15$ requires 16 fourth powers.
$47 = 2^4 + 2^4 + 15$ requires 17 fourth powers.
$63 = 3 \cdot 2^4 + 15$ requires 18 fourth powers.
$79 = 4 \cdot 2^4 + 15$ and requires 19 fourth powers.
$95 = 5 \cdot 2^4 + 15$. But $95 = 3^4 + 14$ and requires only 15 fourth powers.

The method of the preceding example could be used, for any k, to construct an integer requiring more than $2^k - 1$ kth powers. We have only to look for an integer of the form $s \cdot 2^k + 2^k - 1$, where s is chosen as large as possible under the condition that $s \cdot 2^k + 2^k - 1 < 3^k$. Early in the study of Waring's problem, a lower bound for the value of $g(k)$ was determined in this way. Euler stated in 1772 that in order to express every integer as a sum of kth powers, at least

$$I(k) = 2^k + [(\tfrac{3}{2})^k] - 2$$

terms are necessary.

THEOREM 6.5.2. *A Better Lower Bound for $g(k)$.*

$$g(k) \geq I(k) = 2^k + [(\tfrac{3}{2})^k] - 2, \quad k = 1, 2, 3, \ldots.$$

Proof: Let $n = 2^k[(\tfrac{3}{2})^k] - 1$. Since $[x] < x$ when x is not an integer, $n < 2^k(\tfrac{3}{2})^k - 1$; that is, $n < 3^k - 1$. Now, by definition of $g(k)$,

$$n = x_1^k + x_2^k + \cdots + x_{g(k)}^k$$

for some non-negative integers x_i, $i = 1, 2, \ldots, g(k)$. Each of these x_i must be less than 3. Suppose that a of the x_i have the value 2, b have the value 1, and c have the value 0. Then $n = 2^k a + b$, and $g(k) = a + b + c$.

Since $2^k a < n < 2^k[(\tfrac{3}{2})^k]$, we have $a < [(\tfrac{3}{2})^k]$; that is, $a \leq [(\tfrac{3}{2})^k] - 1$. Also, $b = n - 2^k a$, so that $a + b = n - (2^k - 1)a$. Since

$$(2^k - 1)a \leq (2^k - 1)[(\tfrac{3}{2})^k] - (2^k - 1),$$

it follows that

$$\begin{aligned} n - (2^k - 1)a &\geq n - (2^k - 1)[(\tfrac{3}{2})^k] + (2^k - 1) \\ &= 2^k[(\tfrac{3}{2})^k] - 1 - 2^k[(\tfrac{3}{2})^k] + [(\tfrac{3}{2})^k] + 2^k - 1 \\ &= [(\tfrac{3}{2})^k] + 2^k - 2 = I(k). \end{aligned}$$

Since $g(k) \geq a + b$, and $a + b = n - (2^k - 1)a \geq I(k)$, it follows that $g(k) \geq I(k)$. ⬢

From the formula for $I(k)$, we calculate easily $I(2) = 4$, $I(3) = 9$, and $I(4) = 19$, which are exactly the values established for $g(2)$ and $g(3)$ and predicted for $g(4)$. It has been conjectured that $g(k) = I(k)$ for all k, and this prediction is known as the Ideal Waring Theorem. The proof that $g(6) = I(6) = 73$ was finally carried out by Pillai in 1940. In 1933, Dickson established that $g(5) \leq 54$, a bound that was lowered to 40 by Chen in 1959. But, like $g(4)$, the exact value of $g(5)$ is still not determined.

For $k > 6$, it was proved by Dickson that $g(k) = I(k)$, provided that

$$r \leq 2^k - [(\tfrac{3}{2})^k] - 3,$$

where $r = 3^k - 2^k[(\tfrac{3}{2})^k]$. For simplicity, write

$$[(\tfrac{3}{2})^k] = q.$$

By the division algorithm,

$$3^k = 2^k q + r, \qquad 0 < r < 2^k.$$

How much smaller than 2^k is r? Dickson and Rubugunday proved that r cannot equal $2^k - q - 1$ or $2^k - q$. Niven was able to prove that $g(k) = I(k)$ if $r = 2^k - q - 2$. These results combine to show that the Ideal Waring Theorem can be proved for $k > 6$ unless $r > 2^k - q$. Dickson showed that if $r > 2^k - q$,

$$g(k) = I(k) + [(\tfrac{4}{3})^k] - 1.$$

But it seems likely that there does not exist an r such that $r > 2^k - q$. Mahler has proved that if there are any such r, there can be only a finite set of them and hence a greatest one. However, his method does not give an estimate of the size of this greatest one. Using an IBM 7090 computer, Stemmler was able to prove that for $k \leq 200,000$, $r < 2^k - q$.

Thus the determination of $g(k)$ is almost complete. There remain only the cases $g(4)$ and $g(5)$ and the question of whether or not there is a gap between the largest k reached by the computer and the largest k for which r might be greater than $2^k - 1$.

The determination of $g(k)$ will not settle all questions related to Waring's conjecture. Examination of the table of expansion of integers as sums of cubes shows that not many integers require nine or even eight cubes in their representation. Dickson has proved that every integer except 23 and 239 can be expressed as the sum of not more than eight cubes, and it seems that only fifteen integers require eight cubes. This suggests that the smaller numbers that require large numbers of cubes are really exceptional cases and that we might find an integer, $G(k)$, smaller than $g(k)$ with the property that all integers from some point on can be expressed as the sum of not more than $G(k)$ kth powers.

DEFINITION 6.5.1. $G(k)$ is the least integer such that all but a finite number of integers can be represented as the sum of at most $G(k)$ positive kth powers.

What can be said about $G(k)$? Certainly $G(k) \leq g(k)$. Tables show that all integers between 454 and 40,000 require at most seven cubes, and all integers between 8042 and 40,000 require at most six. On this basis, it is conjectured that $G(3) \leq 6$. The only value of $G(k)$ that we know definitely is $G(2)$. Since there is an infinite set of integers that require four squares, $G(2) \geq 4$. But $G(2) \leq g(2) = 4$, so that $G(2) = g(2) = 4$.

If you wish to proceed further in this direction, try the article by W. J. Ellison [6] published in the *American Mathematical Monthly*. This is a most interesting brief discussion of the approaches to Waring's problem, its generalizations, and the direction of current research in the area, together with an extensive bibliography.

EXERCISES 6.5

Checks

1. Seven of the integers requiring eight cubes are less than 200. Can you find those less than 100?

2. Locate an integer requiring $I(5)$ fifth powers.

3. Locate an integer requiring $I(6)$ sixth powers.

4. Is the least number of terms in the expansion always obtained by subtracting the largest possible kth power at each step?

5. Calculate r and $2^k - q$ for $k = 3, 4, 5, 6, 7$.

6. Prove that $G(k) \geq 4$ if k is even.

7. Use the fact that $x^3 \equiv 1, 6 \bmod 7$ to prove that $G(3) \geq 3$.

REFERENCES FOR UNIT 6

1. Chen, Jing-Jung, "Waring's Problem for $g(5)$," *Science Record*, Vol. 3 (1959), pp. 327–330.

2. Dickson, L. E., "Recent Progress on Waring's Theorem and Its Generalizations," *Bulletin of the American Mathematical Society*, Vol. 39 (1933), pp. 701–727.

3. Dickson, L. E., "Proof of the Ideal Waring Theorem for Exponents 7–180," *American Journal of Mathematics*, 58 (1936), pp. 521–529.
4. Dickson, L. E., "Solution of Waring's Problem," *American Journal of Mathematics*, 58 (1936), pp. 530–535.
5. Dickson, L. E., "All Integers Except 23 and 239 Are Sums of Eight Cubes," *Bulletin of the American Mathematical Society*, Vol. XLV (1939), pp. 588–591.
6. Ellison, W. J., "Waring's Problem," *American Mathematical Monthly*, Vol. 78 (1971), pp. 10–36.
7. Gelfond, A. O., and Y. V. Linnik, *Elementary Methods in Analytic Number Theory*, translated by Amie Fienstein, revised and edited by L. J. Morell, (Chicago: Rand McNally, 1965.)
8. Hardy, G. H., *Collected Papers of G. H. Hardy* (Oxford: Clarendon Press, 1966), Vol. 1.
9. Landau, E., *Elementary Number Theory* (New York: Chelsea Publishing Co., 1958.)
10. Niven, I., "An Unsolved Case of the Waring Problem," *American Journal of Mathematics*, Vol. 66 (1944), pp. 137–143.
11. Pillai, S. S., "On Waring's Problem $g(6) = 73$," *Proceedings of the Indian Academy of Science*, (A), Vol. 12 (1940), pp. 30–40.
12. Stemmler, R. M., "The Ideal Waring Theorem for Exponents 401–200,000," *Mathematics of Computation*, Vol. 18 (1964), pp. 144–146.
13. Watson, G. L., "A Simple Proof that All Large Integers Are Sums of at Most Eight Cubes," *Mathematical Gazette*, Vol. 37 (1953), pp. 209–211.

NUMBER-THEORETIC FUNCTIONS— A MEETING OF PATHS

7.1 SOME NUMBER–THEORETIC FUNCTIONS

In our exploration of the integers, we have come across several functions with domain the positive integers. Functions with domain the positive integers and range the real or complex numbers are called *number-theoretic*, or *arithmetic*, functions. That the set of these functions is infinite is clear from the fact that any property of the integers which can be expressed in terms of a real (or complex) number must necessarily lead to a number-theoretic function. Some of these functions have proved sufficiently interesting over the years that a standard notation has been adopted for them. In this section we recall those we have met, we define some simple generalizations of them, and we suggest some new ones. In Section 7.2 we look at the set of number-theoretic functions as a whole and consider some structure on this set. What operations on the set yield elements of the set? Which of these operations have inverses? What is the nature of the ring whose elements are the number-theoretic functions? Section 7.3 deals with the units in the ring of number-theoretic functions for which "convolution product" is the operation corresponding to multiplication. In Section 7.4 we center attention on multiplicative functions and see that the set of nonzero multiplicative functions is a subgroup of the multiplicative group of units. Finally, in Section 7.5, we look briefly at another operation on the ring by which the subgroup of units for which $f(1)$ is positive is mapped onto the additive group of the ring of number-theoretic functions.

We began our exploration with the relation "divides." Many number-theoretic functions are associated with the idea of divisors. For any positive integer n, $\tau(n)$ is the number of positive divisors of n and $\sigma(n)$ is the sum of the positive divisors of n (Section 1.2). We might consider a family of functions of this type,

$$\sigma_k, \quad \text{where} \quad \sigma_k(n) = \sum_{d \mid n} d^k, \quad k = 0, 1, 2, \ldots .$$

This family includes the functions τ and σ, since $\tau = \sigma_0$ and $\sigma = \sigma_1$.

In Unit 2 we met one of the best known of the arithmetic functions, the Euler function, ϕ, defined by $\phi(1) = 1$, $\phi(n) =$ the number of integers in a reduced residue system mod n. Also, in Unit 2, we discussed the Möbius function μ, for which $\mu(1) = 1$, $\mu(n) = (-1)^r$ if n is square-free and has r distinct prime factors; $\mu(n) = 0$ otherwise. This function plays an important role in the structure of the set of number-theoretic functions. This fact is not surprising in view of the important inversion theorem, Theorem 2.5.3. Generalizations of the functions ϕ and μ are suggested in Exercises 7.1, problems 11 and 12. See also [5] and [6].

The functions mentioned in the preceding paragraphs are all *multiplicative*. Recall (Section 1.9) that a function is multiplicative if, when $(m, n) = 1$, $f(mn) = f(m)f(n)$. The multiplicative property of these functions makes it possible to develop formulas for them in terms of the prime factorization of n. In particular, if $n = p_1^{a_1} p_2^{a_2} \cdots p_r^{a_r}$, we have

$$\tau(n) = \prod_{i=1}^{r} (a_i + 1);$$

$$\sigma(n) = \prod_{i=1}^{r} (1 + p_i + \cdots + p_i^{a_i});$$

$$\sigma_k(n) = \prod_{i=1}^{r} (1 + p_i^k + \cdots + p_i^{ka_i});$$

$$\phi(n) = n \prod_{i=1}^{r} \left(1 - \frac{1}{p_i}\right);$$

$\mu(n) = (-1)^r$ if $a_i = 1$ for each i, 0 if $a_i > 1$ for some i, 1 if each $a_i = 0$.

The multiplicative property seems a natural one, since so much of our study of number theory has been concerned with divisibility. A somewhat similar property is the additive property, defined as follows:

DEFINITION 7.1.1. *Additive Function.* A number-theoretic function is *additive* if for every pair of positive integers m, n such that $(m, n) = 1$, $f(mn) = f(m) + f(n)$. The function f is *completely additive* if $f(mn) = f(m) + f(n)$ for every pair m and n.

EXAMPLE 7.1.1. Let f be defined by $f(1) = 0$, $f(n) =$ the number of distinct primes that divide n, if $n > 1$. Thus $f(12) = 2$, $f(24) = 2$, $f(5) = f(25) = 1$, $f(60) = 3$. The function f so defined is additive but not completely additive (Exercises 7.1, problem 3).

EXAMPLE 7.1.2. An example of a function that is completely additive is the function recently defined by L. M. Chawla as follows:

$$f(1) = 0, \quad f(n) = a_1 p_1 + a_2 p_2 + \cdots + a_r p_r, \quad \text{for} \quad n = p_1^{a_1} p_2^{a_2} \cdots p_r^{a_r}.$$

Thus $f(2) = 2$, $f(6) = 5$, $f(5) = 5$, $f(8) = f(9) = 6$, $f(n) \neq 1$ for every n. See Exercises 7.1, problems 1 and 2; also [4].

The additive property, like the multiplicative property, makes it possible to express $f(n)$ in terms of the values of f at prime powers.

THEOREM 7.1.1. *Formula for Additive Functions.* If f is additive, then $f(1) = 0$ and

$$f(n) = \sum_{i=1}^{r} f(p_i^{a_i}), \quad \text{when} \quad n = \prod_{i=1}^{r} p_i^{a_i}.$$

Proof: See Exercises 7.1, problem 6.   ●

Somewhat different in nature is the function π introduced in Section 1.2 and discussed in more detail in Unit 4. This function counts the positive integers that are prime and less than or equal to n. It is not multiplicative or additive, and no simple expression is available as a formula, although the behavior of $\pi(n)$ as n becomes large is similar to that of $n/\log n$. Dirichlet's theorem states that there is an infinite set of primes of the form $a + bk$, provided that $(a, b) = 1$. The number of primes of this form and less than or equal to n is usually denoted $\pi(n; a, b)$. Many other counting functions can be defined. For example, $S_2(n) =$ the number of squares less than or equal to n, and $\pi_2(n) =$ the number of primes less than or equal to n and members of a prime pair. Counting functions of this type are nondecreasing; $S_2(n)$ becomes infinite as n becomes infinite, as does $\pi(n; a, b)$, but the behavior of $\pi_2(n)$ as n becomes infinite is not known.

A useful class of functions with quite different behavior is the set of functions that are characteristic functions of sets of integers.

$$\boxed{E_S}$$

DEFINITION 7.1.2. *Characteristic Function.* Let S be a set of integers. The characteristic function of S, denoted E_S, is defined as follows: $E_S(n) = 1$ if n belongs to S, $E_S(n) = 0$ if n does not belong to S.

EXAMPLE 7.1.3. Let $S = \{n: n \text{ is odd}\}$. $E_S(1) = 1$, $E_S(2) = 0$, $E_S(3) = 1, \ldots$. Certainly E_S is not monotone. Is it multiplicative? Consider $E_S(mn)$. The product of two integers is odd if and only if both the integers are odd; that is, $E_S(mn) = 1$ if and only if $E_S(m) = E_S(n) = 1$. On the other hand, mn is even if and only if at least one of the integers m and n is even. This means that $E_S(mn) = 0$ if and only if $E_S(m) = 0$ or $E_S(n) = 0$, or both. In either case, $E_S(mn) = E_S(m)E_S(n)$. Is the characteristic function of the even integers multiplicative? Is it additive?

EXAMPLE 7.1.4. Let $S = \{n: n \,|\, 24\}$. E_S is multiplicative but not completely multiplicative. For example, $E_S(6)E_S(8) = 1$, but $E_S(48) = 0$.

If $(m, n) = 1$ and m and n both belong to S, then $mn \mid 24$, so that $E_S(mn) = E_S(m)E_S(n)$. If $mn \mid 24$, then $m \mid 24$ and $n \mid 24$, so that $E_S(mn) = 1$ implies $E_S(m)E_S(n) = 1$. If mn does not divide 24 and $(m, n) = 1$, then at least one of m and n does not divide 24, so that $E_S(mn) = 0$ implies $E_S(m)E_S(n) = 0$ if $(m, n) = 1$.

It is clear from Examples 7.1.3 and 7.1.4 that the characteristic function of a set S may or may not be multiplicative. Which is the case depends on the properties of the set S. The following theorem gives a necessary and sufficient condition for a characteristic function to be multiplicative.

THEOREM 7.1.2. *Multiplicative Characteristic Functions.* If S is not empty, the function E_S is multiplicative if and only if S has the following properties:
    1. 1 belongs to S.
    2. If m and n belong to S and $(m, n) = 1$, then mn belongs to S.
    3. If mn belongs to S and $(m, n) = 1$, then m and n both belong to S.

Proof: Suppose that S satisfies the conditions. By property (1), $E_S(1) = 1$; and by properties (2) and (3), if $(m, n) = 1$, $E_S(mn) = 1$ if and only if $E_S(m) = 1$ and $E_S(n) = 1$, so E_S is multiplicative.

Suppose that E_S is multiplicative. Since it is not identically zero, $E_S(1)$ must be 1, so that 1 belongs to S. Since $E_S(mn) = E_S(m)E_S(n)$ when $(m, n) = 1$, and since the only function values are 0 or 1, if mn belongs to S both m and n must, which is property (3), and if mn does not belong to S, then at least one of m and n does not belong to S, which is the contrapositive of (2). ●

One way in which a characteristic function is useful is in defining a function that agrees with a certain function f on a specified set S and that is zero outside S. Let f be a number-theoretic function. The pointwise product $E_S f$, defined by $E_S f(n) = E_S(n)f(n)$, has the property $E_S f(n) = f(n)$ if n belongs to S, and $E_S f(n) = 0$ if n does not belong to S. The function $E_S f$ is called the *restriction of f to S*.

EXAMPLE 7.1.5. Let S be the set of square-free integers, and let f be the function defined in Example 7.1.1. Consider the function F defined by $F(n) = (-1)^{f(n)}$. Since f is additive, F is multiplicative (see Exercises 7.1, problem 9). $F(1) = 1$ and $F(n) = (-1)^r$ if n has r distinct prime divisors. The function $E_S F = \mu$.

EXERCISES 7.1

Checks

1. For Chawla's function (Example 7.1.2) prove that
   (a) $f(mn) = f(m) + f(n)$,
   (b) $f(p) = p$ if p is prime,
   (c) $f(1) = 0$.

2. If g is an arithmetic function satisfying (a), (b), and (c) of problem 1, prove that g is Chawla's function.

3. Prove that the function f defined in Example 7.1.1 is additive but not completely additive. What about the function $v(n) = a_1 + a_2 + \cdots + a_k$ for $n = p_1^{a_1} p_2^{a_2} \cdots p_k^{a_k}$?

4. Let $P = \{n: n \text{ is prime}\}$. Show that

$$\pi(n) = \sum_{k=1}^{n} E_P(k).$$

5. Let $S = \{n: n = 2^i 3^j, i, j = 0, 1, 2, \ldots\}$. Show that E_S is completely multiplicative. Let $T = \{n: n = 1 \text{ or a product of distinct primes}\}$. Show that E_T is multiplicative but not completely multiplicative.

Challenges

6. Use induction and the additive property to prove Theorem 7.1.1.

7. If f is any completely additive function, prove that

$$\sum_{d|n} f(d) = \tfrac{1}{2}\tau(n)f(n).$$

8. Show that E_S is completely multiplicative if and only if S is closed under multiplication, and n belongs to S and $d|n$ implies d belongs to S.

9. Let f be an additive function and k a real number. Define F by the formula $F(n) = k^{f(n)}$. Prove that F is multiplicative.

10. If f is multiplicative and not identically zero, prove that, for $n > 1$,

$$\sum_{d|n} \mu(d)f(d) = \prod_{p|n} [1 - f(p)]. \qquad \text{(Jim Lawrence)}$$

11. Define μ_k as follows: $\mu_k(1) = 1$, and if $n = \prod_{i=1}^{r} p_i^{a_i}$, $\mu_k(n) = (-1)^r$ if $a_1 = a_2 = \cdots = a_r = k$, and $\mu_k(n) = 0$ if $a_i \neq k$ for some i. Prove that

(a) μ_k is multiplicative.

(b) $\mu_k(n^k) = \mu(n)$.

(c) $\sum_{d|n} \mu_k(d) = 1$ or 0, depending on whether n is or is not kth power free.

12. Define ϕ_k by $\phi_k(1) = 1$, $\phi_k(n) =$ the number of integers a in a complete residue system mod n such that (a, n) is not divisible by a kth power. Show that

(a) ϕ_k is multiplicative,

(b) $\phi_k(p^a) = p^a - p^{a-k}$ if $k \leq a$,

(c) $\phi_k(n) = n \prod_{p^k|n} (1 - p^{-k})$,

(d) $\phi_k(n^k) = n^k \sum_{d|n} \mu(d) d^{-k}$.

7.2 BINARY OPERATIONS ON THE SET OF NUMBER-THEORETIC FUNCTIONS

Let us use the notation $\mathscr{F}$ to designate the set of number-theoretic functions. What operations can be defined on this set so that, given two number-theoretic functions, a number-theoretic function is generated? Such an operation is called a *binary* operation because it maps a pair of elements of the set $\mathscr{F}$ into an element of $\mathscr{F}$. The simplest operation of this type is addition, defined pointwise.

$$f + g$$

DEFINITION 7.2.1. *Sum $f + g$.* Let f and g belong to $\mathscr{F}$. The sum of f and g is the function designated $f + g$ and defined by

$$(f + g)(n) = f(n) + g(n)$$

for n a positive integer.

The domain of the function $f + g$ is the positive integers, so $f + g$ is an element of the set $\mathscr{F}$. This is the way we have always added functions. The definition is simply making formal an idea that is intuitively quite familiar.

What sort of algebraic structure is the set $\mathscr{F}$ with the operation $+$? (You may wish to refer to Appendix A to refresh your memory about algebraic structures.) The set $\mathscr{F}$ is closed under the operation $+$. Does $f + g = g + f$? This property, and also the associative property of addition, follows immediately from the corresponding properties of addition of real numbers. Does the set $\mathscr{F}$ have an additive identity? We need a number-theoretic function, z with the property $f + z = f$ for every f in $\mathscr{F}$.

$$\boxed{ z }$$

Clearly z must be the function defined by $z(n) = 0$ for every positive integer n. Every function in $\mathscr{F}$ possesses an additive inverse. For what function g is $f + g = z$? The set $\mathscr{F}$ with the operation $+$ forms a commutative group. We will use the notation $(\mathscr{F}, +)$ to designate this group.

The situation allows more alternatives when we consider how we might define multiplication in a suitable way. The most obvious way is pointwise multiplication, which is just what we ordinarily do when we multiply functions.

$$\boxed{ f \cdot g }$$

DEFINITION 7.2.2. *Pointwise Product, $f \cdot g$.* Let f and g belong to $\mathscr{F}$. The pointwise product of f and g is the function designated $f \cdot g$ and defined by

$$(f \cdot g)(n) = f(n)g(n)$$

for n a positive integer.

Again, the set $\mathscr{F}$ is closed under this operation. The operation is clearly commutative, associative, and distributive over addition, because each of these properties is true for a fixed value of n. The multiplicative identity is the function having the value 1 for each n. Let us represent this function by u: $u(n) = 1$, $n = 1, 2, 3, \ldots$.

$$\boxed{ u }$$

For some functions, multiplicative inverses exist. To find a function g such that $g \cdot f = u$, we must, for every positive integer n, satisfy the equation $g(n)f(n) = 1$. If $f(n) \neq 0$, we can do this easily. But if there is an n such that $f(n) = 0$, a multiplicative inverse for f does not exist.

The set of number-theoretic functions, $\mathscr{F}$, with the operations $+$ and $\cdot$ defined above does form a commutative ring. We use the notation $(\mathscr{F}, +, \cdot)$ to designate this ring. Like the rings Z_m discussed in Section 2.3, the ring $(\mathscr{F}, +, \cdot)$ contains zero divisors.

EXAMPLE 7.2.1. Let f be the characteristic function of the set of odd integers and g be the characteristic function of the set of even integers. Since $(f \cdot g)(n) = f(n)g(n) = 0$ for every n, these functions have the property $f \cdot g = z$, but $f \neq z$ and $g \neq z$. They are, therefore, zero divisors in the ring $(\mathscr{F}, +, \cdot)$. Many more such examples can be found.

When the operations of addition and multiplication are defined pointwise, the ring of number-theoretic functions, as far as its algebraic structure is concerned, is much the same as the ring formed by many other sets of functions with the same operations. Let us look for a binary operation that is more relevant to the set of number-theoretic functions.

Recall that in Section 1.11 we discussed the function $F(n) = \sum_{d|n} f(d)$. In Section 2.6 we derived the Möbius inversion formula: if

$$F(n) = \sum_{d|n} f(d),$$

then

$$f(n) = \sum_{d|n} \mu(d) F\left(\frac{n}{d}\right).$$

Both situations suggest defining a function at n in terms of the values of another function or functions at the divisors of n. In the case of the inversion formula, the values of μ at d are multiplied by the values of F at n/d. This sum could be written

$$\sum_{rs=n} \mu(r) F(s),$$

since the set of pairs of divisors d and n/d is simply the set of pairs r and s such that $rs = n$. A sum of products taken in this way is usually called the convolution product. The word convolution means winding or twisting together and describes the pairing of the divisors d and n/d. You may have met the idea of convolution in differential equations when you considered integrals of the form $\int_0^t f(t-s)g(s)\,ds$.

$$\boxed{f * g}$$

DEFINITION 7.2.3. *Convolution Product, $f * g$*. The convolution product of two functions f and g that belong to $\mathscr{F}$ is the function designated $f * g$ and defined by

$$(f * g)(n) = \sum_{d|n} f(d)g\left(\frac{n}{d}\right) = \sum_{rs=n} f(r)g(s).$$

EXAMPLE 7.2.2

$$(\tau * \sigma)(6) = \tau(1)\sigma(6) + \tau(2)\sigma(3) + \tau(3)\sigma(2) + \tau(6)\sigma(1)$$
$$= 1(12) + 2(4) + 2(3) + 4(1) = 30.$$

In the notation of convolution product, the summation $\sum_{d|n} \mu(d)\, F(n/d)$ is simply $(\mu * F)(n)$. The summation $\sum_{d|n} f(d)$ also falls into this form if we write it as $\sum_{d|n} f(d)\, u(n/d)$, where $u(n) = 1$ for every n. This sum is then $(f * u)(n)$. The Möbius inversion formula in this notation becomes:

if $F = f * u$, then $f = \mu * F$.

We must now examine the properties of the operation $*$. From the symmetry of the definition, especially when the sum is written

$$\sum_{rs=n} f(r)g(s),$$

it is clear that the operation is commutative; that is, $f * g = g * f$. Is it also associative? Is there a multiplicative identity, and, if so, what is it? Is the operation distributive over addition? The answers to these questions are not as obvious as they were in the case of pointwise multiplication, but they are needed in order to examine whether $\mathscr{F}$ with the operations $+$ and $*$ is a ring.

EXAMPLE 7.2.3. Consider $(f * g) + (f * h)$ and $f * (g + h)$ at the point $n = 6$.

$$[(f * g) + (f * h)](6) = (f * g)(6) + (f * h)(6) \quad \text{(definition of } +)$$
$$= f(1)g(6) + f(2)g(3) + f(3)g(2) + f(6)g(1)$$
$$+ f(1)h(6) + f(2)h(3) + f(3)h(2) + f(6)h(1)$$
$$\text{(definition of } *)$$
$$= f(1)[g(6) + h(6)] + f(2)[g(3) + h(3)]$$
$$+ f(3)[g(2) + h(2)] + f(6)[g(1) + h(1)]$$
$$\text{(associative, commutative, and distributive property of real numbers.)}$$
$$= f(1)[g + h](6) + f(2)[g + h](3)$$
$$+ f(3)[g + h](2) + f(6)[g + h](1)$$
$$\text{(definition of } +)$$
$$= [f * (g + h)](6) \quad \text{(definition of } *).$$

Can a similar argument be used to show that

$$[(f * g) + (f * h)](n) = [f * (g + h)](n)$$

for every n and therefore that $(f * g) + (f * h) = f * (g + h)$ and the operation $*$ is distributive over addition?

What should the multiplicative identity be? Let e represent this function. We want $f * e = f$ for every f in $\mathscr{F}$. For $n = 1$, this means $f(1)e(1) = f(1)$, which implies $e(1) = 1$. For $n = 2$, we need $f(1)e(2) + f(2)e(1) = f(2)$ for every f; that is, $f(1)e(2) = 0$ or $e(2) = 0$. Similarly, $f(1)e(3) + f(3)e(1) = f(3)$ implies $e(3) = 0$. Can you guess what the multiplicative identity must be?

THEOREM 7.2.1. *The Ring* $(\mathscr{F}, +, *)$. The set of number-theoretic functions, $\mathscr{F}$, with the operations pointwise addition and convolution product, is a commutative ring with unity. The function e defined by $e(n) = 0$ if $n > 1$, $e(1) = 1$, is the *multiplicative identity*.

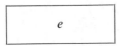

Proof: We have already seen that the set $\mathscr{F}$ with the operation $+$ is a commutative group.

The operation $*$ is commutative. For any f and g in $\mathscr{F}$, we have

$$(f * g)(n) = \sum_{rs=n} f(r)g(s) = \sum_{sr=n} g(s)f(r) = (g * f)(n).$$

The operation $*$ is associative. For any f, g, and h in $\mathscr{F}$, we have

$$[(f * g) * h](n) = \sum_{rs=n} (f * g)(r)h(s)$$

$$= \sum_{rs=n} \left[\sum_{uv=r} f(u)g(v) \right] h(s)$$

$$= \sum_{uvs=n} f(u)g(v)h(s).$$

In the same way, $[f * (g * h)](n)$ is given by the same summation. The operation $*$ is distributive over addition. We have

$$[f * (g + h)](n) = \sum_{rs=n} f(r)[(g + h)](s)$$

$$= \sum_{rs=n} f(r)[g(s) + h(s)]$$

$$= \sum_{rs=n} f(r)g(s) + \sum_{rs=n} f(r)h(s)$$

$$= (f * g)(n) + (f * h)(n)$$

$$= [(f * g) + (f * h)](n).$$

The function e defined by $e(1) = 1$, $e(n) = 0$ for $n > 1$, is the identity for convolution multiplication:

$$(e * f)(n) = \sum_{rs=n} e(r)f(s) = e(1)f(n),$$

since the only nonzero term in this sum occurs when $r = 1$ and therefore $s = n$. Thus $(e * f)(n) = f(n)$ for every n, and $e * f = f$. ⬢

EXERCISES 7.2

Checks

1. Define g such that $f + g = z$.
2. Verify that $h \cdot (f + g) = h \cdot f + h \cdot g$.
3. Prove that μ is a zero divisor in the ring $(\mathscr{F}, +, \cdot)$ by finding a function f such that $f \neq z$ and $\mu \cdot f = z$. Can you characterize the zero divisors in this ring?

4. Calculate $(\tau * \phi)(8)$, $(\tau * \phi)(p^3)$, $(\tau * \phi)(p^k)$, where p is a prime.

5. Calculate $(\mu * \phi)(12)$. Show that

$$(\mu * f)(p^k) = f(p^k) - f(p^{k-1}).$$

6. Write in $*$ notation the formulas

$$\sum_{d|n} \phi(d) = n \quad \text{and} \quad \sum_{d|n} \mu(d) = 0$$

for $n > 1$, obtained in Section 2.5.

Challenges

7. In the ring $(\mathscr{F}, +, \cdot)$ are there any *nilpotent* elements—that is, functions f such that $f \neq z$ and $f^k = z$, where f^k means a product $f \cdot f \cdots \cdot f$ with k factors? Do there exist nilpotent elements in Z_6? In Z_8?

8. The convolution product is sometimes called the Dirichlet product. A Dirichlet series is a series of the form

$$\frac{a_1}{1^s} + \frac{a_2}{2^s} + \frac{a_3}{3^s} + \cdots + \frac{a_n}{n^s} + \cdots \quad \text{or} \quad \sum_{n=1}^{\infty} \frac{a_n}{n^s}.$$

Write two such series

$$\sum \frac{a_n}{n^s} \quad \text{and} \quad \sum \frac{b_n}{n^s},$$

multiply them formally, and collect the terms that will have a denominator n^s so that

$$\sum \frac{a_n}{n^s} \sum \frac{b_n}{n^s} = \sum \frac{c_n}{n^s}.$$

Verify that

$$c_n = \sum_{d|n} a_d b_{n/d}.$$

9. Define the binary operation $\otimes$ which might be called "continuous convolution" by the formula

$$(f \otimes g)(n) = \sum_{k=1}^{n} f(k)g(n+1-k) = \sum_{x+y=n+1} f(x)g(y).$$

Convince yourself that this is commutative, associative, and distributive over addition. What is the multiplicative identity in the ring $(\mathscr{F}, +, \otimes)$?

10. If h is completely multiplicative, prove that

$$(f \cdot h) * (g \cdot h) = (f * g) \cdot h. \quad \text{(Lee Hill)}$$

11. Let the set $S = \{p^i, i = 0, 1, 2, \ldots, p \text{ a prime}\}$. Prove that the set of functions consisting of the functions $f_s = E_s \cdot f$, for f in $\mathscr{F}$, is a subring of the ring $(\mathscr{F}, +, *)$, using the operation $+$ and $*$. Show that $E_s \cdot [g * h] = (E_s \cdot g) * (E_s \cdot h)$ and $E_s \cdot [g - h] = E_s \cdot g - E_s \cdot h$.

12. Prove that the ring $(\mathscr{F}_s, +, *)$, where $\mathscr{F}_s$ is the set of functions $f_s = E_s \cdot f$ for f in $\mathscr{F}$, is isomorphic to the ring $(\mathscr{F}, +, \otimes)$.

7.3   PROPERTIES OF THE RING $(\mathscr{F}, +, *)$

Now that the set of number-theoretic functions, with the operations pointwise addition and convolution multiplication, is known to be a ring, two additional questions arise naturally. Does the ring have any divisors of zero? Which elements, if any, have multiplicative inverses?

EXAMPLE 7.3.1. Can we find nonzero functions f and g such that $f * g = z$? For $n = 1$, this implies $f(1)g(1) = 0$. Thus either $f(1)$ or $g(1) = 0$. Suppose that $f(1) = 0$. Consider $n = 2$. Since $f(1)g(2) + f(2)g(1) = 0$ and $f(1) = 0$, we have $f(2)g(1) = 0$. In the same way, for $n = 3$, $f(1)g(3) + f(3)g(1) = 0$, while for $n = 4$, $f(1)g(4) + f(2)g(2) + f(4)g(1) = 0$. Now if $g(1) \neq 0$, it is easy to see that these equations imply $f(2) = 0$, $f(3) = 0$, and $f(4) = 0$. But suppose that $g(1) = 0$. Can we satisfy all possible equations of this type without having $f(n) = 0$ for all n or $g(n) = 0$ for all n? Direct attack seems to be leading to a cumbersome number of choices. Let us try an indirect approach. Suppose that neither f nor g is the function z, and $f * g = z$. There must be some integer n such that $f(n) \neq 0$ and some integer m such that $g(m) \neq 0$. Can we use these integers to find an integer for which $f * g$ is not zero and thus reach a contradiction? Suppose that $f(4) \neq 0$ but $f(1) = f(2) = f(3) = 0$. Also, suppose that $g(6)$ is not zero but $g(1)$, $g(2)$, $g(3)$, $g(4)$, $g(5)$ are all zero. What about $(f * g)(24)$? We have

$$(f * g)(24) = f(1)g(24) + f(2)g(12) + f(3)g(8) + f(4)g(6) + f(6)g(4) + f(8)g(3) + f(12)g(2) + f(24)g(1).$$

Now the first three terms of this sum are zero because the value of f is zero, and the last four terms are zero because the value of g is zero. This means $(f * g)(24) = f(4)g(6) \neq 0$, which contradicts the assumption that $f * g = z$.

The proof of the following theorem follows the technique of this example.

THEOREM 7.3.1. *Divisors of Zero in* $(\mathscr{F}, +, *)$. The ring $(\mathscr{F}, +, *)$ has no divisors of zero.

Proof: Suppose that $(\mathscr{F}, +, *)$ has divisors of zero. Then there exist functions f and g such that $f \neq z$, $g \neq z$, and $f * g = z$. Since $f \neq z$, there exists some integer n such that $f(n) \neq 0$. Let n_0 be the smallest integer with this property. (The well-ordering principle ensures that n_0 exists.) Similarly, there is an integer m_0 that is the smallest integer such that $g(m_0) \neq 0$.

Consider $(f * g)(n_0 m_0)$. By definition,

$$(f * g)(n_0 m_0) = \sum_{rs = n_0 m_0} f(r)g(s).$$

The terms of this sum fall into three groups: $r < n_0$, $r = n_0$, and $r > n_0$. Note that $r > n_0$ implies $s < m_0$. We can write

$$(f * g)(n_0 m_0) = \sum_{r < n_0} f(r)g(s) + f(n_0)g(m_0) + \sum_{s < m_0} f(r)g(s)$$

where $rs = n_0 m_0$ in each case. The first sum is zero, since $f(r) = 0$ when $r < n_0$. The last sum is zero, since $g(s) = 0$ when $s < m_0$. Therefore

$$(f * g)(n_0 m_0) = f(n_0)g(m_0) \neq 0$$

by hypothesis. This means $f * g \neq z$, which is a contradiction.   ⬡

The ring $(\mathscr{F}, +, *)$ has a multiplicative identity, the function e defined by $e(1) = 1$, $e(n) = 0$ for $n > 1$. We can now ask whether there are any units in the ring—that is, elements possessing multiplicative inverses. This involves a study of the equation $f * g = e$.

EXAMPLE 7.3.2. Let f be the characteristic function of the set of powers of 2; that is, $f(2^k) = 1$ for $k = 0, 1, 2, \ldots$, $f(n) = 0$ otherwise. The equation $f * g = e$ implies

$$f(1)g(1) = 1, \quad \text{and} \quad \sum_{rs = n} f(r)g(s) = 0 \qquad n > 1.$$

From $n = 1$, we get $g(1) = 1$. From $n = 2$, $1 + g(2) = 0$ or $g(2) = -1$. From $n = 3$, $0 + g(3) = 0$; that is, $g(3) = 0$. For $n = 4$, $1 - 1 + g(4) = 0$ implies $g(4) = 0$. In general, we have

$$\sum_{\substack{rs=n \\ s<n}} f(r)g(s) + f(1)g(n) = 0.$$

If we know $g(s)$ for $s < n$, this equation determines $g(n)$ provided that $f(1) \neq 0$. In this case, it seems a reasonable guess that $g(1) = 1$, $g(2) = -1$, and $g(n) = 0$ for $n > 2$. Suppose that we have established this formula for $n < k$. We wish to show that $g(k) = 0$. The formula tells us that

$$g(k) = -\sum_{\substack{rs=k \\ s<k}} f(r)g(s).$$

Since the only nonzero values of g occur for $s = 1$ and $s = 2$, we have at most two terms

$$-f(k)g(1) - f(k/2)g(2).$$

Now if $k = 2^s m$, where m is odd and greater than 1, then $f(k)$ and $f(k/2)$ are both zero, so that the sum is zero. If 2 does not divide k, the term $f(k/2)$ does not occur and $f(k) = 0$, so that the sum is again zero. Now suppose that $k = 2^s$. Then

$$g(k) = -f(2^s)g(1) - f(2^{s-1})g(2) = -1 + 1 = 0.$$

Thus, our guess is correct, and the function g such that $g(1) = 1$, $g(2) = -1$, $g(n) = 0$ for $n > 2$, has the property $f * g = e$.

The particularly easy formula obtained in this example is, of course, the result of the fact that the function f is a rather simple one. Note, however, that what makes it possible to find $g(k)$ for any k is that the coefficient of $g(k)$ in the equation $\sum_{rs=k} f(r)g(s) = 0$ is not zero and that the remaining terms of the sum involve only values of g at integers smaller than k. Since the value of $g(k)$ is defined in terms of values of $g(s)$ for $s < k$, we say that g is defined *recursively*.

THEOREM 7.3.2. *Units in the Ring* $(\mathscr{F}, +, *)$. The function f in $(\mathscr{F}, +, *)$ is a unit if and only if $f(1) \neq 0$. In this case, the multiplicative inverse of f is unique and is defined recursively by the formula

$$f^{-1}(1) = \frac{1}{f(1)}, \quad f^{-1}(n) = -\frac{1}{f(1)} \sum_{\substack{rs=n \\ s<n}} f(r)f^{-1}(s) \qquad \text{if } n > 1.$$

Proof: The function f is a unit if and only if there exists a number-theoretic function g such that $f * g = e$. That is,

$$\sum_{rs=n} f(r)g(s) = 0 \qquad \text{if} \quad n > 1 \quad \text{and} \quad f(1)g(1) = 1.$$

If $f(1) = 0$, there is no real number such that $f(1)g(1) = 1$, and, therefore, $f(1) \neq 0$ is necessary for the existence of the function g.

Suppose that $f(1) \neq 0$. Then

$$g(1) = \frac{1}{f(1)}.$$

For $n = 2$, we obtain

$$f(1)g(2) + f(2)g(1) = 0,$$

that is,

$$g(2) = -\frac{f(2)}{f(1)} g(1).$$

This implies that g is uniquely defined at $n = 2$. Suppose that the value of g is uniquely determined for all integers less than k. Consider $(f * g)(k)$. We obtain

$$\sum_{rs=k} f(r)g(s) = 0,$$

that is,

$$\sum_{\substack{rs=k \\ s<k}} f(r)g(s) + f(1)g(k) = 0.$$

This implies that

$$g(k) = -\frac{1}{f(1)} \sum_{\substack{rs=k \\ s<k}} f(r)g(s).$$

Since every value of s occurring in the summation is less than k, the value of the summation on the right is known, and the function f has a unique inverse. ◆

The calculation of f^{-1} for a particular function f may or may not be a simple process. In the case of multiplicative functions, it can be approached by calculating $f^{-1}(p^k)$ and using the property that the inverse of a multiplicative function is multiplicative. This step will be discussed in the next section. In some cases, the information obtained about the inverses of certain functions can be used to determine the inverses of other functions. A case of particular importance is included in the next theorem.

THEOREM 7.3.3. *The Inverse of u.* The inverse of u is μ.

Proof: This is a direct consequence of Theorem 2.5.2. The statement $\mu(1) = 1$ and $\sum_{d|n}\mu(d) = 0$ if $n > 1$, written in the notation of the present section, becomes $u * \mu = e$, which says that the inverse of u is μ. ●

COROLLARY 7.3.3. The Möbius inversion formula: If $F = u * f$, then $f = \mu * F$.

Proof: Since $F = u * f$, $\mu * F = \mu * u * f = e * f = f$; that is, $f = \mu * F$. ●

EXERCISES 7.3

Checks

1. Calculate $\tau^{-1}(n)$ for $n = 1, 2, 3, 4, 5, 6, 7, 8, 9, 10$.

2. Calculate $\phi^{-1}(n)$ for $n = 1, 2, 3, 4, 5, 6, 7, 8, 9, 10$.

3. Use the commutative and associative properties of $*$ to show that if an inverse exists it must be unique.

4. If f and g have inverses, prove that $f * g$ has an inverse and $(f * g)^{-1} = f^{-1} * g^{-1}$. Does $f + g$ necessarily have an inverse?

5. Prove that $f * f = e$ if and only if $f = \pm e$.

Challenges

6. Calculate $f * g$ if f is the characteristic function of the set of odd integers and g is the characteristic function of the set of even integers.

7. Calculate f^{-1} if f is the characteristic function of the set

$$S = \{p^k, k = 0, 1, 2, \ldots, p \text{ a prime}\}.$$

Calculate f^{-1} if f is the characteristic function of the set

$$S = \{p^j q^k, j, k = 0, 1, 2, \ldots, p \text{ and } q \text{ distinct primes}\}.$$

State a general result of this nature.

8. Let $\mathscr{P}$ be the set of functions f of $\mathscr{F}$ such that $f(1) > 0$. Prove that $(\mathscr{P}, *)$ is a subgroup of the group of units.

9. Let $\mathscr{U}$ be the set of functions f of $\mathscr{F}$ such that $f(1) = 1$. Prove that $(\mathscr{U}, *)$ is a subgroup of the group of units.

10. In the ring $(\mathscr{F}, +, \otimes)$ defined in Exercises 7.2, problem 9, check the counterparts of Theorems 7.3.1 and 7.3.2.

11. Let $\mathscr{I}$ be the set of arithmetic functions for which the range is a subset of Z. Show that $\mathscr{I}$ is closed under $+$ and $*$ and that $(\mathscr{I}, +, *)$ is a subring of $(\mathscr{F}, +, *)$. Prove that the units of $(\mathscr{I}, +, *)$ are the functions in $\mathscr{I}$ for which $f(1) = \pm 1$.   (Jim Lawrence)

7.4 THE SUBGROUP OF MULTIPLICATIVE FUNCTIONS

In the ring $(\mathscr{F}, +, *)$, the units are those functions for which $f(1) \neq 0$. In any ring, the set of units forms a group under the multiplication of the ring. Many of the functions that arose naturally in our explorations have been multiplicative functions. In this section we return to the multiplicative functions and consider which of them are units in $(\mathscr{F}, +, *)$.

THEOREM 7.4.1. *The Multiplicative Units, $\mathscr{M}$*. If f is multiplicative and $f \neq z$, then f is a unit in $(\mathscr{F}, +, *)$.

Proof: Since $f \neq z$, there is an integer n such that $f(n) \neq 0$. If we write $n = n \cdot 1$, the multiplicative property tells us that $f(n) = f(n)f(1)$. Since $f(n) \neq 0$, this implies $f(1) = 1$. Because $f(1) \neq 0$, f is a unit. ●

COROLLARY 7.4.1. If f is multiplicative, $f \neq z$, then $f(1) = 1$.

Not all units are multiplicative functions. For example, the function f which is defined to be 2 for each value of n is a unit in $(\mathscr{F}, +, *)$

but is certainly not multiplicative. However, since the convolution product involves the divisors of n, we expect that the set of multiplicative functions must have some special properties relative to the operation $*$.

EXAMPLE 7.4.1. Is the function $\tau * \sigma$ multiplicative? By definition,

$$(\tau * \sigma)(6) = \tau(1)\sigma(6) + \tau(2)\sigma(3) + \tau(3)\sigma(2) + \tau(6)\sigma(1) = 30,$$
$$(\tau * \sigma)(5) = \tau(1)\sigma(5) + \tau(5)\sigma(1) = 8,$$
$$(\tau * \sigma)(30) = \tau(1)\sigma(30) + \tau(2)\sigma(15) + \tau(3)\sigma(10) + \tau(5)\sigma(6)$$
$$+ \tau(6)\sigma(5) + \tau(10)\sigma(3) + \tau(15)\sigma(2) + \tau(30)\sigma(1) = 240.$$

$(\tau * \sigma)(30) = (\tau * \sigma)(6) \cdot (\tau * \sigma)(5)$ is exactly what we would expect if $\tau * \sigma$ is multiplicative, but it is certainly not enough evidence even to make this conjecture. Let us investigate the result in detail. Recall that the divisors of 30 are of the form $d'd''$, where $d' | 6$ and $d'' | 5$. The following table displays the structure of the divisors.

d'' \ d'	1	2	3	6
1	1	2	3	6
5	5	10	15	30

Each term of the sum for $(\tau * \sigma)(30)$ has the form $\tau(d'd'')\sigma(30/d'd'')$. Since $(6, 5) = 1$, $(d', d'') = 1$. Since $d' | 6$ and $d'' | 5$, we can write

$$\frac{30}{d'd''} = \frac{6}{d'} \cdot \frac{5}{d''}, \quad \text{where} \quad \left(\frac{6}{d'}, \frac{5}{d''}\right) = 1.$$

The multiplicative property of τ and σ allows us to write

$$\tau(d'd'')\sigma\left(\frac{6}{d'} \frac{5}{d''}\right) = \tau(d')\tau(d'')\sigma\left(\frac{6}{d'}\right)\sigma\left(\frac{5}{d''}\right).$$

Thus

$$(\tau * \sigma)(30) = \tau(1)\tau(1)\sigma(6)\sigma(5) + \tau(2)\tau(1)\sigma(3)\sigma(5) + \tau(3)\tau(1)\sigma(2)\sigma(5)$$
$$+ \tau(1)\tau(5)\sigma(6)\sigma(1) + \tau(6)\tau(1)\sigma(1)\sigma(5) + \tau(2)\tau(5)\sigma(3)\sigma(1)$$
$$+ \tau(3)\tau(5)\sigma(2)\sigma(1) + \tau(6)\tau(5)\sigma(1)\sigma(1).$$

If the terms on the right are regrouped, the right-hand side is seen to be the product of

$$[\tau(1)\sigma(6) + \tau(2)\sigma(3) + \tau(3)\sigma(2) + \tau(6)\sigma(1)]$$

and

$$[\tau(1)\sigma(5) + \tau(5)\sigma(1)].$$

That is,

$$(\tau * \sigma)(30) = (\tau * \sigma)(6) \cdot (\tau * \sigma)(5).$$

The structure of the sums in this example gives a guide for the proof in the general situation that the convolution product of two multiplicative functions is a multiplicative function.

THEOREM 7.4.2. *The Multiplicative Property of* $f * g$. If f and g are multiplicative, so is $f * g$.

Proof: Let $(m, n) = 1$. By definition,

$$(f * g)(mn) = \sum_{d \mid mn} f(d)g\left(\frac{mn}{d}\right).$$

The divisors of mn can be written uniquely in the form $d'd''$, where $d' \mid m$ and $d'' \mid n$. (See Section 1.9.) Since $(m, n) = 1$, $(d', d'') = 1$, and also $(m/d', n/d'') = 1$. Thus

$$\sum_{d \mid mn} f(d)g\left(\frac{mn}{d}\right) = \sum_{d' \mid m, \, d'' \mid n} f(d'd'')g\left(\frac{m}{d'}\frac{n}{d''}\right)$$

$$= \sum_{d' \mid m} \sum_{d'' \mid n} f(d')f(d'')g\left(\frac{m}{d'}\right)g\left(\frac{n}{d''}\right).$$

This summation is the product of two summations, so that

$$(f * g)(mn) = \sum_{d' \mid m} f(d')g\left(\frac{m}{d'}\right) \cdot \sum_{d'' \mid n} f(d'')g\left(\frac{n}{d''}\right)$$

$$= (f * g)(m) \cdot (f * g)(n).$$

This proves that $f * g$ is a multiplicative function. ⬢

Theorem 7.4.2 shows that the set of multiplicative functions is closed under the operation $*$. It also gives us a means of calculating easily the convolution product of two multiplicative functions, since it is necessary only to calculate $(f * g)(p^k)$ (see Theorem 1.11.1).

EXAMPLE 7.4.2

$$(\tau * \sigma)(p^k) = \sum_{d \mid p^k} \tau(d)\sigma\left(\frac{p^k}{d}\right) = \sum_{i=0}^{k} \tau(p^i)\sigma(p^{k-i}).$$

But we know that $\tau(p^i) = (i + 1)$ and $\sigma(p^{k-i}) = 1 + p + \cdots + p^{k-i}$, so that

$$(\tau * \sigma)(p^k) = \sum_{i=0}^{k} (1 + i)(1 + p + \cdots + p^{k-i}).$$

If we rearrange this summation and collect terms in powers of p, we obtain

$$(\tau * \sigma)(p^k) = \sum_{i=0}^{k} \frac{(k + 1 - i)(k + 2 - i)}{2} p^i.$$

Now, given any $n = p_1^{k_1} p_2^{k_2} \cdots p_r^{k_r}$, the multiplicative property of $\tau * \sigma$ implies that

$$(\tau * \sigma)(n) = \prod_{j=1}^{r} (\tau * \sigma)(p_j^{k_j}) = \prod_{j=1}^{r} \sum_{i=0}^{k_j} \tfrac{1}{2}(k_j + 1 - i)(k_j + 2 - i)p_j^i.$$

The function $\tau * \sigma$, for which we obtained a formula in Example 7.4.2, is not one of the functions that we studied earlier. In some cases, the result of taking a convolution product may well be a function that is already familiar. This is the case with the function $\phi * \tau$. (See Exercises 7.4, problem 8.)

We have seen that the set of multiplicative functions is closed under convolution product. If we could show that the inverse of a multiplicative function is also multiplicative, it would follow that the set of multiplicative units, $\mathcal{M}$, is a subgroup of the group of units. This result can be proved directly from the definition of the inverse, but it is easier to derive it from the fact that the inverse of a function is unique.

EXAMPLE 7.4.3. What is the inverse of τ? To calculate the value of $\tau^{-1}(p^k)$ from the general formula (Theorem 7.3.2) is not difficult:

$$\tau^{-1}(p^k) = -\sum_{i=0}^{k-1} \tau^{-1}(p^i)\tau(p^{k-i}), \quad \tau^{-1}(1) = 1.$$

Let $k = 1.$ $\tau^{-1}(p) = -\tau(p) = -2.$
Let $k = 2.$ $\tau^{-1}(p^2) = -[\tau(p^2) + \tau^{-1}(p)\tau(p)] = -[3 - 4] = 1.$
Let $k = 3.$ $\tau^{-1}(p^3) = -[4 - 6 + 2] = 0.$
Let $k = 4.$ $\tau^{-1}(p^4) = -[5 - 8 + 3 + 0] = 0.$

Is $\tau^{-1}(p^k) = 0$ if $k > 2$? This conjecture can be established by induction. Now construct the multiplicative function g by the formula

$$g(1) = 1, \quad g(n) = \prod_{p|n} \tau^{-1}(p^k) \quad \text{if} \quad n = \prod_{p|n} p^k.$$

From the definition, g is clearly multiplicative. Is $\tau * g = e$? Since τ and g are multiplicative, $\tau * g$ is multiplicative (Theorem 7.4.2). Therefore

$$(\tau * g)(n) = \prod_{p|n} (\tau * g)(p^k) \quad \text{when} \quad n = \prod_{p|n} p^k.$$

But by the construction of g, $(\tau * g)(p^k) = e(p^k)$. Therefore $\tau * g = e$ and the multiplicative function g is actually τ^{-1}. Since g is multiplicative, τ^{-1} must also be multiplicative.

THEOREM 7.4.3. *Inverses of Multiplicative Functions.* If f is a multiplicative function and $f \neq z$, the inverse of f is the multiplicative function g defined by

$$g(1) = 1, \quad g(n) = \prod_{p|n} g(p^k) \quad \text{for} \quad n = \prod_{p|n} p^k,$$

where

$$g(p^k) = -\sum_{i=0}^{k-1} f(p^{k-i})g(p^i).$$

Proof: The function g has the property that $(f * g)(p^k) = 0$ if $k > 0$, since

$$(f * g)(p^k) = \sum_{i=0}^{k} f(p^{k-i})g(p^i) = g(p^k) + \sum_{i=0}^{k-1} f(p^{k-i})g(p^i) = 0.$$

The function g is multiplicative. Let $(m, n) = 1$. Then

$$m = p_1^{a_1} p_2^{a_2} \cdots p_r^{a_r}$$

and

$$n = q_1^{b_1} q_2^{b_2} \cdots q_s^{b_s},$$

where $p_i \neq q_j$ for any i, j. Therefore

$$g(mn) = g(p_1^{a_1})g(p_2^{a_2}) \cdots g(p_r^{a_r})g(q_1^{b_1})g(q_2^{b_2}) \cdots g(q_s^{b_s})$$
$$= g(m)g(n).$$

Since f and g are multiplicative, so is $f * g$, and

$$(f * g)(n) = \prod_{p|n} (f * g)(p^k) = \prod_{p|n} e(p^k) = e(n).$$

Since the inverse of f is unique, this implies that $g = f^{-1}$ and that f^{-1} is multiplicative. ●

The preceding theorem makes it easier to calculate the inverse of multiplicative functions, although even in this case the inverse is defined recursively. We have already seen that $u^{-1} = \mu$. Some additional examples are given below and in Exercises 7.4.

EXAMPLE 7.4.4. The function u_1 is defined by: $u_1(n) = n$ for every n. What is u_1^{-1}? We first find $u_1^{-1}(p^k)$, for $k = 1, 2, 3, \ldots$. We have

$$u_1^{-1}(p) = -u_1(p)u_1^{-1}(1) = -p,$$
$$u_1^{-1}(p^2) = -u_1(p^2)u_1^{-1}(1) - u_1(p)u_1^{-1}(p) = -p^2 - p(-p) = 0,$$
$$u_1^{-1}(p^3) = -u_1(p^3)u_1^{-1}(1) - u_1(p^2)u_1^{-1}(p) - u_1(p)u_1^{-1}(p^2)$$
$$= -p^3 + p^3 = 0.$$

Is $u_1^{-1}(p^k) = 0$ if $k > 1$? This can be established by a simple induction argument. We can now write

$$u_1^{-1}(n) = \prod_{p|n} (-p)$$

if n is square free, $u_1^{-1}(n) = 0$ otherwise. This can also be written $u_1^{-1}(n) = n\mu(n)$.

EXAMPLE 7.4.5. Find a formula for $\sigma^{-1}(n)$. Again we begin by calculating $\sigma^{-1}(p^k)$. We obtain

$$\sigma^{-1}(p) = -(1 + p), \quad \sigma^{-1}(p^2) = p, \quad \sigma^{-1}(p^3) = 0, \ldots.$$

The fact that $\sigma^{-1}(p^k) = 0$ for $k > 2$, is easily established by induction when we note that

$$\sigma^{-1}(p^k) = \sigma^{-1}(p^{k-1}) + p^{k-2}[-p^2 + p(1 + p) - p].$$

Again,

$$\sigma^{-1}(n) = \prod_{p|n} \sigma^{-1}(p^k), \qquad \text{when } n = \prod_{p|n} p^k.$$

EXAMPLE 7.4.6. Since $\sigma(n) = \sum_{d|n} d$, we can write $\sigma = u * u_1$. This relation can be used, along with Example 7.4.4, to establish the result of Example 7.4.5. We have

$$\sigma^{-1} = u^{-1} * u_1^{-1} = \mu * u_1^{-1}.$$

As we have seen (Exercises 7.2, problem 5), convolution products involving μ are especially easy to evaluate. We have

$$(\mu * u_1^{-1})(p^k) = u_1^{-1}(p^k) - u_1^{-1}(p^{k-1})$$

from which the result of Example 7.4.5 follows.

EXERCISES 7.4

Checks

1. Calculate $(\phi * \tau)(24)$, $(\phi * \tau)(8)$, and $(\phi * \tau)(3)$ as in Example 7.4.1.

2. Show that

$$\sum_{i=0}^{k} (1 + i)(1 + p + \cdots + p^{k-i}) = \sum_{i=0}^{k} \frac{(k + 1 - i)(k + 2 - i)}{2} p^i.$$

    Recall the formula

$$1 + 2 + \cdots + n = \frac{n(n + 1)}{2}.$$

3. Carry out the induction argument in Example 7.4.4.

4. Find an expression for f^{-1} if f is the characteristic function of the set of odd integers.

5. Write τ, σ, σ_k as convolutions, using the set of functions u, u_1, u_k, where $u(n) = 1$ for every n and $u_k(n) = n^k$ for every n, $k = 1, 2, \ldots$.

6. Calculate τ^{-1} from $\tau = u * u$.

7. Prove that the groups $(\mathcal{M}, *)$, $(\mathcal{U}, *)$, $(\mathcal{P}, *)$ are each proper subgroups of the group $(\mathcal{E}, *)$, where $\mathcal{E}$ represents the set of units in $\mathcal{F}$.

Challenges

8. Prove that $\phi * \tau = \sigma$. (Compare Exercises 7.2, problems 4 and 5.)

9. Write ϕ as a convolution of μ and a u-function.

10. Prove that $u_k^{-1}(n) = \mu(n)n^k$.

11. Find a formula for $\sigma_k^{-1}(p^s)$.

12. Derive the results of problem 11 from problem 10 and the fact that $\sigma_k = u * u_k$.

13. If $f * g$ is multiplicative and f is multiplicative, prove that g is multiplicative. Is it possible for $f * g$ to be multiplicative without f and g being multiplicative? In this problem $f \neq z$.

7.5   AN OPERATOR ON THE SET $\mathcal{P}$

During the past decade, many people have investigated the set of number-theoretic functions and its structure from various points of view. A selection of articles is listed in the references. One of the most interesting suggestions is included in this section, as representative of the exploration that might be undertaken in this area with little in the way of tools except that most important one, a creative mind. By defining an operation similar to the logarithm of a positive number, David Rearick [8] has demonstrated an isomorphism between the group $(\mathcal{P}, *)$ and the additive group of the ring of number-theoretic functions. In this mapping, the group $(\mathcal{M}, *)$ maps into the subgroup of the additive group, which consists of those functions f that are zero when n is not a power of a prime. As we did in Unit 4, we use the notation log a to mean the natural logarithm of a—that is, the base used is the transcendental number e.

$$\boxed{Lf}$$

DEFINITION 7.5.1. *The Logarithm Operator L.* If f is a number-theoretic function for which $f(1) > 0$, then Lf is defined as follows:

$$Lf(1) = \log f(1),$$

$$Lf(n) = \sum_{d|n} f(d) f^{-1}\left(\frac{n}{d}\right) \log d, \quad n > 1.$$

EXAMPLE 7.5.1. We have already seen that if f is the characteristic function of the set $S = \{2^k, k = 0, 1, 2, \ldots\}$, then $f^{-1}(1) = 1, f^{-1}(2) = -1$, and $f^{-1}(n) = 0$ for $n > 2$. For this function, the calculation of Lf is rather simple.

$$Lf(1) = 0 \quad \text{and} \quad Lf(n) = \sum_{d|n} f(d) f^{-1}\left(\frac{n}{d}\right) \log d.$$

Since $f(d) = 0$ unless $d = 2^k$, it is immediate that we need consider only $n = 2^k m$. But $f^{-1}(n/d) = 0$ unless $n/d = 1$ or 2. Thus the only n for which nonzero terms can occur in the sum is $n = 2^k$. In this case,

$$Lf(2^k) = f(2^k) f^{-1}(1) \log 2^k + f(2^{k-1}) f^{-1}(2) \log 2^{k-1} + 0$$
$$= \log 2^k - \log 2^{k-1} = \log 2.$$

Thus $Lf(2^k) = \log 2$, $k = 1, 2, 3, \ldots$; $Lf(n) = 0$ otherwise.

EXAMPLE 7.5.2. By analogy with the logarithm function for real numbers, we might expect a simple result for the logarithm of the multiplicative identity. Since $e^{-1} = e$, the definition of Lf gives

$$Le(1) = \log 1 = 0, \quad Le(n) = \sum_{d|n} e(d) e\left(\frac{n}{d}\right) \log d.$$

But every term in this summation is zero, since $e(n) = 0$ for $n > 1$, so that $Le(n) = 0$ for $n = 1, 2, 3, \ldots$; that is, $Le = z$. The operator L is not defined on the function z, since z does not belong to $\mathscr{P}$.

The summation that defines Lf is so similar to a convolution product that it seems desirable to introduce a notation that would make it possible to write it using a convolution product.

$$\boxed{Df}$$

DEFINITION 7.5.2. *The Operator D.* For any number-theoretic function f, the function Df is defined as follows:

$$Df(n) = f(n) \log n, \quad n \geq 1.$$

The notation Df suggests the derivative of elementary calculus. Surprisingly, the operation that Df describes has many properties in common with the derivative, as the following theorem shows.

THEOREM 7.5.1. *The Properties of D.* The operation D has the following properties:
1. $D(f + g) = Df + Dg$.
2. $D(af) = aDf$, for any real number a.
3. $De = z$.
4. $D(f * g) = (f * Dg) + (Df * g)$.

Proof: Properties (1), (2), and (3) follow immediately from the definition; their verification is left as an exercise.

To establish (4), we consider $(f * Dg)(n)$ and $(Df * g)(n)$.

$$(f * Dg)(n) = \sum_{d|n} f(d)g\left(\frac{n}{d}\right) \log \left(\frac{n}{d}\right)$$

$$= \sum_{d|n} f(d)g\left(\frac{n}{d}\right) \log n - \sum_{d|n} f(d)g\left(\frac{n}{d}\right) \log d.$$

$$(Df * g)(n) = \sum_{d|n} [f(d) \log d]g\left(\frac{n}{d}\right).$$

It is now clear that $(f * Dg)(n) + (Df * g)(n) = (f * g)(n) \log n$, which is equal to $D(f * g)(n)$.   ●

The definition of the operator L, expressed in terms of the operator D and a convolution product, becomes

$$Lf(n) = (f^{-1} * Df)(n) \qquad \text{when } n > 1.$$

Note that $Lf(1) \neq (f^{-1} * Df)(1)$, in general, since $Lf(1) = \log f(1)$ and $(f^{-1} * Df)(1) = f^{-1}(1)Df(1) = 0$. However, if $f(1) = 1$, $\log f(1) = 0$. For these functions, $Lf = f^{-1} * Df$. In particular, if f is multiplicative, $f(1) = 1$ and $Lf = f^{-1} * Df$.

What properties would we expect the operator L to possess? We have seen in Example 7.5.2 that $Le = z$. One of the best-known properties of the log function for real numbers is $\log ab = \log a + \log b$. By analogy, we might hope that $L(f * g) = Lf + Lg$.

EXAMPLE 7.5.3. Let f be the characteristic function of the set $S = \{2^k, k = 0, 1, 2, \ldots\}$. We know (Example 7.3.2) that the inverse of f is the function g defined by $g(1) = 1$, $g(2) = -1$, $g(n) = 0$, $n > 2$. Thus we have $f * g = e$. Is $Lf + Lg = Le$? We have calculated Lf in Example 7.5.1. In the same way, we see that $Lg(1) = 0$, $Lg(n) = 0$ unless $n = 2^k$, and

$$Lg(2^k) = g(2^k)g^{-1}(1) \log 2^k + \cdots + g(2)g^{-1}(2^{k-1}) \log 2 + g(1)g^{-1}(2^k) \log 1.$$

Since $g^{-1}(2^i) = f(2^i) = 1$, this becomes $Lg(2^k) = -\log 2$. Thus

$$Lf(n) + Lg(n) = 0 \quad \text{and} \quad Lf + Lg = z = Le.$$

THEOREM 7.5.2. *The Logarithm of a Product.* For all f and g in $\mathscr{P}$, $L(f * g) = Lf + Lg$.

Proof:

$$L(f * g)(1) = \log\,[(f * g)(1)] = \log f(1)g(1) = \log f(1) + \log g(1)$$
$$= Lf(1) + Lg(1).$$

For $n \neq 1$, $Lf(n) = (f^{-1} * Df)(n)$. Recall that $(f * g)^{-1} = f^{-1} * g^{-1}$. Then, with the exception of $n = 1$, we have

$$
\begin{aligned}
L(f * g) &= (f * g)^{-1} * D(f * g) \\
&= (f^{-1} * g^{-1}) * [f * Dg + Df * g] \\
&= (f^{-1} * g^{-1}) * (f * Dg) + (f^{-1} * g^{-1}) * (g * Df) \\
&= (f^{-1} * f * g^{-1} * Dg) + (g^{-1} * g * f^{-1} * Df) \\
&= (g^{-1} * Dg) + (f^{-1} * Df) \\
&= Lg + Lf.
\end{aligned}
$$

In this development we used the properties of D, the distributive property of $*$ over $+$, and the commutative and associative properties of $*$. ●

The preceding theorem gives us some information about the mapping defined by L. We see that this operation maps the functions in $\mathscr{P}$

into the set of functions $\mathscr{F}$ in such a way that $L(f * g) = Lf + Lg$. The multiplicative identity in $(\mathscr{P}, *)$ maps into the additive identity in $(\mathscr{F}, +)$. Thus L is a homomorphism from $(\mathscr{P}, *)$ into $(\mathscr{F}, +)$. To see whether L is an isomorphism, we must see whether the mapping is *onto*; that is, given an arithmetic function h, is there a function f in $\mathscr{P}$ such that $h = Lf$? Also, we must show that f is uniquely defined when h is given, so that the mapping is one to one.

EXAMPLE 7.5.4. How would we look for f such that $Lf = h$, where h is a known function? Let us experiment with a function for which we know the answer. Define h as follows: $h(1) = 0$, $h(2^k) = \log 2$ for $k = 1, 2, 3, \ldots$, $h(n) = 0$ otherwise. (Compare with Example 7.5.1.) To find f we must use the equation

$$h(n) = \sum_{d \mid n} f(d) f^{-1}\left(\frac{n}{d}\right) \log d, \quad \text{if} \quad n > 1, \quad h(1) = \log f(1).$$

Since $h(1) = 0$, we see that $\log f(1) = 0$ and $f(1) = 1$. Take $n = 2$;

$$f(2) f^{-1}(1) \log 2 + f(1) f^{-1}(2) \log 1 = h(2).$$

Here we can calculate $f^{-1}(1)$ because we have already found $f(1)$. We do not need $f^{-1}(2)$, since $\log 1 = 0$ and this term is zero. Thus $f(2) \log 2 = \log 2$, which implies $f(2) = 1$.

Take $n = 3$; $f(3) f^{-1}(1) \log 3 + 0 = h(3)$ implies $f(3) = 0$.

Take $n = 4$; $f(4) f^{-1}(1) \log 4 + f(2) f^{-1}(2) \log 2 + 0 = h(4)$. Since we know $f(1)$ and $f(2)$, we can determine $f^{-1}(2)$, which is -1 in this case. This gives $f(4) \log 4 - \log 2 = \log 2$; that is, $f(4) = 1$. Suppose that we have already determined f for $n < m$; that is, $f(n) = 1$ if $n = 2^k$, and $f(n) = 0$ otherwise. From these values of f, we can determine $f^{-1}(n)$ for $n < m$. We find $f^{-1}(1) = 1$, $f^{-1}(2) = -1$, $f^{-1}(n) = 0$ otherwise. Now consider $n = m$. We have

$$h(m) = \sum_{d \mid m} f(d) f^{-1}\left(\frac{m}{d}\right) \log d.$$

The term involving $f^{-1}(m)$ is not present, since $\log 1 = 0$. The other values of f and f^{-1} are known. In this case, all terms are zero unless $m = 2^k$. For this m, we have

$$\log 2 = f(2^k) \log 2^k - f(2^{k-1}) \log 2^{k-1};$$

that is,

$$\log 2 = f(2^k) \log 2^k - \log 2^{k-1},$$

which implies $f(2^k) = 1$. As anticipated for the h defined in this example, if $Lf = h, f$ is the characteristic function of the set $S = \{2^k, k = 0, 1, 2...\}$.

In a particular case, the calculations may be quite cumbersome. However, in principle, the situation is clear and the values of $f(n)$ can be determined uniquely by the process outlined in Example 7.5.4.

THEOREM 7.5.3. *The Inverse of the Operator L.* For each h in $\mathscr{F}$, there is a unique f in $\mathscr{P}$ such that $h = Lf$.

Proof: Since $h(1) = \log f(1)$, $f(1) = e^{h(1)}$. Assume that $f(k)$ has been determined for all $k < n$. The values of $f^{-1}(k)$ can be determined for all $k < n$ from the formula

$$\sum_{rs=k} f^{-1}(r)f(s) = e(k).$$

The value of $f(n)$ can be found from the equation

$$h(n) = \sum_{uv=n} f(u)f^{-1}(v) \log u.$$

Here $h(n)$ is known, $f^{-1}(v)$ is known for $v < n$, and $f(u)$ is known for $u < n$. The value of $f^{-1}(n)$ is not needed, since it is multiplied by $\log 1$, which is zero. The value of $f(n)$ can be determined uniquely from this equation, since the coefficient of $f(n)$ is $f^{-1}(1) \log n$, which is not zero.
    Since $e^{h(1)} \neq 0$, the function f so determined is in $\mathscr{P}$. ⬢

Theorem 7.5.3 gives the additional information necessary to conclude that the operator L defines an isomorphism between the group $(\mathscr{P}, *)$ and the group $(\mathscr{F}, +)$. Since the subgroup $(\mathscr{M}, *)$ is one of particular interest, it is desirable to investigate the image of this subgroup under the isomorphism L.

EXAMPLE 7.5.5. A simple multiplicative function is the function u, the inverse of which is μ. Because u is multiplicative,

$$Lu = u^{-1} * Du = \mu * Du.$$

Thus

$$Lu(p^k) = Du(p^k) - Du(p^{k-1}) = \log p^k - \log p^{k-1} = \log p.$$

What about $Lu(n)$ if n is not a power of a prime? Since $Lu(1) = 0$, we know that Lu cannot be multiplicative; thus we are not able to write the values of $Lu(n)$ in terms of $Lu(p^k)$ as we did in Theorem 1.11.1. Consider $Lu(n)$, where $n = pq$, the product of two distinct primes.

$$Lu(pq) = \mu(1) \log pq + \mu(p) \log q + \mu(q) \log p = 0.$$

In general, if n is composite and not a power of a prime, there exist integers a and b, $(a, b) = 1$, a and b both greater than 1, such that $n = ab$. The divisors of n can be written in the form $d = d'd''$ where $d'\,|\,a$ and $d''\,|\,b$. Also,

$$(d', d'') = 1 \quad \text{and} \quad \left(\frac{a}{d'}, \frac{b}{d''}\right) = 1.$$

Then

$$Lu(n) = \sum_{d\,|\,ab} u(d)\mu\left(\frac{ab}{d}\right) \log d$$

$$= \sum_{d'\,|\,a,\, d''\,|\,b} u(d'd'')\mu\left(\frac{a}{d'}\frac{b}{d''}\right) \log (d'd'')$$

$$= \sum_{d'\,|\,a} \sum_{d''\,|\,b} u(d')u(d'')\mu\left(\frac{a}{d'}\right)\mu\left(\frac{b}{d''}\right)[\log d' + \log d'']$$

$$= \sum_{d'\,|\,a} u(d')\mu\left(\frac{a}{d'}\right) \log d' \sum_{d''\,|\,b} u(d'')\mu\left(\frac{b}{d''}\right)$$

$$+ \sum_{d'\,|\,a} u(d')\mu\left(\frac{a}{d'}\right) \sum_{d''\,|\,b} u(d'')\mu\left(\frac{b}{d''}\right) \log d''$$

$$= [(\mu * Du)(a)][(\mu * u)(b)] + [(\mu * u)(a)][(\mu * Du)(b)]$$

$$= Lu(a)e(b) + e(a)Lu(b)$$

$$= 0,$$

since $e(a)$ and $e(b)$ are both zero. We now see that $Lu(1) = 0$, $Lu(p^k) = \log p$, $Lu(n) = 0$ if n is not a power of a prime. This function is called the von Mangoldt function and is designated by Λ. It is used in the study of the prime number theorem.

Note that in the discussion of Lu for composite n in the preceding example, no use was made of special properties of μ and u. The only property needed was the multiplicative property of the function and its inverse. This fact leads to the conjecture that Lf for any multiplicative function is zero if $n \neq p^k$. This is, in fact, the property that characterizes the set $\{Lf, f \text{ multiplicative}\}$. In the following theorem, n is a power of a prime if $n = p^k$, $k \geq 1$. Thus 1 is not a power of a prime.

THEOREM 7.5.4. *The Image of $\mathcal{M}$ under L.* A function f of $\mathcal{P}$ is multiplicative if and only if $Lf(n) = 0$ whenever n is not a power of a prime.

Proof: Suppose that f belongs to $\mathcal{M}$. Then $f \neq z$ and $f(1) = 1$, so that $Lf(1) = 0$. Let n be an integer not a power of a prime. There exist integers a and b greater than 1 and relatively prime such that $n = ab$. The divisors of n are of the form $d'd''$, where $d'|a$, $d''|b$, so that $(d', d'') = 1$ and $(a/d', b/d'') = 1$. Since f is multiplicative, f^{-1} is also multiplicative.

$$
\begin{aligned}
Lf(n) &= \sum_{d|ab} f(d)f^{-1}\left(\frac{ab}{d}\right) \log d \\
&= \sum_{d'|a,\, d''|b} f(d'd'')f^{-1}\left(\frac{a}{d'}\frac{b}{d''}\right) \log d'd'' \\
&= \sum_{d'|a} \sum_{d''|b} f(d')f(d'')f^{-1}\left(\frac{a}{d'}\right)f^{-1}\left(\frac{b}{d''}\right)[\log d' + \log d''] \\
&= Lf(a)e(b) + Lf(b)e(a) \\
&= 0,
\end{aligned}
$$

since both a and b are greater than 1.

Now suppose that $Lf = h$, where $h = 0$ when n is not a power of a prime. We wish to prove that f is multiplicative. We can do so by creating a function g that is multiplicative and that has the property $Lg = h$. Then, since the inverse operation defines a unique function, we can conclude that $f = g$ and hence that f is multiplicative. We first calculate $f(p^k)$ from the equation $Lf(p^k) = h(p^k)$; that is,

$$
h(p^k) = \sum_{i=1}^{k} f(p^i)f^{-1}(p^{k-i}) \log p^i.
$$

In this equation $h(p^k)$ is known, the coefficient of $f(p^k)$ is $f^{-1}(1) \log p^k$, which is not zero, and the values of $f(p^i)$ and $f^{-1}(p^i)$ are determined recursively, $i = 0, 1, 2, \ldots, k - 1$. The term $i = 0$ in the sum is not present, since $\log 1 = 0$.

Now define $g(p^k) = f(p^k)$, $k = 0, 1, 2, \ldots$. It is clear that this definition also implies $g^{-1}(p^k) = f^{-1}(p^k)$, since the value of $f^{-1}(p^k)$ depends only on the values of $f(p^j)$, $j \leq k$. Since

$$f(p^k) = g(p^k) \quad \text{and} \quad f^{-1}(p^k) = g^{-1}(p^k),$$

we have

$$Lf(p^k) = Lg(p^k) = h(p^k).$$

We can now obtain a multiplicative function g by the definition

$$g(n) = g(p_1^{a_1})g(p_2^{a_2})\cdots g(p_r^{a_r}) \quad \text{for} \quad n = p_1^{a_1}p_2^{a_2}\cdots p_r^{a_r}.$$

It follows that $Lg(n) = 0$ if n is not a power of a prime, by the proof of the first half of this theorem. Thus we have created a function g that is multiplicative and that has the property $Lg = h$. Because the inverse operation is unique, f and g must be identical and f must be multiplicative. ●

The task of calculating Lf for a multiplicative function is much simpler now that we know the nature of Lf when f is multiplicative. We have only to calculate $Lf(p^k)$.

EXAMPLE 7.5.6. Calculate $L\tau$.

In Example 7.4.3 we saw that

$$\tau^{-1}(p) = -2, \quad \tau^{-1}(p^2) = 1, \quad \tau^{-1}(p^k) = 0, \qquad k > 2.$$

Since $L\tau = \tau^{-1} * D\tau$, we obtain

$$L\tau(1) = 0,$$
$$L\tau(p) = 2 \log p,$$
$$L\tau(p^2) = 3 \log p^2 - 4 \log p = 2 \log p,$$
$$L\tau(p^3) = 4 \log p^3 - 6 \log p^2 + 2 \log p = 2 \log p.$$

In general, when $k > 2$,

$$L\tau(p^k) = (k + 1)k \log p - 2k(k - 1) \log p + (k - 1)(k - 2) \log p$$
$$= 2 \log p.$$

$L\tau(n) = 0$ if n is not a power of a prime.

This result also follows rather more quickly from the relation $\tau = u * u$.

$$L\tau = Lu + Lu = \Lambda + \Lambda,$$

from Example 7.5.5.

EXERCISES 7.5

Checks

1. Calculate $Lu(p_1^2 p_2)$, using Definition 7.5.1.

2. Verify properties (1), (2), and (3) of the operator D.

3. Prove that $Df(mn) = f(n)Df(m) + f(m)Df(n)$ if f is multiplicative and $(m, n) = 1$.

4. Prove that the set $\{f: f$ in $\mathscr{F}$ and $f(n) = 0$ when n is not a power of a prime$\}$ is a subgroup of the group $(\mathscr{F}, +)$.

5. Calculate $Lu_1(p^k)$, $Lu_2(p^k)$. Can you predict $Lu_s(p^k)$?

Challenges

6. Identify the functions $L\sigma$ and $L\phi$.

7. Prove that $D(f^{-1}) = -(f^{-1} * f^{-1}) * Df$.

8. Define the *operator* E as follows: if h belongs to $\mathscr{F}$, $Eh = f$ if and only if $Lf = h$. Prove that $E(f + g) = Ef * Eg$ for any function f and g in $\mathscr{F}$.

9. If r is an integer, we define f^r by $f * f * \cdots * f^r$ where there are r factors of f. Show that $f^r = E(rLf)$.

10. For any real number r, define $f^r = E(rLf)$. Prove that

$$(f^r)^s = f^{rs}, \quad f^{r+s} = f^r * f^s, \quad (f * g)^r = f^r * g^r.$$

11. If f is multiplicative, prove that f^r is multiplicative for every real number r. *Hint*: Use Theorem 7.5.4.

12. Consider the Dirichlet series

$$\sum_{n=1}^{\infty} \frac{f(n)}{n^s}.$$

Show that the series obtained by formally differentiating this Dirichlet series, with respect to s, term by term is

$$-\sum_{n=1}^{\infty} \frac{Df(n)}{n^s}.$$

REFERENCES FOR UNIT 7

1. Apostol, T. M., "Some Properties of Completely Multiplicative Arithmetical Functions," *American Mathematical Monthly*, Vol. 78 (1971), pp. 266–271.
2. Carlitz, L., "Rings of Arithmetic Functions," *Pacific Journal of Mathematics*, Vol. 14, No. 4 (Winter, 1964), pp. 1165–1171.
3. Cashwell, E. D., and C. J. Everett, "The Ring of Number-Theoretic Functions," *Pacific Journal of Mathematics*, Vol. 9 (1959), pp. 975–985.
4. Chawla, L. M., "On a Pair of Arithmetic Functions," *Journal of Natural Sciences and Mathematics (Lahore)* Vol. 8 (1968), pp. 262–269.
5. Cohen, E., "Some Totient Functions," *Duke Mathematical Journal*, Vol. 23 (1956), pp. 515–522.
6. Klee, V. L., Jr., "A Generalization of Euler's ϕ Function," *American Mathematical Monthly*, Vol. 55 (1948), pp. 358–359.
7. Le Van, M. O., "An Induction Formula for $q(n)$," *Journal of Natural Sciences and Mathematics (Lahore)* Vol. 9 (1969), pp. 47–49.
8. Rearick, D., "Operators on Algebras of Arithmetic Functions," *Duke Mathematical Journal*, Vol. 35 (1968), pp. 761–766.
9. Rearick, D., "The Trigonometry of Numbers," *Duke Mathematical Journal*, Vol. 35 (1968), pp. 767–776.

THE p-ADIC INTEGERS—
JOURNEY IN SPACE

8.1 p-ADIC SEQUENCES

The fundamental theorem of arithmetic says that an integer can be written uniquely as a product of primes in a specified order. If we focus attention on a particular prime p, this theorem leads directly to the conclusion that every integer, except zero, can be expressed uniquely in the form $p^k n_0$, where $k \geq 0$, and n_0 is a positive or negative integer not divisible by p. In Section 1.12 a number-theoretic function called the p-value of n was defined, based on this representation of the integer n. With the ideas of congruence, we can look at this representation in more detail. Since k is the highest power of p that divides the integer n, we have

$$n \equiv 0 \bmod p^i, \qquad i = 1, 2, 3, \ldots, k,$$

263

and

$$n \not\equiv 0 \bmod p^i, \qquad i \geq k + 1.$$

Let us examine the sequence of residue classes to which a given integer belongs mod p^i, $i = 1, 2, 3, \ldots$.

EXAMPLE 8.1.1. Let $p = 3$. For convenience, use the least non-negative element to represent the residue class mod 3^i to which the integer a belongs. The table shown summarizes this information for $a = 33$, 60, 9, -54, -48.

	$a =$	33	60	9	-54	-48
$a \equiv a_0 \bmod 3$	$a_0 =$	0	0	0	0	0
$a \equiv a_1 \bmod 3^2$	$a_1 =$	6	6	0	0	6
$a \equiv a_2 \bmod 3^3$	$a_2 =$	6	6	9	0	6
$a \equiv a_3 \bmod 3^4$	$a_3 =$	33	60	9	27	33
$a \equiv a_4 \bmod 3^5$	$a_4 =$	33	60	9	189	195
$\vdots$	$\vdots$	$\vdots$	$\vdots$	$\vdots$	$\vdots$	$\vdots$
$a \equiv a_i \bmod 3^{i+1}$	$a_i =$	33	60	9	$-54 + 3^{i+1}$	$-48 + 3^{i+1}$
$\vdots$	$\vdots$	$\vdots$	$\vdots$	$\vdots$	$\vdots$	$\vdots$

Examine the table vertically. In any given column, to what residue class do all the entries belong? To what residue class do all the entries below a_2 belong? We can make the general statement that $a_i \equiv a_{i+1} \bmod 3^{i+1}$ for each i, indeed that $a_i \equiv a_{i+n} \bmod 3^{i+n}$ for $n \geq 1$. This statement simply reflects the fact that each residue class mod 3^i is the union of three residue classes mod 3^{i+1}, each of which is in turn the union of three residue classes mod 3^{i+2} (see Section 2.1).

Now examine the table horizontally. The five integers 33, 60, 9, -54, -48 all belong to the same residue class mod 3. For what values of i are 33 and 60 in the same residue class mod 3^i? What is the smallest i such that 33 and 60 are in distinct residue classes mod 3^i? How is this value of i related to $60 - 33$?

The diagram shown represents the situation described in the table. Here an interval is used to represent a residue class.

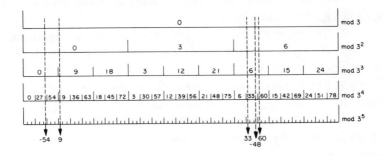

If we associate with the integer 33 the sequence 0, 6, 6, 33, 33, ... and with the integer -48 the sequence 0, 6, 6, 33, 195, ..., $-48 + 3^{n+1}$, ... we are associating with each integer a nested sequence of residue classes mod p^n. Before discussing the general situation, we note two things about these sequences. Let $a_0, a_1, a_2, ..., a_n, ...$ designate the sequence. We have $a_{n-1} \equiv a_n$ mod p^n. The second observation is that the choice of integers to represent the residue classes is not unique. For convenience, in the table of Example 8.1.1, we used the least non-negative element, but instead of $-54 \equiv 27$ mod 81, we might have written $-54 \equiv -135$ mod 81, or $-54 \equiv 109$ mod 81, or $-54 \equiv -54$ mod 81. All possible choices must be congruent to -54 and hence congruent to each other mod 81.

DEFINITION 8.1.1. *p-adic Sequences*. Let p be a fixed prime. A sequence of integers $\langle x_n \rangle$ with the property that $x_{n-1} \equiv x_n$ mod p^n, $n = 1, 2, 3, ...$, is called a *p-adic sequence*. Two *p*-adic sequences $\langle x_n \rangle$ and $\langle y_n \rangle$ are equivalent, written $\langle x_n \rangle = \langle y_n \rangle$, if for every n, $x_n \equiv y_n$ mod p^{n+1}. If there exists an integer b such that for every n, $b \equiv x_n$ mod p^{n+1}, then the *p*-adic sequence $\langle x_n \rangle$ is *associated with* the integer b.

In this unit, our exploration will center on sets of equivalent *p*-adic sequences. The letter p will represent a fixed prime throughout. From the sets of sequences, we will construct the ring of *p*-adic integers. The investigation of the properties of the *p*-adic integers will take us outside our accustomed Euclidean space into a non-Archimedean metric space and will provide some interesting and perhaps surprising insight into the structure of mathematical systems.

THEOREM 8.1.1. *Sums and Products of p-adic Sequences.* Let $\langle x_n \rangle$ and $\langle y_n \rangle$ be *p*-adic sequences.

1. The sequences $\langle x_n + y_n \rangle$ and $\langle x_n y_n \rangle$ are *p*-adic sequences.

2. If $\langle x'_n \rangle = \langle x_n \rangle$ and $\langle y'_n \rangle = \langle y_n \rangle$, then $\langle x'_n + y'_n \rangle = \langle x_n + y_n \rangle$ and $\langle x'_n y'_n \rangle = \langle x_n y_n \rangle$.

3. There exists a *p*-adic sequence associated with b, where b is an integer. If $\langle x_n \rangle$ and $\langle y_n \rangle$ are both associated with b, then $\langle x_n \rangle = \langle y_n \rangle$.

4. If $\langle x_n \rangle$ is associated with the integer a and $\langle y_n \rangle$ is associated with the integer b, then $\langle x_n + y_n \rangle$ is associated with $a + b$, and $\langle x_n y_n \rangle$ is associated with ab.

Proof: The statements of the theorem follow immediately from the properties of congruence. For example, $x_{n-1} \equiv x_n \bmod p^n$ and $y_{n-1} \equiv y_n \bmod p^n$ imply $x_{n-1} + y_{n-1} \equiv x_n + y_n \bmod p^n$, yielding $\langle x_n + y_n \rangle$ as a *p*-adic sequence. Also, $x'_n \equiv x_n \bmod p^{n+1}$ and $y'_n \equiv y_n \bmod p^{n+1}$ imply $x'_n + y'_n \equiv x_n + y_n \bmod p^{n+1}$. Therefore if $\langle x'_n \rangle = \langle x_n \rangle$ and $\langle y'_n \rangle = \langle y_n \rangle$, then $\langle x'_n + y'_n \rangle = \langle x_n + y_n \rangle$. ◆

EXERCISES 8.1

Checks

1. Write three *p*-adic sequences associated with 534, if $p = 3$. Locate 534 on the diagram in Example 8.1.1.

2. Write three *p*-adic sequences associated with -89, if $p = 3$.

3. Show that $\langle x_n \rangle = 0, 0, 0, 27, 108, 108, 837, 3024, 3024, \ldots$ is a 3-adic sequence associated with 3024, and that $\langle y_n \rangle = 0, 0, 0, 27, 189, 675, \ldots, -54 + 3^{n+1}, \ldots$ is a 3-adic sequence associated with -54. Calculate $\langle x_n + y_n \rangle$ and show that this is a 3-adic sequence associated with 2970.

4. Let $\langle y'_n \rangle = -54, -54, -54, -54, \ldots$. Show that $\langle y_n \rangle = \langle y'_n \rangle$, where $\langle y_n \rangle$ is defined in problem 3.

5. Show that $\langle x_n + y'_n \rangle$ is a 3-adic sequence associated with 2970, where $\langle x_n \rangle$ is defined in problem 3 and $\langle y'_n \rangle$ in problem 4.

Challenges

6. Prove statement (4) of Theorem 8.1.1.

7. If $a \equiv b \bmod p^n$, $n = 1, 2, 3, \ldots$, prove that $a = b$.

8. Use number 7 to prove the following. If $\langle x_n \rangle$ is a *p*-adic sequence associated with a and $\langle y_n \rangle$ is a *p*-adic sequence associated with b, then $\langle x_n \rangle = \langle y_n \rangle$ implies $a = b$.

8.2   THE p-ADIC INTEGERS

Equivalence of sequences is an " equivalence relation." Equivalence is (a) reflexive: $\langle x_n \rangle = \langle x_n \rangle$; (b) symmetric: $\langle x_n \rangle = \langle y_n \rangle$ implies $\langle y_n \rangle =$ $\langle x_n \rangle$; (c) transitive: $\langle x_n \rangle = \langle y_n \rangle$ and $\langle y_n \rangle = \langle z_n \rangle$ imply $\langle x_n \rangle = \langle z_n \rangle$. By this relation, the set of p-adic sequences is divided into a collection of disjoint subsets called equivalence classes, each of which consists of a collection of equivalent sequences. Let $[\langle x_n \rangle]$ denote the set of p-adic sequences, each of which is equivalent to the p-adic sequence$\langle x_n \rangle$. This equivalence class of p-adic sequences is called a *p-adic integer*. Any sequence $\langle x_n \rangle$ from the equivalence class determines the equivalence class and therefore is said to *determine* the p-adic integer. In general, Greek letters will be used to represent p-adic integers. Since it will be necessary to speak of p-adic integers that are determined by p-adic sequences of integers, it will be helpful, in order to avoid confusion, to use the term " rational integers " to describe the usual integers of arithmetic. Lowercase Arabic letters will be used to denote rational integers.

In order to construct a ring whose elements are p-adic integers, we need to define two operations with appropriate properties. We use addition and multiplication of sequences to define addition and multiplication of equivalence classes of sequences.

DEFINITION 8.2.1. *p-adic Integers.* A *p-adic integer* is an equivalence class of p-adic sequences. The *sum* and *product* of p-adic integers are defined as follows: Let $\alpha = [\langle x_n \rangle]$ and $\beta = [\langle y_n \rangle]$. Then $\alpha + \beta = [\langle x_n + y_n \rangle]$ and $\alpha \cdot \beta = [\langle x_n y_n \rangle]$.

THEOREM 8.2.1. *The Ring of p-adic Integers,* O_p. The set of p-adic integers, with the operations $+$ and $\cdot$, is a commutative ring with unity. It contains a subring isomorphic to the ring of rational integers.

Proof: Theorem 8.1.1 shows that the operations $+$ and $\cdot$ are well defined, since $[\langle x_n + y_n \rangle]$ and $[\langle x_n y_n \rangle]$ are equivalence classes of p-adic sequences and are not dependent on the particular sequence chosen to represent the p-adic integers α and β.

The verification of the following ring properties requires only Theorem 8.1.1, the corresponding properties of the rational integers, and a reasonable amount of patience: Addition is commutative and associative, multiplication is commutative and associative, multiplication is distributive over addition; the equivalence class that contains the sequence $0, 0, 0, \ldots$ is the additive identity; the equivalence class that contains the sequence $1, 1, 1, \ldots$ is the multiplicative identity; every p-adic integer α has an additive inverse, $-\alpha$; if $\alpha = [\langle x_n \rangle]$, then $-\alpha = [\langle -x_n \rangle]$.

Let the rational integer b correspond to the p-adic integer determined by $b, b, b, \ldots$. This correspondence is one to one and preserves sums and products [Theorem 8.1.1, part (4)]. Also, 0 corresponds to the additive identity and 1 corresponds to the multiplicative identity. A one-to-one correspondence with these properties is an isomorphism. ●

$$\boxed{\quad O_p \quad Z \quad}$$

The notation O_p is used for the ring of p-adic integers, and Z is used for the ring of rational integers. We will say that Z *is* a subring of O_p, instead of using the more accurate and cumbersome phrase "is isomorphic to a subring of O_p." One of the first questions that might be asked about O_p is: Does it contain any elements that are not in Z? Before we consider this question, it will help to select, from all the possible sequences that determine α, a standard or *canonical* sequence. We will then see that the canonical sequences that represent rational integers have special properties (Theorem 8.2.2 and Exercises 8.3). As a result, we conclude that Z is a proper subset of O_p.

DEFINITION 8.2.2. *Canonical Sequence.* Let $\alpha = [\langle x_n \rangle]$ and let $\bar{x}_n$ be the unique rational integer such that $\bar{x}_n \equiv x_n \bmod p^{n+1}$ and $0 \le \bar{x}_n < p^{n+1}$. Then $\alpha = [\langle \bar{x}_n \rangle]$ and $\langle \bar{x}_n \rangle$ is the canonical sequence that determines α.

EXAMPLE 8.2.1. Let $p = 5$, and $\alpha = [\langle x_n \rangle]$, where $\langle x_n \rangle = 26$, $-19, 256, 756, -1744, \ldots$. Since $26 \equiv 1 \bmod 5$, $-19 \equiv 6 \bmod 5^2$, $256 \equiv 6 \bmod 5^3, 756 \equiv 131 \bmod 5^4, -1744 \equiv 1381 \bmod 5^5, \ldots$, the canonical sequence that determines α is

$$\langle \bar{x}_n \rangle = 1, 6, 6, 131, 1381, \ldots.$$

Let $\beta = [\langle y_n \rangle]$, where $\langle y_n \rangle = 5, 5^2, 5^3, 5^4, \ldots$; the canonical sequence that determines β is $0, 0, 0, \ldots$.

Let $\gamma = [\langle z_n \rangle]$, where $\langle z_n \rangle = -1, -1, -1, \ldots;$ the canonical sequence that determines γ is 4, 24, 124, 624, ...

In the preceding example, the p-adic integer γ is identified with the rational integer -1. It might seem that the sequence $\langle z_n \rangle$, in which each term is -1, is a more natural choice than the canonical sequence. The importance of the canonical sequence lies in its relation to a unique representation for a p-adic integer as a series.

It is frequently convenient to express a sequence in series form. This is a common technique in analysis as well as in other areas of mathematics. Let $\langle s_n \rangle$ be a sequence and let $c_0 = s_0$, $c_1 = s_1 - s_0$, $c_2 = s_2 - s_1$, $\ldots, c_n = s_n - s_{n-1}, \ldots$. The symbol $\sum_{i=0}^{\infty} c_i$ or $c_0 + c_1 + c_2 + \cdots$ is called an *infinite series* and is defined to be identical to the sequence $\langle s_n \rangle$. Since $s_n = c_0 + c_1 + \cdots + c_n$, s_n is called the nth *partial sum* of the infinite series $\sum_{i=0}^{\infty} c_i$. All this, of course, agrees with your calculus experience, but notice that we are not concerned here with "convergence."

We now write the canonical sequence associated with α in series form. Since $\bar{x}_n \equiv \bar{x}_{n-1} \bmod p^n$, $\bar{x}_n = \bar{x}_{n-1} + a_n p^n$, and because $0 \leq \bar{x}_n < p^{n+1}$, the integer a_n must satisfy the inequality $0 \leq a_n < p$. Thus

$$\bar{x}_0 = a_0, \quad \bar{x}_1 = a_0 + a_1 p, \quad \bar{x}_2 = a_0 + a_1 p + a_2 p^2, \ldots, \quad \bar{x}_n = \sum_{i=0}^{n} a_i p^i, \ldots$$

and we have

$$\langle \bar{x}_n \rangle = \sum_{i=0}^{\infty} a_i p^i,$$

where $0 \leq a_i < p$ for every i.

Conversely, if we consider a series of the form

$$\sum_{i=0}^{\infty} a_i p^i, \quad 0 \leq a_i < p,$$

its sequence of partial sums is a canonical sequence for some p-adic integer. Because $s_{n+1} - s_n = a_{n+1} p^{n+1}$, we have $s_{n+1} \equiv s_n \bmod p^{n+1}$, so that $\langle s_n \rangle$ is a p-adic sequence. Also,

$$s_n = a_0 + a_1 p + \cdots + a_n p^n < p^{n+1}$$

(see Exercises 8.2, problem 5), so that s_n is the least non-negative residue mod p^{n+1} and $\langle s_n \rangle$ is a canonical sequence.

DEFINITION 8.2.3. *p-series*. A series of the form $\sum_{i=0}^{\infty} a_i p^i$, where $0 \le a_i < p$ for $i = 0, 1, 2, \ldots$, is called a *p-series*. Let $\alpha = [\langle x_n \rangle]$, let $\langle \bar{x}_n \rangle$ be the canonical sequence that determines α, and let $\sum_{i=0}^{\infty} a_i p^i$ be the *p-series* equal to $\langle \bar{x}_n \rangle$. We write

$$\alpha = \sum_{i=0}^{\infty} a_i p^i = \langle \bar{x}_n \rangle.$$

EXAMPLE 8.2.2. For the integers α, β, γ defined in Example 8.2.1, we have

$$\alpha = 1 + 5 + 0 \cdot 5^2 + 1 \cdot 5^3 + 2 \cdot 5^4 + \cdots,$$
$$\beta = 0 + 0 \cdot 5 + 0 \cdot 5^2 + 0 \cdot 5^3 + \cdots,$$
$$\gamma = 4 + 4 \cdot 5 + 4 \cdot 5^2 + 4 \cdot 5^3 + \cdots.$$

EXAMPLE 8.2.3. Find the 5-series corresponding to 25821. The canonical sequence for 25821 is found by computing the remainders on dividing by powers of 5 and is 1, 21, 71, 196, 821, 10196, 25821, 25821, ... In series form, we have

$$25821 = 1 + 4 \cdot 5 + 2 \cdot 5^2 + 1 \cdot 5^3 + 1 \cdot 5^4 + 3 \cdot 5^5 + 1 \cdot 5^6 + 0 \cdot 5^7 + \cdots.$$

The coefficients from a_7 on are all zero. These terms can be omitted and the 5-series is finite. In this case, the $=$ between 25821 and the 5-series can be thought of as an ordinary $=$ sign, since, if the finite sum is calculated, it is equal to 25821.

THEOREM 8.2.2. *p-series for a Positive Rational Integer*. Every positive rational integer has a finite *p-series*.

Proof: Let b be a positive rational integer. Then $0 \le b < p^n$ for some n. The canonical sequence will have $\bar{x}_{n-1} = b$, $\bar{x}_n = b, \ldots$ since the least non-negative residue of b mod p^k for $k \ge n$ is b itself. This fact implies that $a_n = 0$, $a_{n+1} = 0, \ldots$ and the *p-series* is finite. ●

Anyone who has studied the representation of integers in base p will recognize that the expansion of a positive integer in a *p-series* is equivalent to expressing that integer in expanded form in base p.

EXERCISES 8.2

Checks

1. Find the 3-series for the integers 318, 24, 2213.
2. Find the 5-series for the integers 318, 24, 2213.
3. Verify that multiplication is distributive over addition, Theorem 8.2.1.
4. Prove that $\bar{x}_{n-1} \leq \bar{x}_n$ for all n in the canonical representation of any p-adic sequence.
5. Prove that

$$a_0 + a_1 p + a_2 p^2 + \cdots + a_n p^n < p^{n+1},$$

   where $0 \leq a_i \leq p - 1$ for every i. *Hint:* Set $a_i = p - 1$.

Challenges

6. Show that the coefficients of the p-series for the integer b can be found by successive applications of the division algorithm:

$$b = pq_0 + a_0, \quad q_0 = pq_1 + a_1, \quad q_1 = pq_2 + a_2, \ldots.$$

7. Given $[\langle x_n \rangle] = \alpha$ in O_p, define $\langle y_n \rangle$ as follows:

$$y_0 = 0, \quad y_n = px_{n-1}, \quad \text{for } n = 1, 2, 3, \ldots.$$

   (a) Prove that $[\langle y_n \rangle]$ is an element of O_p.
   (b) If $\langle x_n \rangle$ is a canonical sequence, prove that $\langle y_n \rangle$ is a canonical sequence.
   (c) If $\langle x_n \rangle$ is associated with a rational integer b, what rational integer is associated with $\langle y_n \rangle$?   (Lee Hill)

8. Let $S = \{[\langle x_n \rangle]$ such that $\bar{x}_0 = 0, \bar{x}_1 = 0, [\langle x_n \rangle]$ in $O_5\}$. Prove that the set S is an ideal in O_5. Characterize the rational integers that are associated with elements of S.   (Lee Hill)

8.3 THE FUNDAMENTAL THEOREM

Addition and multiplication of p-adic integers has already been defined. If the operation is carried out on canonical sequences, the resulting sequence is not necessarily canonical but can be made into a canonical sequence by calculating the termwise sum (or product) mod p^{n+1}. If the operations are carried out on the p series, the series are added or multiplied in the usual way, each coefficient must then be replaced by its p-series, and the terms must be regrouped. An example will illustrate the technique. You can devise shortcuts for yourself.

EXAMPLE 8.3.1. $p = 5$.

$221 = 1, 21, 96, 221, 221, 221, \ldots$.
$113 = 3, 13, 113, 113, 113, 113, \ldots$.
$334 = 4, 34, 209, 334, 334, 334, \ldots$ (add termwise),
$\quad = 4, 9, 84, 334, 334, 334, \ldots$, the canonical sequence.

In series form, the same operations are

$221 = 1 + 4 \cdot 5 + 3 \cdot 5^2 + 5^3$.
$113 = 3 + 2 \cdot 5 + 4 \cdot 5^2$.
$334 = 4 + 6 \cdot 5 + 7 \cdot 5^2 + 5^3$ (add termwise),
$\quad = 4 + (1 + 1 \cdot 5)5 + (2 + 1 \cdot 5)5^2 + 5^3$ (replace coefficients by p-series),
$\quad = 4 + 1 \cdot 5 + 3 \cdot 5^2 + 2 \cdot 5^3$ (regroup).

For products, the computation may be more lengthy.

$113 \cdot 221 = 3, 273, 10848, 24973, 24973, 24973, \ldots$ (termwise product),
$\quad = 3, 23, 98, 598, 3098, 9348, 24973, \ldots$ (canonical).
$113 \cdot 221 = 3 + 14 \cdot 5 + 21 \cdot 5^2 + 25 \cdot 5^3 + 14 \cdot 5^4 + 4 \cdot 5^5$
$\quad = 3 + (4 + 2 \cdot 5)5 + (1 + 4 \cdot 5)5^2 + (5^2)5^3 + (4 + 2 \cdot 5)5^4 + 4 \cdot 5^5$
$\quad = 3 + 4 \cdot 5 + 3 \cdot 5^3 + 4 \cdot 5^2 + 4 \cdot 5^4 + 7 \cdot 5^5$
$\quad = 3 + 4 \cdot 5 + 3 \cdot 5^2 + 4 \cdot 5^3 + 4 \cdot 5^5 + 2 \cdot 5^5 + 5^6$.

Since O_p is a ring, we can ask whether division is possible within the system. Can we define a relation "divides" in O_p just as we did in Z? We wish this definition to have the property that if $a|b$ in Z, then $a|b$ in O_p.

DEFINITION 8.3.1. *Divides.* Let α and β belong to O_p. Then α divides β if there exists a p-adic integer γ such that $\beta = \alpha \cdot \gamma$.

It is not difficult to see that those properties of "divides" that depend only on the definition and not on the ordering of the integers or the concept of absolute value can be established for O_p in exactly the same way they were established for Z. For example, if $\alpha|\beta$, then α divides $\gamma\beta + \delta\alpha$ for any γ and δ in O_p. The p-adic integers divisible by p^k are easy to identify by using the p series for the p-adic integer.

THEOREM 8.3.1. *Division by* p^k. If

$$\alpha = \sum_{i=0}^{\infty} a_i p^i, \quad \text{then} \quad p^k\alpha = \sum_{i=0}^{\infty} a_i p^{k+i}.$$

If $\beta = \sum_{i=0}^{\infty} b_i p^{k+i}$, then $\beta = p^k\gamma$ where $\gamma = \sum_{i=0}^{\infty} b_i p^i$.

Proof: Since $\alpha = \sum_{i=0}^{\infty} a_i p^i$, the canonical sequence for α is $\langle x_n \rangle$, where

$$x_n = a_0 + a_1 p + \cdots + a_n p^n.$$

By the definition of multiplication, a sequence for $p^k\alpha$ is $\langle y_n \rangle$, where

$$y_n = a_0 p^k + a_1 p^{k+1} + \cdots + a_n p^{k+n}.$$

The canonical sequence for $p^k\alpha$ is $\langle z_n \rangle$, where

$$z_n = 0, \quad n = 0, 1, 2, \ldots, k - 1, \quad \text{and} \quad z_n = y_{n-k}, \quad n \geq k.$$

The corresponding p-series is

$$p^k\alpha = \sum_{i=0}^{\infty} a_i p^{k+i}.$$

The second statement of the theorem is a direct consequence of the first, because of the uniqueness of the p series representation. ◆

COROLLARY 8.3.1a. $p^k|\alpha$, where $\alpha = \sum_{i=0}^{\infty} a_i p^i$, if and only if $a_0 = a_1 = \cdots = a_{k-1} = 0$.

COROLLARY 8.3.1b. $p^k|\alpha$ if and only if $\alpha = \langle x_n \rangle$, where $x_n \equiv 0$ mod p^{n+1} for $n = 0, 1, 2, \ldots, k - 1$.

Questions of divisibility immediately suggest the search for units in the ring. We have found that in Z the only units are 1 and -1. In the ring of residues mod m, Z_m, the number of units depends on the nature of m. What about the units in O_p?

EXAMPLE 8.3.2. $p = 5$. Is 3 a unit? Can we find x such that $3x = 1$? Let $x = \langle x_n \rangle$. A sequence for $3x$ is $\langle 3x_n \rangle$. Thus $3x = 1$ if and only if $3x_n \equiv 1 \bmod 5^{n+1}$ for every n. Since $(3, 5) = 1$, each of the congruences $3x_n \equiv 1 \bmod 5^{n+1}$ can be solved for x_n. We find that $x_0 = 2$, $x_1 = 17$, $x_2 = 42$, $x_3 = 317$, ... Is the sequence so determined a p-adic sequence?

$$17 \equiv 2 \bmod 5, \quad 42 \equiv 17 \bmod 5^2, \quad 317 \equiv 42 \bmod 5^3.$$

In general, since $3x_n \equiv 1 \bmod 5^{n+1}$, we have $3x_n \equiv 1 \bmod 5^n$. Also, $3x_{n-1} \equiv 1 \bmod 5^n$. Hence,

$$3x_n \equiv 3x_{n-1} \bmod 5^n \quad \text{and} \quad x_n \equiv x_{n-1} \bmod 5^n$$

since $(3, 5) = 1$.

Is 5 a unit in O_5?

THEOREM 8.3.2. *Identification of a Unit.* A p-adic integer $\alpha = \langle x_n \rangle$ is a unit if and only if $x_0 \not\equiv 0 \bmod p$.

Proof: Suppose that α is a unit. There exists $\beta = \langle y_n \rangle$ such that $\alpha\beta = 1$. That is, $x_n y_n \equiv 1 \bmod p^{n+1}$, for each $n \geq 0$. In particular, $x_0 y_0 \equiv 1 \bmod p$, which implies that $x_0 \not\equiv 0 \bmod p$ (Section 2.3).

Conversely, suppose that $x_0 \not\equiv 0 \bmod p$. Since $x_n \equiv x_{n-1} \bmod p^n$, for each $n \geq 1$, $x_n \equiv x_0 \bmod p$ and hence $x_n \not\equiv 0 \bmod p$ for each n. This implies that there exists a y_n such that $x_n y_n \equiv 1 \bmod p^{n+1}$. Since $x_n y_n \equiv 1 \bmod p^{n+1}$, $x_n y_n \equiv 1 \bmod p^n$. Also, $x_{n-1} y_{n-1} \equiv 1 \bmod p^n$. But $\langle x_n \rangle$ is a p-adic sequence, so $x_n \equiv x_{n-1} \bmod p^n$. This implies that $y_n \equiv y_{n-1} \bmod p^n$. Thus $\langle y_n \rangle$ is also a p-adic sequence and determines a p-adic integer β such that $\alpha\beta = 1$. Hence α is a unit. ●

COROLLARY 8.3.2a. A p-adic integer $\alpha = \sum_{i=0}^{\infty} a_i p^i$ is a unit if and only if $a_0 \neq 0$.

COROLLARY 8.3.2b. A rational integer b is a unit in O_p if and only if $(b, p) = 1$.

COROLLARY 8.3.2c. A rational number of the form r/s, $(r, s) = 1$, is a unit if and only if $(s, p) = 1$ and $(r, p) = 1$.

COROLLARY 8.3.2d. The set of units in O_p is infinite.

What sort of factorization is possible in O_p? The fact that the set of units is so large makes it less surprising that every p-adic integer can be expressed in terms of a unit and a power of p, as the next theorem shows.

THEOREM 8.3.3. *The Fundamental Theorem in O_p.* Every p-adic integer that is not zero has a unique representation in the form $\alpha = p^m \varepsilon$, where ε is a unit in O_p and m is a non-negative rational integer.

Proof: If α is a unit, then $m = 0$ and $\alpha = \varepsilon$. Let

$$\alpha = \sum_{i=0}^{\infty} a_i p^i.$$

Since $\alpha \neq 0$, not all the coefficients a_i are zero. Let m be the smallest integer such that a_m is not zero. Then

$$\alpha = a_m p^m + a_{m+1} p^{m+1} + \cdots;$$

that is, $\alpha = p^m \varepsilon$, where $\varepsilon = \sum_{i=0}^{\infty} a_{m+i} p^i$ (Theorem 8.3.1). Since $a_m \neq 0$, ε is a unit. Hence $\alpha = p^m \varepsilon$, the form required in the theorem.

The uniqueness of the representation follows immediately from the uniqueness of the p-series representation for α. ●

COROLLARY 8.3.3. Let $\alpha = p^k \varepsilon$ and $\beta = p^m \eta$. Then α divides β if and only if $k \leq m$.

Proof: If $p^m \eta = p^k \varepsilon \cdot p^n \delta$, then $m = k + n$ and $m \geq k$. If $m \geq k$, then $p^m \eta = p^k \varepsilon \cdot p^{m-k} \eta \varepsilon^{-1}$. ●

Let S_p denote the set of rational numbers of the form a/b, where a and b are rational integers, $(a, b) = 1$ and $(b, p) = 1$. The corollaries of Theorem 8.3.2 show that the set O_p contains not only the rational integers but also all numbers in S_p. Again, the word "contains" is an abbreviation for "contains a subset isomorphic to." The following theorem, the proof of which is outlined in the exercises, shows how to identify the elements of S_p.

THEOREM 8.3.4. *The Set* S_p. The set S_p is a subset of O_p. If α is an element of S_p, the p-series for α is either finite or periodic. In particular, if α is a negative integer, its p-series is infinite and, from some point on, the coefficients are $p - 1$.

There are rational numbers that are not in O_p—namely, those of the form a/b, $(a, b) = 1$, where $p \mid b$. There are elements of O_p which are not rational numbers—namely, those whose p-series is not periodic. The diagram shown illustrates the situation.

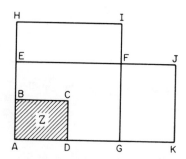

$ABCD = Z$, the set of rational integers.
$AEFG = S_P$.
$AHIG = Q$, the rational numbers.
$AEJK = O_P$.
$Z \subset S_P \subset O_P$ and $Z \subset S_P \subset Q$.
$FGKJ = O_P - Q$ is not empty.
$HEIF = Q - O_P$ is not empty.

EXERCISES 8.3

Checks

1. Prove that the sum of two units in O_p is not necessarily a unit.

2. Prove that S_p is a subring of the rational numbers.

3. Find the 5-series for $\frac{1}{3}$. Find the 5-series for $\frac{2}{3}$.

4. Prove that $1 - p^m$ is a unit. Prove that the inverse of $1 - p^m$ is $1 + p^m + p^{2m} + \cdots$.

5. Prove that the canonical sequence for -1 is $\langle p^{n+1} - 1 \rangle$ and that the p series for -1 is

$$\sum_{i=0}^{\infty} (p - 1)p^i.$$

Challenges

6. Prove that the set of units in O_p is uncountably infinite. *Hint:* The set is in one-to-one correspondence with the set of sequences $(a_0, a_1, a_2, \ldots)$, $a_0 \neq 0$.

7. Prove that if $\alpha = \sum_{i=0}^{\infty} a_i p^i$ is a finite or a periodic p-series, then α belongs to S_p. *Hint:* If the series for α is periodic, it can be written in the form

$$\alpha = A + p^k B + p^{k+m} B + p^{k+2m} B + \cdots,$$

   where

$$A = a_0 + a_1 p + \cdots + a_{k-1} p^{k-1} \quad \text{and} \quad B = b_0 + b_1 p + \cdots + b_{m-1} p^{m-1}.$$

   Then

$$\alpha - A = p^k B + p^m (\alpha - A).$$

8. Let $\alpha = r/s$, where $(r, s) = 1$ and $(s, p) = 1$, $s > 0$, and $-1 < r/s < 0$. Let m be the smallest rational integer such that $p^m \equiv 1 \bmod s$; that is, $1 - p^m = zs$. Prove that

$$zr = b_0 + b_1 p + \cdots + b_{m-1} p^{m-1}.$$

   From

$$\frac{r}{s} = \frac{rz}{1 - p^m}$$

   derive the fact that r/s has a periodic p-series

$$\frac{r}{s} = B + Bp^m + Bp^{2m} + \cdots.$$

9. If α is positive and in S_p, use $\alpha = N + r/s$ and the preceding problem to prove that α has a finite or a periodic p-series.

10. If $\alpha = \sum_{i=0}^{\infty} a_i p^i$, prove that·

$$-\alpha = p - a_0 + \sum_{i=1}^{\infty} (p - 1 - a_i) p^i.$$

11. Use problems 9 and 10 to prove that if α is negative and in S_p, then α has a periodic p-series.

12. Prove that there exists a 7-adic integer x such that $x^2 = 2$.

13. Prove that there does not exist a 5-adic integer x such that $x^2 = 2$.

14. For which primes p does there exist a p-adic integer x such that $x^2 = -1$?

15.   Define $\alpha \equiv \beta$ mod γ if and only if $\gamma \mid \alpha - \beta$. Prove that $\alpha \equiv \beta$ mod γ if and
only if $\alpha \equiv \beta$ mod p^k, where $\gamma = p^k \varepsilon$. Prove that, given α, there exists a
rational integer b such that $\alpha \equiv b$ mod p^m. Thus show that there are only
p^m distinct residue classes mod p^m in O_p.

8.4   DISTANCE

So far we have been concerned with algebraic properties of the
p-adic integers (addition, multiplication, divisibility). When we think about
the rational integers, there are other properties that concern us. We say,
for example, 3 and -3 have the same absolute value; the integers 3 and
-3 are the same distance from 0; the distance between the integers 4 and
-6 is ten units.

The meaning of the word "distance" depends on the context.
A father and his son in the living room may be a short distance apart
physically but a great distance from each other ideologically. The distance
between two cities may be much longer by car than by plane. Yet there are
certain properties that we feel a good distance function ought to possess.

Let S be a set. For every pair a and b in S, let $d(a, b)$ be a real
number, which we will call "the distance from a to b." Then $d(a, b)$ ought
have the properties to:

1. $d(a, b) \geq 0$, and $d(a, b) = 0$ if and only if $a = b$,
2. $d(a, b) = d(b, a)$,
3. $d(a, c) \leq d(a, b) + d(b, c)$.

For the rational integers, we frequently use the idea of absolute
value to define distance. If a and b are in Z, we define $d(a, b) = |a - b|$.
The function d so defined has the three properties listed (see Exercises 8.4,
problem 1). By analogy, we define a distance function on O_p by first
defining a value function, called p-value.

$$\boxed{|\alpha|_p; \qquad d(\alpha, \beta)}$$

DEFINITION 8.4.1. p-value, Distance. Let $\alpha = p^k \varepsilon$ be a p-adic
integer, ε a unit.

The p-value of α, written $|\alpha|_p$, is equal to $1/p^k$.

The p-value of 0, $|0|_p = 0$.

If α and β belong to O_p, the distance between α and β, written $d(\alpha, \beta)$, is defined as $d(\alpha, \beta) = |\alpha - \beta|_p$.

In Section 1.12, Definition 1.12.1 the p-value of a rational integer was defined. Definition 8.4.1 is a natural extension of this definition of p-value to the ring O_p. The properties of p-value established in Theorem 1.12.3 are equally true of p-value in its more general setting.

THEOREM 8.4.1. *Basic Properties of p-Value.*

1. $|\alpha|_p \geq 0$ and equals zero only if $\alpha = 0$.
2. $|\alpha\beta|_p = |\alpha|_p |\beta|_p$.
3. $|\alpha + \beta|_p \leq \max(|\alpha|_p, |\beta|_p)$, non-Archimedean property.
4. $|\alpha + \beta|_p \leq |\alpha|_p + |\beta|_p$, triangle inequality.

Proof: See Exercises 8.4, problem 2. ●

The properties of p-value imply certain basic properties of the distance function.

THEOREM 8.4.2. *Properties of Distance.*

1. $d(\alpha, \beta) \geq 0$, and $d(\alpha, \beta) = 0$ if and only if $\alpha = \beta$.
2. $d(\alpha, \beta) = d(\beta, \alpha)$.
3. $d(\alpha, \gamma) \leq d(\alpha, \beta) + d(\beta, \gamma)$.
4. $d(\alpha, \gamma) \leq \max[d(\alpha, \beta), d(\beta, \gamma)]$.

Proof. Properties (1) and (2) are an immediate consequence of the definition and the properties of p-value. Property (4) implies property (3). Property (4) is a consequence of the non-Archimedean property of p-value. We have

$$d(\alpha, \gamma) = |\alpha - \gamma|_p = |(\alpha - \beta) + (\beta - \gamma)|_p$$

which, by the non-Archimedean property, is less than or equal to $\max[|\alpha - \beta|_p, |\beta - \gamma|_p]$. Thus we have

$$d(\alpha, \gamma) \leq \max[d(\alpha, \beta), d(\beta, \gamma)]. ●$$

EXAMPLE 8.4.1. Let $p = 5$. Let

$$\alpha = -1 = 4 + 4 \cdot 5 + 4 \cdot 5^2 + 4 \cdot 5^3 + \cdots \quad \text{and} \quad \beta = 99 = 4 + 4 \cdot 5 + 3 \cdot 5^2.$$

What is $d(\alpha, \beta)$? Since

$$-1 - 99 = 1 \cdot 5^2 + 4 \cdot 5^3 + \cdots$$

we see that

$$|-1 - 99|_5 = \frac{1}{5^2},$$

so that

$$d(-1, 99) = \frac{1}{5^2}.$$

What other 5-adic integers x have the property that $d(-1, x) = 1/5^2$? If $|-1 - x|_5 = 1/5^2$, then

$$5^2 | -1 - x \quad \text{and} \quad 5^3 \nmid -1 - x;$$

that is, $x = -1 - 5^2 \varepsilon$, where ε is a unit. This means that

$$x = 4 + 4 \cdot 5 + a_2 5^2 + a_3 5^3 + \cdots,$$

where $a_2 \neq 4$. The set $\{x, \text{ with } d(-1, x) = 1/5^2\}$ could be thought of as a circle of center -1 and radius $1/5^2$.

In plane geometry, a circle of radius r is defined to be the set of points a distance r from a fixed point called the center. The circle, together with its interior, is called a *closed disk* and is defined as the set of points whose distance from a fixed point is less than or equal to r. If "less than or equal to" is replaced by "less than," the resulting set of points is an *open disk*. Since these definitions are expressed in terms of points and distances, they can be applied to the set of p-adic integers with the integers considered as points and distance defined by Definition 8.4.1.

$$\boxed{S(x, r); \quad S[x, r]}$$

DEFINITION 8.4.2. *Disks in O_p.* The set $S(x, r) = \{y: d(x, y) < r\}$ is called an *open disk* with center x and radius r. The set $S[x, r] = \{y: d(x, y) \leq r\}$ is called a *closed disk* with center x and radius r.

EXAMPLE 8.4.2. $S(0, 1)$ is the set $\{y: d(0, y) < 1\}$; that is, $\{y: |y|_p < 1\}$. This implies that $y = a_1 p + a_2 p^2 + \cdots$.

$S[0, 1]$ is the set $\{y: d(0, y) \leq 1\}$; that is, $\{y: |y|_p \leq 1\}$. This set is O_p. $S[0, 1/p]$ is the set $\{y: d(0, y) \leq 1/p\}$; that is, $\{y: |y|_p \leq 1/p\}$. This is exactly the set $S(0, 1)$.

We see that the same set of points can be thought of as an open disk $S(0, 1)$ or as a closed disk $S[0, 1/p]$. This property is different from the properties of disks in ordinary plane geometry. It is a consequence of the fact that the values which the distance function can assume are discrete—that is, they are separated real numbers and not a continuous interval of real numbers. The technical description of this property is that the range of the distance function has only one accumulation point, in our case 0.

THEOREM 8.4.3. *Disks Both Open and Closed.* Every disk in the p-adic integers can be interpreted both as an open disk and a closed disk.

Proof: Let $S(x, r) = \{y: d(x, y) < r\}$. If $r > 1$, choose $r_1 = 1$. If $r < 1$, there is a k such that $1/p^k \geq r > 1/p^{k+1}$. Choose $r_1 = 1/p^{k+1}$. Then $S(x, r) = S[x, r_1]$. Similarly, given the closed disk $S[x, r]$, choose r_1 greater than r and less than or equal to the next larger value of the distance function. The set $S[x, r] = S(x, r_1)$. ●

In the following discussion, theorems are stated for closed disks. Because of the property exhibited in Theorem 8.4.3, the theorems for closed disks are equally true for open disks.

EXAMPLE 8.4.3. $S[1, 1/5^2] = \{x: d(1, x) \leq 1/5^2\}$—that is, $\{x: x = 1 + a_2 5^2 + a_3 5^3 + \cdots\}$. What about $S[1 + 5^2, 1/5^2]$? This is the set $\{x: x = 1 + 5^2 + b_2 5^2 + b_3 5^3 + \cdots\}$. Since there are no restrictions on the values of a_i and b_i except for $0 \leq a_i, b_i \leq 4$, it is clear that the sets are the same. What about $S[x_0, 1/5^2]$, where $x_0 = 1 + c_2 5^2 + c_3 5^3 + \cdots$?

This example indicates that although a disk is defined in terms of center and radius, any point in the disk may be taken as a center. This is quite different from Euclidean plane geometry, in which the center of a disk is unique. The unusual properties of disks in O_p are a consequence of the non-Archimedean property of the distance function. Algebraically, the key to understanding these properties is the fact that the distance function is defined in terms of divisibility and hence the disks are residue classes mod some power of p. These remarks are stated formally in the following theorems.

THEOREM 8.4.4. *Nonuniqueness of Center.* Let y be a point in a disk with center x and radius r. The disk with center y and radius r is the same set of points as the disk with center x and radius r.

Proof: Since y is in $S[x, r]$, then $d(y, x) \leq r$. For any z, $d(z, x) \leq$ max $\{d(z, y), d(y, x)\}$. Therefore if $d(z, y) \leq r$, $d(z, x)$ is also $\leq r$ and hence $S[y, r] \subset S[x, r]$.

Also, $d(z, y) \leq$ max $\{d(z, x), d(y, x)\}$. Hence if $d(z, x) \leq r$, then $d(z, y) \leq r$ as well, and $S[x, r] \subset S[y, r]$. This implies $S[x, r] = S[y, r]$. ●

THEOREM 8.4.5. *Overlapping Disks.* Let S_1 be a disk of radius r_1 and S_2 a disk of radius r_2, where $r_1 \geq r_2$. If $S_1 \cap S_2 \neq \phi$, then $S_1 \supset S_2$.

Proof. Let x belong to $S_1 \cap S_2$. Then $S_1 = S[x, r_1]$ and $S_2 = S[x, r_2]$. If y belongs to S_2, $d(x, y) \leq r_2 \leq r_1$. Hence y belongs to S_1 and $S_2 \subset S_1$. ●

THEOREM 8.4.6. *The Set of p-adic Disks.* There are exactly p^k closed disks of radius $1/p^k$. These disks are disjoint and their union is O_p.

Proof: Let x and y belong to O_p. Then x and y belong to the same disk of radius $1/p^k$ if and only if $|x - y|_p \leq 1/p^k$; that is, if and only if $p^k | x - y$. But if $x = \Sigma a_i p^i$ and $y = \Sigma b_i p^i$, $p^k | x - y$ if and only if $a_i = b_i$, $i = 0, 1, 2, \ldots, k - 1$. The set of disks of radius $1/p^k$ is thus in one-to-one correspondence with the p-adic integers of the form $a_0 + a_1 p + \cdots + a_{k-1} p^{k-1}$. There are p^k of these integers.

Let $z = \Sigma c_i p^i$. The p-adic integer z belongs to the disk that contains $c_0 + c_1 p + \cdots + c_{k-1} p^{k-1}$. Hence the union of the disks of radius $1/p^k$ is O_p. The fact that the disks are disjoint follows from Theorem 8.4.5. ●

EXERCISES 8.4

Checks

1. Using the properties of absolute value, show that the distance function defined by $d(a, b) = |a - b|$ satisfies the three properties listed for a distance function. Show that it does not satisfy the non-Archimedean property.

2. Verify the properties of p-value stated in Theorem 8.4.1.

3. Identify the p-adic integers in $S[p^2, 1/p]$; in $S[p^2, 1/p^2]$; in $S[p^2, 1/p^3]$. Which of these disks contain 0?

4. List the nine disjoint disks of radius $1/3^2$ in O_3.

5. What open disk is equal to $S[p, 1/p^3]$? What closed disk is equal to $S(p^2, 1/p^3)$?

6. Consider the disk $S[p^2, r]$ in O_p. For what values of r does 0 belong to this disk?

Challenges

7. Show that the disks $S[0, 1/5^2]$ and $S[1, 1/5^2]$ are disjoint. Let x belong to $S[0, 1/5^2]$ and y belong to $S[1, 1/5^2]$. What is $d(x, y)$? Let S_1 and S_2 be any two disjoint disks of radius $1/p^k$ in O_p. What are the possible values of $d(x, y)$ if x is in S_1 and y is in S_2?

8. Prove that every disk of radius $r > 0$ contains at least one rational integer. Is there a disk that contains exactly one?

9. Prove that every closed disk of radius $1/p^k$ is the union of p disjoint closed disks of radius $1/p^{k+1}$.   (Verlin Koper)

10. Prove that $\alpha | \beta$ if and only if $|\alpha|_p \geq |\beta|_p$.

11. Prove that $\alpha \equiv \beta \bmod \gamma$, where $\gamma = p^k \varepsilon$, if and only if $d(\alpha, \beta) \leq 1/p^k$ (see Exercises 8.3, problem 15).

8.5   A GEOMETRIC REPRESENTATION OF O_p

What picture do we have in mind when we speak of disks in O_p? A mathematical system can sometimes be better understood if a geometric model can be constructed which demonstrates its characteristic properties. A basic consideration for such a construction is the concept of distance.

When we think of the rational integers with distance defined in terms of absolute value, we usually think of points equally spaced on a line extending indefinitely in both directions. The distance between a and b

$$\begin{array}{cccccccccccccccc} -6 & -5 & -4 & -3 & -2 & -1 & 0 & 1 & 2 & 3 & 4 & 5 & 6 & 7 & 8 \end{array}$$

is $|a - b|$. Thus the distance between 4 and -6 is ten units. For a proper choice of a and b, $|a - b|$ can be made as large as desired. The set Z, with this definition of distance, is an *unbounded* set.

The concept of distance generated by the p-value function suggests a different geometric picture. Since $|\alpha - \beta|_p \le 1$ for every α, β in O_p, the set of p-adic integers is a *bounded* set relative to the distance function determined by p-value. This property will be reflected in our geometric picture if we represent the elements of O_p by a set of points in a finite portion of the plane. We choose here to explore the representation of p-adic integers by points on the line segment $[0, 1)$. The points on this line segment are in a one-to-one correspondence with the set of real numbers $\{r: 0 \le r < 1\}$. We obtain the representation of O_p by defining a subset G of the set of real numbers $\{r: 0 \le r < 1\}$ and setting up a one-to-one correspondence between O_p and G.

Each real number r, $0 \le r < 1$, can be represented by using base $p + 1$ in the form

$$r = \frac{a_0}{p + 1} + \frac{a_1}{(p + 1)^2} + \frac{a_2}{(p + 1)^3} + \cdots = \sum_{i=0}^{\infty} \frac{a_i}{(p + 1)^{i+1}}, \qquad 0 \le a_i \le p.$$

As is customary, we write $r = \cdot a_0 a_1 a_2 \cdots$ to represent this series, recalling that the base of the representation is $p + 1$ in place of the usual 10. The representation $r = \cdot a_0 a_1 a_2 \cdots$ is unique unless there is a k such that $a_i = p$ for all $i > k$. If this happens, r has two representations, $r = \cdot a_0 a_1 \cdots a_k pppp \cdots = \cdot a_0 a_1 \cdots (a_k + 1)000 \cdots$. Define the set G as follows: $G = \{r: r = \cdot a_0 a_1 a_2 \cdots, 0 \le a_i \le p - 1\}$. Note that for the real numbers r in G the representation is unique.

EXAMPLE 8.5.1. Let $p = 3$. Then $G = \{r: r = \cdot a_0 a_1 a_2 \cdots,$ where $a_i = 0$, 1, or 2}. What points of the half-open interval $[0, 1)$ correspond to real numbers in G? Certainly no points of G lie in the interval $[\frac{3}{4}, 1)$, since in this interval the real numbers are represented by $\cdot 3 a_1 a_2 a_3 \cdots$. Let $E_1 = [\frac{3}{4}, 1)$. The set G is a subset of $[0, 1) - E_1$. Since a_1 cannot equal 3, there are no points of G in the four intervals

$$\left[\frac{3}{4^2}, \frac{4}{4^2}\right), \quad \left[\frac{7}{4^2}, \frac{8}{4^2}\right), \quad \left[\frac{11}{4^2}, \frac{12}{4^2}\right), \quad \left[\frac{15}{4^2}, 1\right).$$

Call the union of these four intervals E_2. Then G is a subset of $[0, 1)$ $- E_1 - E_2$. The last interval listed in E_2 is a subset of E_1, but no problem arises from including it in both. In the same way, let

$$E_3 = \bigcup_{i=1}^{16} \left[\frac{4i-1}{4^3}, \frac{4i.}{4^3} \right).$$

These intervals contain the real numbers for which $a_2 = 3$ and therefore contain no points of G. Continue this process. The set G consists of the

$$[0,1) - E_1 = AG \quad [0,1) - E_1 - E_2 = AB \cup CD \cup EF$$

real numbers corresponding to points in $[0, 1) - \bigcup_{n=1}^{\infty} E_n$. Every real number not in G must have $a_k = 3$ for some k. The point corresponding to this number is eliminated in E_{k+1}. Therefore the points corresponding to the set G are exactly $[0, 1) - \bigcup_{n=1}^{\infty} E_n$.

The construction of the set G for a general p is carried out along the lines of Example 8.5.1. First the interval is divided into $p + 1$ equal half-open intervals and the last one eliminated. Each of the remaining intervals is divided into $p + 1$ parts and the last one eliminated. Thus

$$G = [0, 1) - \bigcup_{n=1}^{\infty} E_n, \quad \text{where} \quad E_n = \bigcup_{j=1}^{(p+1)^{n-1}} \left[\frac{(p+1)j - 1}{(p+1)^n}, \frac{(p+1)j}{(p+1)^n} \right).$$

You may have read of a famous example in set theory called the Cantor set, which is constructed in a similar way.

Now that the set G and its image on the interval $[0, 1)$ has been examined, we can set up a correspondence between $O_{p.}$ and G. Each p-adic integer is uniquely determined by its p-series,

$$\alpha = \sum_{i=0}^{\infty} a_i p^i,$$

and therefore by a sequence of integers $a_0, a_1, a_2, \ldots$ such that $0 \le a_i \le p - 1$. Let α belong to O_p and r belong to G; then α corresponds to r if and only if $\alpha = \sum_{i=0}^{\infty} a_i p^i$ and $r = \cdot a_0 a_1 a_2 \cdots$, where the representation of r is base $p + 1$.

EXAMPLE 8.5.2. Let $p = 3$. The table shown illustrates this correspondence for some simple 3-adic integers α. The corresponding points are shown on the diagram.

α in 0_3	r in G	Point in $[0,1)$
1	.1	A
2	.2	B
3	.01	C
$4 = 1 + 1 \cdot 3$	.11	D
$5 = 2 + 1 \cdot 3$	.21	E
$-1 = \sum_{i=0}^{\infty} 2 \cdot 3^i$	.222...	F
$1/2 = 2 + \sum_{i=1}^{\infty} 3^i$	.2111...	G
$33 = 0 + 2 \cdot 3 + 0 \cdot 3^2 + 3^3$	.0201	H
$60 = 0 + 2 \cdot 3 + 0 \cdot 3^2 + 2 \cdot 3^3$	.0202	I
$-48 = 0 + 2 \cdot 3 + 0 \cdot 3^2 + 1 \cdot 3^3 + \sum_{i=4}^{\infty} 2 \cdot 3^i$	.0201222...	J

In what part of $[0,1)$ do the units fall? Where are the nonunits? What is the 3-adic distance from A to 0? Note that the 3-adic distance from E to F is $|5 + 1|_3 = \frac{1}{3}$. What other points lie in $S[F, \frac{1}{3}]$?

What do the disks look like in this geometric picture? A disk with 0 in its interior is the intersection of G and one of the intervals for which 0 is the left endpoint. Thus $S[0, 1/p^k]$ is the set of points that are in G and also in the interval $[0, \cdot 00 \cdots 1)$, where 1 occurs in the kth position, since $|\alpha|_p \leq 1/p^k$ if and only if $a_0 = a_1 = \cdots = a_{k-1} = 0$. Let

$$\beta = \sum_{i=0}^{\infty} b_i p^i.$$

If x is in $S[\beta, 1/p^k]$, $|x - \beta|_p \leq 1/p^k$; that is,

$$x = \sum_{i=0}^{\infty} x_i p^i, \quad \text{where} \quad b_i = x_i \quad \text{for} \quad i = 0, 1, 2, \ldots, k - 1.$$

The corresponding real numbers must satisfy the inequality

$$\cdot b_0 b_1 b_2 \cdots b_{k-1} \leq r < \cdot b_0 b_1 b_2 \cdots (b_{k-1} + 1).$$

The real numbers satisfying this inequality lie in one of the p^k nonempty intervals of equal length that were constructed at the kth step in the construction of the image of G on the interval $[0, 1)$. We have already seen (Theorem 8.4.6) that O_p is the union of p^k closed disjoint disks of radius $1/p^k$. The geometric representation of these disks is the set of p^k point sets $G \cap I_j$, where I_j is one of the p^k intervals that remain when $[0, 1)$ is divided into $(p + 1)^k$ equal half-open intervals and those that contain no points of G are discarded.

EXAMPLE 8.5.3. Let $p = 3$. The nine disks of radius $\frac{1}{9}$ are the intersections of G and the nine half-open intervals: $[0, .01)$, $[.01, .02)$, $[.02, .03)$, $[.1, .11)$, $[.11, .12)$, $[.12, .13)$, $[.2, .21)$, $[.21, .22)$, $[.22, .23)$. When each of these intervals is divided into four parts and the fourth is eliminated, 27 half-open intervals result. The 27 closed disks of radius $\frac{1}{27}$ are the intersections of G with each of these 27 half-open intervals.

Disks of radius $\frac{1}{3}$:

Disks of radius $\frac{1}{9}$:

Disks of radius $\frac{1}{27}$:

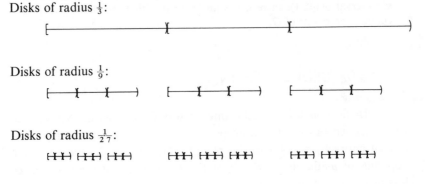

EXERCISES 8.5

Checks

1. Let $p = 3$. Locate on $[0, 1)$ the points corresponding to the following integers in O_3: 10, 30, 90, 1, 4, 22, 49, 130.

2. Locate on $[0, 1)$ the disk of radius $\frac{1}{9}$ that contains the 3-adic integer 34. To what integer does the left endpoint of this disk correspond?

3. In Example 8.5.3 a representation of the nine disks of radius $\frac{1}{9}$ is given. Let x belong to the first of these disks and y belong to the second. Find $d(x, y)$. Suppose that y belongs to the third, what is $d(x, y)$? What is $d(x, y)$ if y belongs to the fourth? If y belongs to the ninth?

4. Find a sequence of disks of decreasing radius such that 5 belongs to each disk of the sequence.

5. Let α be a p-adic integer. Show that there is a rational integer in $S[\alpha, 1/p^k]$. Is there more than one such rational integer? Is there a negative rational integer in this disk? Is there an element of S_p in this disk?

Challenges

6. Show that the disks have the following property characteristic of intervals: If x and y belong to D, then $\alpha x + (1 - \alpha)y$ belongs to D, for α any integer in O_p.

7. A linear ordering is induced on O_p by the one-to-one correspondence exhibited between O_p and a subset of the rationals. In terms of this ordering, prove that the smallest p-adic integer in any disk is a positive rational integer. Is there a largest? If so, what is it?

8. Prove that every disk of radius greater than 0 contains at least one rational integer. Is there a disk with $r > 0$ which contains exactly one?

9. Prove that the set of points in G is *nowhere dense* in $[0, 1)$; that is, given any subinterval of $[0, 1)$, there exists an interval within the subinterval which contains no points of G.

8.6   SEQUENCES IN O_p

In Section 8.2 a p-adic integer was defined as an equivalence class of sequences with special properties. Does it make sense to represent such an entity by a point in the plane? The distance function and the properties of p-adic disks shed light on the meaning of the equivalence class of p-adic sequences, the canonical sequence, and the statement

$$\alpha = \langle x_n \rangle = \sum_{i=0}^{\infty} a_i p^i.$$

EXAMPLE 8.6.1. Let $p = 5$, $\alpha = 1 + 2 \cdot 5 + 1 \cdot 5^2 + 2 \cdot 5^3 + 1 \cdot 5^4 + \cdots$. The canonical sequence for α is 1, 11, 36, 286, 911, ... Let $\langle x_n \rangle$ be any other sequence that determines α. Since $x_0 \equiv 1 \bmod 5$, x_0 belongs to $S[1, 1/5]$. Since $x_1 \equiv 11 \bmod 5^2$, x_1 belongs to $S[11, 1/5^2]$. Since 11 belongs both to $S[1, 1/5]$ and to $S[11, 1/5^2]$, the disk $S[11, 1/5^2] \subset S[1, 1/5]$.

A = 1, B = 11, C = 36.

The illustration shows the geometric picture. If we proceed in this way, we see that

$$S\left[1,\frac{1}{5}\right], \quad S\left[11,\frac{1}{5^2}\right], \quad S\left[36,\frac{1}{5^3}\right], \quad S\left[286,\frac{1}{5^4}\right], \dots$$

is a sequence of disks, each contained in the preceding one. The intersection of this set of disks contains a single point, α. Any sequence that determines α determines the same nested sequence of disks.

THEOREM 8.6.1. p-adic Integers and Nested Sequences of Disks. The p-adic integer α is associated with a unique nested sequence of p-adic disks:

$$D_n = S\left[x_n, \frac{1}{p^{n+1}}\right], \qquad n = 0, 1, 2, \dots.$$

The canonical sequence chooses as center the smallest non-negative rational integer in the disk.

$$\bigcap_{n=0}^{\infty} D_n = \alpha.$$

Proof: Let $\alpha = \langle x_n \rangle = \langle \bar{x}_n \rangle$, where $\langle \bar{x}_n \rangle$ is the canonical sequence. Since $x_n \equiv \bar{x}_n \bmod p^{n+1}$, x_n belongs to $S[\bar{x}_n, 1/p^{n+1}] = D_n$. Since $\bar{x}_n$ is the least non-negative residue mod p^{n+1}, $\bar{x}_n$ is the least non-negative rational integer in D_n. The disks are nested, since $x_n \equiv x_{n+1} \bmod p^{n+1}$ implies x_{n+1} belongs to D_n. But x_{n+1} also belongs to D_{n+1}. Since the disks D_n and D_{n+1} have a point in common, $D_{n+1} \subset D_n$.

Since $\alpha \equiv x_n \bmod p^{n+1}$ for each n, α belongs to D_n for each n, and hence α belongs to $\bigcap_{n=0}^{\infty} D_n$. Suppose that $\beta = \langle y_n \rangle \neq \alpha$. There exists an m such that $y_m \not\equiv x_m \bmod p^{m+1}$; that is, y_m does not belong to D_m. This means that $S[y_m, 1/p^{m+1}]$ and $S[x_m, 1/p^{m+1}]$ are disjoint. Since β is in $S[y_m, 1/p^{m+1}]$, β is not in D_m and hence β is not in $\bigcap_{n=0}^{\infty} D_n$. ●

The way in which the sequence $\langle x_n \rangle$ or the nested sequence of p-adic disks D_n determines the p-adic integer α corresponds to the way in which a sequence "converges to" a limit in analysis or topology. We say that the sequence $\langle s_n \rangle$ converges to s if the distance between s_n and s can be made arbitrarily small by choosing n large enough. In the present setting, we must remember that distance is determined by the p-adic distance function.

DEFINITION 8.6.1. *Convergence.* The sequence of p-adic integers α_n converges to α if $d(\alpha_n, \alpha)$ approaches 0 as n becomes infinite.

EXAMPLE 8.6.2. Let $p = 5$ and define the sequence x_n as follows:
If n is odd, $n = 2k + 1$, $x_{2k+1} = 1 + 2 \cdot 5 + \cdots + 1 \cdot 5^{2k} + 2 \cdot 5^{2k+1}$.
If n is even, $n = 2k$, $x_{2k} = 1 + 2 \cdot 5 + 1 \cdot 5^2 + \cdots + 1 \cdot 5^{2k}$.

Now group the powers of 5 that have coefficient 1 and the powers of 5 that have coefficient 2. We get

$$x_{2k+1} = (1 + 5^2 + \cdots + 5^{2k}) + 2 \cdot 5(1 + 5^2 + \cdots + 5^{2k});$$

that is,

$$x_{2k+1} = 11(1 + 5^2 + \cdots + 5^{2k}) = 11 \cdot \frac{1 - 5^{2k+2}}{1 - 5^2} = -\frac{11}{24} + \frac{11 \cdot 5^{2k+2}}{24}.$$

Also,

$$x_{2k} = 11 \cdot \frac{1 - 5^{2k}}{1 - 5^2} + 1 \cdot 5^{2k} = -\frac{11}{24} + 5^{2k}\left(\frac{11}{24} + 1\right) = -\frac{11}{24} + \frac{7}{24} \cdot 5^{2k+1}.$$

It is now easy to calculate $d(x_n, -11/24)$, since

$$d\left(x_{2k+1}, -\frac{11}{24}\right) = \left|\frac{11}{24} \cdot 5^{2k+2}\right|_5 = \frac{1}{5^{2k+2}}$$

and

$$d\left(x_{2k}, -\frac{11}{24}\right) = \left|\frac{7}{24} \cdot 5^{2k+1}\right|_5 = \frac{1}{5^{2k+1}}.$$

From these expressions for $d(x_n, -11/24)$, we see that $d(x_n, -11/24)$ approaches zero as n (and therefore k) becomes infinite. Thus the sequence $\langle x_n \rangle$ converges to $-11/24$.

The preceding calculations represent one method of calculating the rational number to which the p-series $1 + 2 \cdot 5 + 1 \cdot 5^2 + \cdots$ converges. Another argument goes as follows: Let $\alpha = 1 + 2 \cdot 5 + 1 \cdot 5^2 + \cdots$; then

$5^2\alpha = 1 \cdot 5^2 + 2 \cdot 5^3 + \cdots$ so that $\alpha = 1 + 2 \cdot 5 + 5^2\alpha$. Therefore $(1 - 5^2)\alpha = 1 + 2 \cdot 5$; that is, $\alpha = -11/24$. Do you recall using a similar technique in high school to change a repeating decimal to a fraction?

THEOREM 8.6.2. *The Limit of a p-adic Sequence.* Let $\alpha = \langle x_n \rangle$. The sequence $\langle x_n \rangle$ converges to α.

Proof: Since $\alpha = \langle x_n \rangle$, $\alpha \equiv x_n \bmod p^{n+1}$ for each n. Hence $d(\alpha, x_n) \leq 1/p^{n+1}$ for each n. Since $d(\alpha, x_n)$ approaches zero as n becomes infinite, $\langle x_n \rangle$ converges to α. ◆

A more detailed study of the *p*-adic integers must be left for a future excursion. The references listed at the end of this section are among many that introduce us to additional related information and ideas. For example, the ring of *p*-adic integers can be embedded in a field Q_p in much the same way as the rational integers are embedded in the field of rational numbers. Every element of O_p is of the form $p^k\varepsilon$, where ε is a unit in O_p and k is a rational integer (positive, negative, or zero). This field Q_p contains a subfield isomorphic to the field of rationals, and it can be shown to be the completion of the field of rationals with respect to the *p*-value distance.

We can consider Q_p as a topological space with the topology induced by the *p*-value; that is, the neighborhoods of the points are the disks containing the points. This is a metric space of dimension zero, totally disconnected and spherically complete. The set O_p is a compact subset of the space Q_p.

Does the study of *p*-adic integers seem to be not properly a part of elementary number theory? Certainly it is a natural outgrowth of the ideas of divisibility. In this area, we see an example of the interrelation of many different branches of mathematics. The *p*-adic integers were first introduced by Hensel in the study of Diophantine equations, especially in connection with quadratic forms. But they provide a simple setting for much current research relating to analysis, topology, convexity, and linear vector spaces over non-Archimedean fields.

... some larger way
Where many paths and errands meet.
And whither then?

EXERCISES 8.6

Checks

1. Prove that the sequence

    $$p, 2p, p^2, 2^2p^2, p^3, 3^3p^3, \ldots, p^k, k^kp^k, \ldots$$

    converges to 0. Is this a p-adic sequence (Definition 8.1.1)?

2. Find the rational number associated with the 5-series

    $$3 + 2\cdot5 + 3\cdot5^2 + 4\cdot5^3 + 2\cdot5^4 + 3\cdot5^5 + 4\cdot5^6 + 2\cdot5^7 + \cdots.$$

3. Prove that every p-adic integer is the limit of a sequence of rational integers—that is, Z is *dense* in O_p.

4. Prove that the set of p-adic integers is a *perfect* set—that is, every element of the set is a limit of a sequence of distinct points of the set.

5. Recall that the infinite series $\sum_{k=0}^{\infty} a_k$ is defined to be the limit of the sequence of partial sums $S_n = a_0 + a_1 + \cdots + a_{n-1}$. From what you know about the convergence of a sequence in O_p, prove that the infinite series $\sum_{k=0}^{\infty} a_k$ converges in O_p if and only if $\lim_{k\to\infty} a_k = 0$. (Do not use a theorem like this in your calculus class!)

Challenges

6. A sequence $\langle x_n \rangle$ is said to be Cauchy if $d(x_n, x_m)$ can be made arbitrarily small by choosing n and m large enough. Prove that the sequence of p-adic integers $\langle x_n \rangle$ is Cauchy if $d(x_n, x_{n+1})$ approaches zero as n becomes infinite. *Hint*: Show that $d(x_n, x_m) \le \max [d(x_{n+i}, x_{n+i+1}), i = 0, 1, \ldots, m - n - 1]$.

7. Let q be a prime different from p. Show that the sequence $q, q^2, q^3, \ldots$ does not converge in O_p. Can you choose a convergent subsequence from this sequence?

8. Consider the sequence of rational integers that are prime: $2, 3, 5, 7, 11, \ldots$. Does this sequence converge in O_p? Show how to find a convergent subsequence.

9. Prove that from any sequence of p-adic integers it is possible to select a convergent subsequence. *Hint*: Since the number of residue classes mod p is finite, there are infinitely many terms of the sequence α_n in one of them. These terms are a subsequence $\alpha_n^{(1)}$, each integer of which is congruent to some x_0 mod p. Repeat to determine a subsequence $\alpha_n^{(2)}$ of $\alpha_n^{(1)}$, each element of which is congruent to some x_1 mod p^2. In this way, the p-adic integer $\langle x_n \rangle$ is constructed. What subsequence converges to x_n?

10. Let $\langle \alpha_n \rangle$ be a sequence of p-adic integers which is Cauchy but which does not converge to zero. Prove that from some value of n on, the p-value of α_n is constant.

11. Construct the set of ordered pairs $O_p \times O_p$, defined by

$$O_p \times O_p = \{(x, y) \text{ such that } x \text{ and } y \text{ are elements of } O_p\}.$$

Define: $(a, b) + (c, d) = (a + c, b + d)$ and $k(a, b) = (ka, kb)$. Let d be the function defined as follows:

$$d[(x, y), (a, b)] = \max \left[|x - a|_p, |y - b|_p \right].$$

Prove that d is a distance function on the vector space $O_p \times O_p$. Prove that d is non-Archimedean.

REFERENCES FOR UNIT 8

1. Bachman, G., *Introduction to p-adic Numbers and Valuation Theory* (New York: Academic Press, 1964.)

2. Borevich, Z. I., and I. R. Shafarevich, *Number Theory*, translated by Newcomb Greenleaf (New York: Academic Press, 1966.)

3. Lewis, D. J., "Diophantine Equations: p-Adic Methods," in *Studies in Number Theory*, W. J. LeVeque (Ed.), MAA Studies in Mathematics, Vol. 6 (1969), pp. 25–75.

4. MacDuffee, C. C., "The p-adic Numbers of Hensel," *The American Mathematical Monthly*, Vol. 45 (1938), pp. 500–508.

5. Monna, A. F. "Linear Topological Spaces over non-Archimedean Valued Fields," *Proceedings of a Conference on Local Fields*, T. A. Springer (Ed.) (New York: Springer-Verlag, 1967.)

SOME TERMINOLOGY RELATING TO ALGEBRAIC SYSTEMS

An algebraic system can be described as a set of elements in which certain operations have been defined. The type of system under consideration depends upon the number and properties of the operations.

Let S represent a set of elements. The set $S \times S$ is the set of ordered pairs (a, b) such that a and b are elements of S. A *binary operation* on S is a function from $S \times S$ into S. In other words, a binary operation associates a unique element of S with each ordered pair of elements in S. Such an operation is frequently written in the form $a \circ b = c$, where a, b, c represent elements of S.

The operation $\circ$ is *associative* if, for every set of elements a, b, c of S, $a \circ (b \circ c) = (a \circ b) \circ c$.

The operation $\circ$ is *commutative* if, for every pair of elements a, b of S, $a \circ b = b \circ a$.

An element e of S is an *identity* with respect to the operation $\circ$ if $a \circ e = e \circ a = a$ for each a in S.

An element a of S has an *inverse* with respect to the operation $\circ$ if there is an element a^* of S such that $a \circ a^* = a^* \circ a = e$.

The set S and the binary operation $\circ$, written $(S, \circ)$ is called a *semi-group* if the operation obeys the associative law.

If the semi-group $(S, \circ)$ has an identity element and if each element of S has an inverse, the semi-group is a group.

Thus a *group* is an algebraic system consisting of a set S and an operation $\circ$ with the following properties:

1. Closure. For each a and b in S, $a \circ b$ is an element of S.

2. Associative law. For arbitrary a, b, c in S, $(a \circ b) \circ c = a \circ (b \circ c)$.

3. Existence of an identity element. There is an element e in S such that $a \circ e = e \circ a = a$ for each a in S.

4. Existence of inverses. For each a in S there is an inverse element a^* in S such that $a \circ a^* = a^* \circ a = e$.

The group $(S, \circ)$ is *commutative*, or *abelian*, if the operation satisfies the commutative law.

A nonempty subset H of a group is a subgroup if H is closed under the group operation and satisfies the other requirements of a group in its own right. The simplest condition which ensures that a subset H of S is a subgroup is that $a \circ b^*$ belongs to H for each a and b in H.

The group $(S, \circ)$ and the group $(S', \circ')$ are *homomorphic* if there is a function f from S into S' with the property that, for any a and b in S,

$$f(a \circ b) = f(a) \circ' f(b).$$

The group $(S, \circ)$ and the group $(S', \circ')$ are *isomorphic* if there is a one-to-one function f from S onto the set S' such that, for any a and b in S,

$$f(a \circ b) = f(a) \circ' f(b).$$

A *ring* is an algebraic system consisting of a set S along with two binary operations with properties outlined below. These operations are usually called addition $(+)$ and multiplication $(\times)$. In order to be a ring, the system $(S, +, \times)$ must satisfy the three conditions listed below.

1. $(S, +)$ is an abelian group.
2. $(S, \times)$ is a semi-group.
3. The following distributive laws hold:

$$a \times (b + c) = (a \times b) + (a \times c),$$
$$(b + c) \times a = (b \times a) + (c \times a).$$

If a subset of $(S, +, \times)$ is closed under both operations and also satisfies the conditions (1), (2), (3) as a subsystem, the subset is a *subring*.

A subring E of $(S, +, \times)$ with the property that $y \times a$ and $a \times y$ are in E for every a in S and y in E is called an *ideal*.

The ring $(S, +, \times)$ and the ring $(S', +', \times')$ are *homomorphic* if there is a mapping f from S into S' such that

$$f(a + b) = f(a) +' f(b),$$
$$f(a \times b) = f(a) \times' f(b).$$

If the mapping f is one-to-one and onto S', the rings are *isomorphic*.

A ring is *commutative* if its multiplicative semi-group is commutative. A ring has an *identity* if its multiplicative semi-group has an identity.

An element a in a ring is a *divisor of zero* if $a \neq 0$ and there exists an element $b \neq 0$, such that $a \times b = 0$. (Here 0 represents the additive identity of the ring.) A ring is an *integral domain* if it has no divisor of zero.

In a commutative ring, a *divides* b if there is an element c such that $b = ac$. Two elements are called *associates* if each divides the other. A ring element that has a multiplicative inverse is called a *unit*. Two elements of a commutative integral domain with an identity element are associates if and only if each is the product of the other by a unit of the ring.

DISTRIBUTION OF THE ODD PRIMES IN THE ODD INTEGERS LESS THAN 2000

–	11	–	31	41	–	61	71	–	–
3	13	23	–	43	53	–	73	83	–
5	–	–	–	–	–	–	–	–	–
7	17	–	37	47	–	67	–	–	97
–	19	29	–	–	59	–	79	89	–
101	–	–	131	–	151	–	–	181	191
103	113	–	–	–	–	163	173	–	193
–	–	–	–	–	–	–	–	–	–
107	–	127	137	–	157	167	–	–	197
109	–	–	139	149	–	–	179	–	199

–	211	–	–	241	251	–	271	281	–
–	–	223	233	–	–	263	–	283	293
–	–	–	–	–	–	–	–	–	–
–	–	227	–	–	257	–	277	–	–
–	–	229	239	–	–	269	–	–	–

–	311	–	331	–	–	–	–	–	–
–	313	–	–	–	353	–	373	383	–
–	–	–	–	–	–	–	–	–	–
307	317	–	337	347	–	367	–	–	397
–	–	–	–	349	359	–	379	389	–

401	–	421	431	–	–	461	–	–	491
–	–	–	433	443	–	463	–	–	–
–	–	–	–	–	–	–	–	–	–
–	–	–	–	–	457	467	–	487	–
409	419	–	439	449	–	–	479	–	499

–	–	521	–	541	–	–	571	–	–
503	–	523	–	–	–	563	–	–	593
–	–	–	–	–	–	–	–	–	–
–	–	–	–	547	557	–	577	587	–
509	–	–	–	–	–	569	–	–	599

601	–	–	631	641	–	661	–	–	691
–	613	–	–	643	653	–	673	683	–
–	–	–	–	–	–	–	–	–	–
607	617	–	–	647	–	–	677	–	–
–	619	–	–	–	659	–	–	–	–

701	–	–	–	–	751	761	–	–	–
–	–	–	733	743	–	–	773	–	–
–	–	–	–	–	–	–	–	–	–
–	–	727	–	–	757	–	–	787	797
709	719	–	739	–	–	769	–	–	–

–	811	821	–	–	–	–	–	881	–
–	–	823	–	–	853	863	–	883	–
–	–	–	–	–	–	–	–	–	–
–	–	827	–	–	857	–	877	887	–
809	–	829	839	–	859	–	–	–	–

–	911	–	–	941	–	–	971	–	991
–	–	–	–	–	953	–	–	983	–
–	–	–	–	–	–	–	–	–	–
907	–	–	937	947	–	967	977	–	997
–	919	929	–	–	–	–	–	–	–

–	–	1021	1031	–	1051	1061	–	–	1091
–	1013	–	1033	–	–	1063	–	–	1093
–	–	–	–	–	–	–	–	–	–
–	–	–	–	–	–	–	–	1087	1097
1009	1019	–	1039	1049	–	1069	–	–	–

–	–	–	–	–	1151	–	1171	1181	–
1103	–	1123	–	–	1153	1163	–	–	1193
–	–	–	–	–	–	–	–	–	–
–	1117	–	–	–	–	–	–	1187	–
1109	–	1129	–	–	–	–	–	–	–
1201	–	–	1231	–	–	–	–	–	1291
–	1213	1223	–	–	–	–	–	1283	–
–	–	–	–	–	–	–	–	–	–
–	1217	–	1237	–	–	–	1277	–	1297
–	–	1229	–	1249	1259	–	1279	1289	–
1301	–	1321	–	–	–	1361	–	1381	–
1303	–	–	–	–	–	–	1373	–	–
–	–	–	–	–	–	–	–	–	–
1307	–	1327	–	–	–	1367	–	–	–
–	1319	–	–	–	–	–	–	–	1399
–	–	–	–	–	1451	–	1471	1481	–
–	–	1423	1433	–	1453	–	–	1483	1493
–	–	–	–	–	–	–	–	–	–
–	–	1427	–	1447	–	–	–	1487	–
1409	–	1429	1439	–	1459	–	–	1489	1499
–	1511	–	1513	–	–	–	1571	–	–
–	–	1523	–	1543	1553	–	–	1583	–
–	–	–	–	–	–	–	–	–	–
–	–	–	–	–	–	1567	–	–	1597
–	–	–	–	1549	1559	–	1579	–	–
1601	–	1621	–	–	–	–	–	–	–
–	1613	–	–	–	–	1663	–	–	1693
–	–	–	–	–	–	–	–	–	–
1607	–	1627	1637	–	1657	1667	–	–	1697
1609	1619	–	–	–	–	1669	–	–	1699
–	–	1721	–	1741	–	–	–	–	–
–	–	1723	1733	–	1753	–	–	1783	–
–	–	–	–	–	–	–	–	–	–
–	–	–	–	1747	–	–	1777	1787	–
1709	–	–	–	–	1759	–	–	1789	–
1801	1811	–	1831	–	–	1861	1871	–	–
–	–	1823	–	–	–	–	1873	–	–
–	–	–	–	–	–	–	–	–	–
–	–	–	–	1847	–	1867	1877	–	–
–	–	–	–	–	–	–	1879	1889	–
1901	–	–	1931	–	1951	–	–	–	–
–	1913	–	1933	–	–	–	1973	–	1993
–	–	–	–	–	–	–	–	–	–
1907	–	–	–	–	–	–	–	1987	1997
–	–	–	–	1949	–	–	1979	–	1999

SOME NUMBER-THEORETIC FUNCTIONS, THEIR DEFINITIONS AND FORMULAS, $n = \prod_{i=1}^{r} p^{a_i}$

τ	The number of positive divisors of n	$\tau(n) = \sum_{d\mid n} 1$
		$= \prod_{i=1}^{r} (a_i + 1)$
σ	The sum of the positive divisors of n	$\sigma(n) = \sum_{d\mid n} d$
		$= \prod_{i=1}^{r} \dfrac{p_i^{a_i+1} - 1}{p_i - 1}$
σ_k	The sum of the kth powers of the positive divisors of n	$\sigma_k(n) = \sum_{d\mid n} d^k$
		$= \prod_{i=1}^{r} \dfrac{p_i^{k(a_i+1)} - 1}{p_i^k - 1}$

ϕ The number of positive integers less than n and relatively prime to n

$$\phi(n) = \prod_{i=1}^{r} (p_i^{a_i} - p_i^{a_i - 1})$$

$$= n \prod_{p \mid n} \left(1 - \frac{1}{p}\right)$$

μ The Möbius function

$$\mu(n) = \prod_{i=1}^{r} \mu(p_i^{a_i}) \qquad \text{where}$$

$$\mu(p_i^{a_i}) = 1 \ \text{if} \ a_i = 0$$
$$-1 \ \text{if} \ a_i = 1$$
$$0 \ \text{if} \ a_i > 1.$$

π The number of primes in the first n positive integers

π_2 The number of those primes in the first n positive integers which are members of prime pairs

E_S The characteristic function of the set S

$E_S(n) = 1$ if n belongs to S
$= 0$ if n does not belong to S

z The additive identity in the group $(\mathscr{F}, +)$

$z(n) = 0$ for every n.

e The multiplicative identity in the ring $(\mathscr{F}, +, *)$

$e(n) = 1$ if $n = 1$
$= 0$ if $n > 1$

u Constant function

$u(n) = 1$ for every n

u_k Power function

$u_k(n) = n^k$ for every n

Λ the von Mangoldt function

$\Lambda(n) = \log p$ if $n = p^k$ and $k \geq 1$
$= 0$ otherwise

GENERAL REFERENCES

Out of the vast collection of books related to elementary number theory, the following ones are suggested.

For historical perspective:

1. Dickson, L. E., *History of the Theory of Numbers*, 3 vols. (Washington: Carnegie Institution of Washington, 1919–1923. New York: Chelsea Publishing Co., 1952.)
2. Ore, O., *Number Theory and Its History* (New York: McGraw-Hill, 1948.)

For a collection of more recent results and references:

3. Sierpinski, W., *Elementary Theory of Numbers* (Warszawa, Poland: Panstwowe Wydawnictwo Naukowe, 1965.)

Textbooks selected arbitrarily because they have been adopted at Oklahoma State University at some time during the past fifteen years:

4. LeVeque, W. J., *Topics in Number Theory*, 2 vols. (Reading, Mass.: Addison-Wesley, 1962.)
5. Long, C. T., *Elementary Introduction to Number Theory* (Boston: D. C. Heath & Company, 1965.)
6. Niven, I., and H. S. Zuckerman, *An Introduction to the Theory of Numbers* (New York: Wiley & Sons, 1960.)
7. Schockley, J. E., *Introduction to Number Theory* (New York: Holt, Rinehart & Winston, 1967.)
8. Stewart, B. M., *Theory of Numbers*, 2nd edition. (New York: Macmillan, 1964.)

INDEX OF NOTATION

INDEX